D0076998

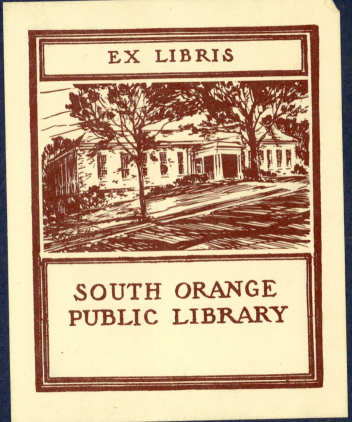

EX LIBRIS

SOUTH ORANGE
PUBLIC LIBRARY

THE
MATHEMATICAL
GARDNER

edited by

David A. Klarner

PRINDLE, WEBER & SCHMIDT
Boston, Massachusetts 02116

WADSWORTH INTERNATIONAL
Belmont, California 94002

793.74
Ma ✓

© Copyright 1981 by Wadsworth International
10 Davis Drive, Belmont, California

All rights reserved. No part of this book may be reproduced or transmitted in
any form or by any means, electronic or mechanical, including photocopying,
recording, or any information storage and retrieval system, without permission, in writing, from the publisher.

Wadsworth International is a division of Wadsworth,Inc.

Library of Congress Cataloging in Publication Data

Main entry under title:

The Mathematical Gardner.

 Bibliography: p.

 1. Mathematical recreations—Addresses, essays,
lectures. 2. Gardner, Martin, 1914
I. Klarner, David A.
QA95.M3676   793.7'4   80-22046
ISBN 0-534-98015-5

Cover image "Bees in Clover" by Marjorie Rice. Used by permission of the artist.

Cover and text design by Susan G. Graham in collaboration with the staff of Prindle,
Weber & Schmidt. Composed in Baskerville by H. Charlesworth & Co., Ltd. Dust
jacket and color insert printed by Bradford & Bigelow. Text and covers printed and
bound by The Alpine Press.

Dedication and Introduction

The articles in this book are dedicated to Martin Gardner, the world's greatest expositor and popularizer of mathematics. While our papers are confined to this single subject, Gardner's interests and accomplishments have a wide range of subjects. Hence, we have entitled the book the *Mathematical Gardner*, and would like to see other volumes such as the *Magical*, the *Literary*, the *Philosophical*, or the *Scientific Gardner* accompany it. Of course, our title is also an appropriate pun, for Martin Gardner's relationship to the mathematical community is similar to a gardener's relationship to a beautiful flower garden. The contributors to this volume comprise only a small part of a large body of mathematicians whose work has been nurtured by its exposition in "Mathematical Games"; Martin's column which appears every month in *Scientific American*. More than just a mathematical journalist, Martin connects his readers by passing along problems and information and stimulating creative activity. Thus, he is a force behind the scenes as well as a public figure.

Two people were particularly helpful in putting this book together. Ronald Graham and Donald Knuth not only made contributions themselves but also helped in the quiet solicitation of manuscripts from the other authors. The project was carried out secretly, although Dr. I. J. Matrix seems to have gotten a hold of the manuscript and he is coming out with a pirated edition. Because of the clandestine nature of the project, the contributors were selected on the basis of personal contact rather than as a consequence of a published appeal for papers. Many who might have wished to contribute have my apologies now. A published call for papers not only would have spoiled the surprise, but it would have resulted in an overwhelming avalanche of material.

In the nuts and bolts of making the book there are others to thank. Foremost among these is Dean Hoffman. Many of his ideas were incorporated in the papers he edited. He also encouraged me to press on and carry out the project in spite of the time it was taking from my research work. I would also like to thank my wife for all of her help and patience. Kara Lynn graciously posed as Dr. Matrix's beautiful Eurasian daughter Eva for the centerfold, but this had to be left out at the publisher's insistence. There is a lot of correspondence involved in making a book of this sort, and my secretary Elizabeth Newton was invaluable in this regard. Finally, I would like to thank Susan Graham and the staff at Prindle, Weber and Schmidt for their sincere interest and help in making this book. In particular, Theron Shreve deserves considerable credit.

David A. Klarner

Table of Contents

GAMES

GEOMETRY

TWO-DIMENSIONAL TILING

THREE-DIMENSIONAL TILING

FUN AND PROBLEMS

NUMBERS AND CODING THEORY

THE
MATHEMATICAL
GARDNER

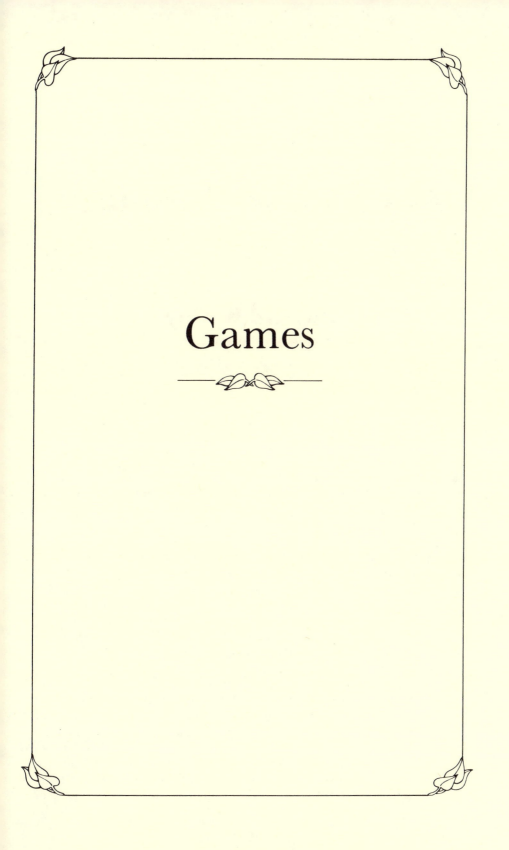

Games

Anyone for Twopins?

Richard K. Guy

UNIVERSITY OF CALGARY

The bowling game of Twopins (pronounced "Tuppins") is played by two people, with columns of pins lined up as in Figure 1. Each column contains one or two pins. The columns are spaced so that the bowler may knock out any one column, or any two neighboring columns. If a column of two is hit, both pins fall. After a shot, the pins are not reset before the opponent takes his turn. The game ends when the last pin falls and the person knocking it down is declared the winner. For a proper shot, at least two pins must fall; you are not allowed to remove a single column when it contains only one pin, and the game may end with some isolated single pins still standing. In Figure 1, for example, you may remove column *d* only, but not columns *b*, *c*, *e* or *h*, unless you remove an adjacent column at the same time. After removal of *d*, the opponent cannot remove columns *c* and *e* because these are not neighboring.

Twopins is considered an *impartial* game because in any position, the available options are the same for each of the two players. In contrast, chess is a *partisan* game because, in any position, Black has a different set of available options from White. The theory of impartial games in which the *last* player is declared the winner, is not as widely known as it deserves to be. It was discovered independently by Sprague [21] and Grundy [12] and by various people since. They found that every position in any impartial game has a *nim-value*; that is, the position is equivalent to a *nim-heap*, or a heap of beans in the game of Nim [4,2,15]. There is a simple rule for finding the nim-value of a position:

Take the mex of the nim-values of the options.

The *mex* (minimum excluded value) of a set of nonnegative integers is the least nonnegative integer *not* in the set. For example, mex $\{5,3,0,7,1\} = 2$ and mex $\varnothing = 0$, therefore, the nim-value for the *Endgame* (when there are no options and the game is finished) is zero.

The importance of the nim-value, or Sprague-Grundy function, derives from the fact that all (positions in) impartial games form an additive Abelian group. Indeed, so do all last-player-winning games, including the partisan ones, but the Sprague-Grundy theory applies only to the subgroup of impartial games.

The sum (or disjunctive combination) of two or more positions, (not necessarily in the same game) is played as follows:

The player whose turn it is to move chooses *one* of the component games and makes a legal move in that component.

The compound game ends when each component has ended, and the last player is again the winner. It is easy to see that this kind of addition is associative and commutative.

The identity of the group is, of course, the Endgame, and the negative of any position is the same position with the opposing player to move. (In impartial games each position is its own negative.) Most people have come across examples of the *Tweedledum and Tweedledee principle*, in which a symmetry strategy, mimicking your opponent's moves, enables you to win

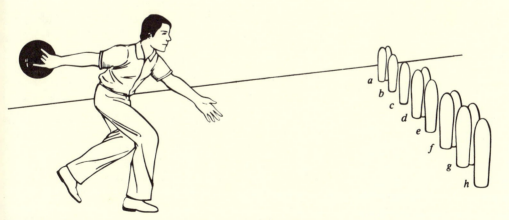

FIGURE 1
Ready for a shot at Twopins.

a game. This additive group is not only mathematically pretty, but is also important practically, since many games break up into sums of separate games in the normal course of play. A typical move in Twopins, for example, breaks a row into two shorter rows, and therefore, the next move must be made in one of the two new rows.

The main result of the Sprague-Grundy theory for impartial games with the last player winning is summarized in the theorem:

> The nim-value of the sum of two games is the nim-sum of their nim-values.

To find the nim-sum of two nonnegative integers, add them in binary without carrying. This is the operation used by Bouton [4] in his original analysis of Nim (see also [2, 15]). Indeed, now that we have the Sprague-Grundy theory, Nim is seen to be the archetype of *all* impartial games: a typical Nim position is the disjunctive sum of games of Nim, each played with one heap.

$\mathbf{d}_r =$	*Game played with rows of beans*	*Game played with heaps of beans*
0	There is no legal move in which r beans may be taken.	
$\mathbf{1 = 2^0}$	r beans may be taken if they comprise a whole row.	A heap of exactly r beans may be removed completely.
$\mathbf{2 = 2^1}$	r beans may be taken from either end of a longer row.	r beans may be taken from a (larger) heap, leaving a non-empty heap.
$\mathbf{3 = 2^1 + 2^0}$	r beans may be taken in either of the last two circumstances, (leaving 0 or 1 rows).	(leaving 0 or 1 heaps).
$\mathbf{4 = 2^2}$	r consecutive beans may be taken strictly from within a longer row, leaving 2 nonempty rows.	r beans may be taken from a heap of $r + 2$ or more, leaving the remainder as two non-empty heaps.
$\mathbf{5 = 2^2 + 2^0}$	r consecutive beans may be taken from a row, if this leaves just 0 or 2 rows.	A heap of r beans may be taken, or r beans may be taken from a heap of $\geqslant r + 2$, leaving the rest as two nonempty heaps.
$\mathbf{6 = 2^2 + 2^1}$	r consecutive beans may be taken from a longer row, leaving 1 or 2 rows.	r beans may be taken from a larger heap, with the rest left as 1 or 2 heaps.
$\mathbf{7 = 2^2 + 2^1}$ $\mathbf{+ 2^0}$	r beans may be taken in any of these circumstances, (leaving 0, 1 or 2 rows).	(leaving 0, 1 or 2 heaps).

TABLE 1

Meaning of the octal code digit $\mathbf{d}_r$.

The game of Twopins was discovered by Elwyn Berlekamp in the course of his ingenious analysis [3, Chapter 16] of the well-known paper-and-pencil game, Dots-and-Boxes, or Dots-and-Squares [10]. It contains, as special cases, the games of Kayles [8, 19, 9] and Dawson's Kayles [6, 7], which we'll soon describe and whose analyses are already known. In fact, Guy and Smith [14] investigated a large class of "take and break" games, played with rows or heaps of beans. These may be called *octal games* because the rules can be described by a code name in the scale of eight:

$$d_0 \cdot d_1 d_2 d_3 \ldots$$

where $d_0 = 0$ or 4 (split a row or heap into two nonempty rows or heaps without removing any beans) and, $0 \leq d_r \leq 7$ for $r \geq 1$; the meaning of the digits is given in Table 1. For example, the code name for the game called *Kayles* by Dudeney [8] and Rip Van Winkle's Game by Loyd [19, 9] is $0 \cdot 77$. It is the special case of Twopins where every column contains two pins, so that the rules can be concisely stated as: take 1 or 2 adjacent columns.

Analysis of octal games was first prompted by a problem proposed by T. R. Dawson, the fairy chess expert [6, 7]. We call it *Dawson's Chess*. It is played on a chessboard with 3 ranks and n files (Figure 2). White and Black pawns occupy the first and third ranks, respectively, and the game is "losing chess' in that the capture is obligatory and the last player loses. Those who know how pawns move and capture will soon see that pairs of pawns become blocked on a file after any pawns in the neighboring files have been swapped. So Dawson's Chess may be played with a row of beans, with the option to take any bean, provided that its immediate neighbors, if any, are removed at the same time. You can check that in octal code, this is the game $0 \cdot 137$.)

The game as Dawson originally proposed it is in *misère form*; that is, the last player *loses*. The analysis of misère games is inordinately more complicated than the *normal form*, where the last player wins. Because misère Nim involves only a small change of strategy near the end, people have often been deceived into thinking that strategies for other impartial games can be similarly modified. For the vast majority of them this is not

FIGURE 2
Ready for a game of Dawson's chess.

true. (See Grundy and Smith [13] or Conway [5, chapter 12], who give an analysis of the first few positions of the misère forms of Dawson's Chess (in the form 0·4 and of Kayles [p. 145]. More extensive analyses will be found in Chapter 16 of [2].)

It is not difficult to show [14] that the games 0·137, 0·07 and 0·4 are closely related. We call the form 0·07 *Dawson's Kayles*. It may be played with a row of beans; a move is defined as the taking of two adjacent beans. Thus, it is the special case of Twopins in which each column contains a single pin. The nim-values for Kayles and Dawson's Kayles, played with a row of n beans, were found [14] to be periodic, apart from some irregular values for small values of n, with periods 12 and 34, respectively.

It is unrealistic to ask for a complete analysis of Twopins, since its positions are too various. How many essentially different positions are there with n columns? Because there are just two kinds of columns, the simple answer is 2^n positions, but we do not need to investigate all of these because Berlekamp has already pointed out various equivalences between positions, which you may easily verify:

1 $0*ijk \cdots = *ijk \cdots = 00ijk \cdots,$

2 $\cdots ijk*0*lmn \cdots = \cdots ijk* + *lmn \cdots,$

3 $\cdots ijk*00*lmn \cdots = \cdots ijk***lmn \cdots,$

where 0 represents a column with one pin (on its own it can be removed by neither player and is the Endgame); * represents a column with two pins (which may be removed by either player, therefore it is equivalent to a nim-heap of 1); the game star, [5, p. 72] and letters represent columns of either sort; and the sum sign on the right of equation **2** is the disjunctive sum we've already described.

We need to analyze then, only those Twopins positions which have a star at either end, as in equation **1**, and in which 0's (columns of 1 pin) do not occur except in blocks of at least three, as in equations **2** and **3**. Binary sequences of this kind were enumerated by Austin and Guy [1], who had 0 and 1 in place of our * and 0. The relevant number, t_n, of such Twopins positions is $a_{n-2}^{(3)}$ in their notation and the difference of 2 in rank is due to the 2 stars at either end of the row. The quantity t_n satisfies the recurrence

$$t_n = 2t_{n-1} - t_{n-2} + t_{n-4}.$$

In fact

$$t_n = \left(\frac{1}{2}\right)f_n + \left(\frac{1}{\sqrt{3}}\right)\sin\left(\frac{n\pi}{3}\right)$$

where f_n is the Fibonacci number

4 $$f_n = \frac{1}{\sqrt{5}}\left\{\left(\frac{1+\sqrt{5}}{2}\right)^n - \left(\frac{1-\sqrt{5}}{2}\right)^n\right\}.$$

However, it is unnecessary to analyze positions which are mere reflections of those already analyzed, so we next ask for the number of *symmetrical* positions, s_n.

The center of a symmetrical position is of one of the 4 types, A, B, C, D, shown on the left of Figure 3, if n is odd, where ? denotes either 0 or $*$. If n is even, replace the central symbol by a pair of equal ones. The central symbol (if n is odd) may be replaced by

(a) $* * *$, (b) $0 * 0$, or (c) $0\ 0\ 0$

to yield symmetrical positions with two more columns, except that (a) may

$$A \qquad \cdots * \,|*|\, * \cdots$$

$$\left\{ \begin{array}{l} \cdots * |* * *| * \cdots \qquad A \\ \cdots * |0\ 0\ 0| * \cdots \qquad B \end{array} \right.$$

$$B \qquad \cdots * 0 |0| 0 * \cdots$$

$$\cdots * 0 |0\ 0\ 0| 0 * \cdots \qquad C$$

$$C \qquad \cdots * 0\ 0 |0| 0\ 0 * \cdots$$

$$\cdots * 0\ 0 |0\ ?\ 0| 0\ 0 * \cdots \qquad D$$

$$D \qquad \cdots 0\ 0\ 0 |?| 0\ 0\ 0 \cdots$$

$$\left\{ \begin{array}{l} \cdots 0\ 0\ 0 |0\ ?\ 0| 0\ 0\ 0 \cdots \qquad D \\ \cdots 0\ 0\ 0 |* * *| 0\ 0\ 0 \cdots \qquad A \end{array} \right.$$

FIGURE 3

The four types of center for a symmetrical position.

not be used in cases B and C, and (b) may not be used in A or B. If n is even, replaces the central pair of symbols by

(a) $* * * *$, (b) $0 * * 0$, or (c) $0\ 0\ 0\ 0$.

Let A_n denote the number of symmetrical n-column positions of type A, etc., so that

$$A_n = A_{n-2} + D_{n-2},$$
$$B_n = A_{n-2},$$
$$C_n = B_{n-2},$$
$$D_n = C_{n-2} + D_{n-2},$$

and insert a coefficient 2 to allow for the ambiguity in D,

$$s_n = A_n + B_n + C_n + 2D_n$$
$$= (A_{n-2} + B_{n-2} + C_{n-2} + 2D_{n-2}) + (A_{n-2} + C_{n-2} + D_{n-2})$$
$$= (A_{n-2} + B_{n-2} + C_{n-2} + 2D_{n-2}) +$$
$$(A_{n-4} + B_{n-4} + C_{n-4} + 2D_{n-4})$$

Thus, s_n satisfies the recurrence

5
$$s_n = s_{n-2} + s_{n-4}$$

and has value

$$s_n = f_{\lfloor(n+1)/2\rfloor}$$

where $\lfloor \ \rfloor$ is the floor function (greatest integer not greater than) and f is the Fibonacci number **4**.

Therefore the number, u_n, of *un*symmetrical Twopins positions, not counting reflections as distinct, is

$$u_n = \frac{1}{2}(t_n - s_n) = \frac{1}{4}f_n - \frac{1}{2}f_{\lfloor(n+1)/2\rfloor} + \frac{1}{2\sqrt{3}}\sin\frac{n\pi}{3},$$

and the total number, not counting reflections, is

$$v_n = \frac{1}{2}(t_n + s_n) = \frac{1}{4}f_n + \frac{1}{2}f_{\lfloor(n+1)/2\rfloor} + \frac{1}{2\sqrt{3}}\sin\frac{n\pi}{3}.$$

The more general case

$$t_n^{(k)} = a_{n-2}^{(k)}$$

was discussed in [1], in which 0 occurs in blocks of length at least k. Here we extend the analysis to obtain the corresponding sequences $s_n^{(k)}$, $u_n^{(k)}$ and $v_n^{(k)}$ for general k. The formulae are generally true for $k \geqslant 1$, but for $k = 1$ no restriction is implied (apart from the requirement of $*$ at each end) and it is easy to see that for $n \geqslant 2$ (and $k = 1$),

$$t_n = 2^{n-2}, \ s_n = 2^{\lfloor(n-1)/2\rfloor}, \ u_n = 2^{n-3} - 2^{\lfloor(n-3)/2\rfloor}, \ v_n = 2^{n-3} + 2^{\lfloor(n-3)/2\rfloor}$$

From now on, we will omit the superscripts (k).

First, we use the fact [1] that

6
$$t_m = 2t_{m-1} - t_{m-2} + t_{m-k-1},$$

so that

$$t_m - t_{m-1} = t_{m-1} - t_{m-2} + t_{m-k-1}$$
$$= t_{m-2} - t_{m-3} + t_{m-k-1} + t_{m-k-2}$$
$$\vdots$$
$$= t_{k+1} - t_k + t_{m-k-1} + t_{m-k-2} + \cdots + t_2 + t_1$$

and since $t_1 = t_2 = \ldots = t_k = t_{k+1} = 1$, we have

7
$$t_m - t_{m-1} = \sum_{i=1}^{m-k-1} t_i,$$

a convenient algorithm for calculating $\{t_m\}$. We may also sum this formula to obtain

8 $$t_m = 1 + \sum_{i=1}^{m-k-1} (m - k - i)t_i.$$

Formulas **7** and **8** were not given in [1].

We next establish formula **5** in the more general form

9 $$s_n = s_{n-2} + s_{n-k-1}.$$

CASE A $k = 2l - 1$ odd, $n = 2m - 1$ or $2m$.

←——— $2m-1$ ———→	←——— $2m$ ———→
* ⋯ ⋯ ⋯ * ⋯ ⋯ ⋯ *	* ⋯ ⋯ ⋯ * * ⋯ ⋯ ⋯ *
←—— m ——→	
* ⋯ * 0 0 0 0 0 * ⋯ *	* ⋯ * 0 0 0 0 0 0 * ⋯ *
← $m-l$ → ←$k=2l-1$→	← $m-l$ → ←$k+1=2l$ +
* ⋯ * 0 0 0 0 0 0 0 * ⋯ *	* ⋯ * 0 0 0 0 0 0 0 0 * ⋯ *
← $m-l-1$ ←— $2l+1$ —→	← $m-l-1$ → ←—$2l+2$—→
⋯ ⋯ . ⋯ ⋯ . ⋯	⋯ ⋯ ⋯
* 0 0 0 ⋯ ⋯ ⋯ 0 0 0 *	* 0 0 0 ⋯ ⋯ ⋯ 0 0 0 *

FIGURE 4

Symmetrical Twopins positions with 0's in blocks of $\geqslant k$.

From Figure 4 we see that the number of symmetrical positions is

$$s_n = s_{2m-1} = s_{2m} = t_m + t_{m-l} + t_{m-l-1} + \cdots + t_1$$

$$s_{n-2} = s_{2m-3} = s_{2m-2} = t_{m-1} + t_{m-l-1} + t_{m-l-2} + \cdots + t_1$$

$$s_n - s_{n-2} = t_m - t_{m-1} + t_{m-l}$$

$$= t_{m-l} + \sum_{i=1}^{m-2l} t_i.$$

by **7**, so that $s_n - s_{n-2} = s_{2(m-l)-1} = s_{2(m-l)} = s_{n-k-1}$, as required.

CASE B $k = 2l$ (even), is similar to Case A, but we have to treat $n = 2m$ and $n = 2m - 1$ separately:

$$s_{2m} = t_m + t_{m-l} + t_{m-l-1} + \cdots + t_1,$$

$$s_{2m-1} = t_m + t_{m-l-1} + t_{m-l-2} + \cdots + t_1,$$

$$s_{2m-2} = t_{m-1} + t_{m-l-1} + t_{m-l-2} + \cdots + t_1,$$

$$s_{2m-3} = t_{m-1} + t_{m-l-2} + t_{m-l-3} + \cdots + t_1,$$

$$s_{2m} - s_{2m-2} = t_m - t_{m-1} + t_{m-l},$$

$$s_{2m-l} - s_{2m-3} = t_m - t_{m-1} + t_{m-l-1},$$

and we obtain **9** in either case by the use of **7**, as before.

The generating functions for t_n and s_n are

$$T(z,k) = \sum_{i=0}^{\infty} t_i^{(k)} z^i = \frac{z(1-z)}{(1-z)^2 - z^{k+1}}, \quad S(z,k) = \sum_{i=0}^{\infty} s_i^{(k)} z^i$$

$$= \frac{z(1+z)}{1 - z^2 - z^{k+1}}.$$

Formulae **7** and **9**, together with

$$u_n = \frac{1}{2}(t_n - s_n), \quad v_n = \frac{1}{2}(t_n + s_n)$$

enable us to calculate the values in Table 2, where dots indicate that the sequences are constant for earlier positive values of n.

Of these sequences, the only ones to appear in Sloane's Handbook [20] are

the powers of two, $t_n^{(1)} = 2^{n-2}$,
the Fibonacci numbers, $s_n^{(3)}$, and
sequence #102, $s_n^{(2)}$.

This last appeared in [11] as an example of a sum, having taken over the generalized diagonals

$$3x + 2y = n - 1$$

of entries in Pascal's triangle. It is also given in [16,17,18] with factorizations and a discussion of divisibility properties. For example,

$s_n^{(2)}$ is even just if $n = 7m - 3$, $7m - 2$ or $7m$.
The highest power of 2 which divides $s_{7m-3}^{(2)}$ is the "ruler function"; the highest power of 2 in $2m$.
3 divides $s_n^{(2)}$ just if $n = 13m - 3$, $13m - 2$, $13m$ or $13m + 6$.

We have however, wandered away from the game of Twopins. What is the best hit to make in Figure 1? Berlekamp's equivalence equation **1** tells us that column h can be ignored, and equation **2** that we can remove e without affecting the position. Equivalence equation **3** then enables us to put b and c together and the position is

$$*{*}* + *{*}.$$

n	...3	4	5	6	7	8	9	10	11	12	13	14	15	16	17	18	19
t_n	...1	2	4	7	12	21	37	65	114	200	351	616	1081	1897	3329	5842	10252
s_n	...1	2	2	3	4	5	7	9	12	16	21	28	37	49	65	86	114
u_n	...0	0	1	2	4	8	15	28	51	92	165	294	522	924	1632	2878	5069
v_n	...1	2	3	5	8	13	22	37	63	108	186	322	559	973	1697	2964	5183

$k = 2$

20	21	22	23	24	25	26	27	28	29	30
17991	31572	55405	97229	170625	299426	525456	922111	1618192	2839729	4983377
151	200	265	351	465	616	816	1081	1432	1897	2513
8920	15686	27570	48439	85080	149405	262320	460515	808380	1418916	2490432
9071	15886	27835	48790	85545	150021	263136	461596	809812	1420813	2492945

	...4	5	6	7	8	9	10	11	12	13	14	15	16	17	18	19	20	21
	...1	2	4	7	11	17	27	44	72	117	189	305	493	798	1292	2091	3383	5473
	...1	2	2	3	3	5	5	8	8	13	13	21	21	34	34	55	55	89
	...0	0	1	2	4	6	11	18	32	52	88	142	236	382	629	1018	1664	2692
	...1	2	3	5	7	11	16	26	40	65	101	163	257	416	663	1073	1719	2781

$k = 3$

22	23	24	25	26	27	28	29	30	31	32
8855	14328	23184	37513	60697	98209	158905	257114	416020	673135	1089155
89	144	144	233	233	377	377	610	610	987	987
4383	7092	11520	18640	30232	48916	79264	128252	207705	336074	544084
4472	7236	11664	18873	30465	49293	79641	128862	208315	337061	545071

	...5	6	7	8	9	10	11	12	13	14	15	16	17	18	19	20	21	22	23
	...1	2	4	7	11	16	23	34	52	81	126	194	296	450	685	1046	1601	2452	3753
	...1	2	2	3	3	4	5	6	8	9	12	14	18	22	27	34	41	52	63
	...0	0	1	2	4	6	9	14	22	36	57	90	139	214	329	506	780	1200	1845
	...1	2	3	5	7	10	14	20	30	45	69	104	157	236	356	540	821	1252	1908

$k = 4$

24	25	26	27	28	29	30	31	32	33	34	35
5739	8771	13404	20489	31327	47904	73252	112004	171245	261813	400285	612009
79	97	120	149	183	228	280	348	429	531	657	811
2830	4337	6642	10170	15572	23838	36486	55828	85408	130641	199814	305599
2909	4434	6762	10319	15755	24066	36766	56176	85837	131172	200471	306410

TABLE 2 (continued on next 2 pages)

Values of $r_n^{(k)}$, $s_n^{(k)}$, $u_n^{(k)}$, $v_n^{(k)}$ for $k = 2, 3, \cdots, 9$.

k = 5

...6	7	8	9	10	11	12	13	14	15	16	17	18	19	20	21	22	23	24
...1	2	4	7	11	16	22	30	42	61	91	137	205	303	443	644	936	1365	1999
...1	2	2	3	3	4	4	6	6	9	9	13	13	19	19	28	28	41	41
...0	0	1	2	4	6	9	12	18	26	41	62	96	142	212	308	454	662	979
...1	2	3	5	7	10	13	18	24	35	50	75	109	161	231	336	482	703	1020

25	26	27	28	29	30	31	32	33	34	35	36	37
2936	4316	6340	9300	13625	19949	29209	42785	62701	91917	134758	197548	289547
60	60	88	88	129	129	189	189	277	277	406	406	595
1438	2128	3126	4606	6748	9910	14510	21298	31212	45820	67176	98571	134476
1498	2188	3214	4694	6877	10039	14699	21487	31489	46097	67582	98977	145071

k = 6

...7	8	9	10	11	12	13	14	15	16	17	18	19	20	21	22	23	24	25
...1	2	4	7	11	16	22	29	38	51	71	102	149	218	316	452	639	897	1257
...1	2	2	3	3	4	4	5	6	7	9	10	13	14	18	20	25	29	35
...0	0	1	2	4	6	9	12	16	22	31	46	68	102	149	216	307	434	611
...1	2	3	5	7	10	13	17	22	29	40	56	81	116	167	236	332	463	646

26	27	28	29	30	31	32	33	34	35	36	37	38
1766	2493	3536	5031	7165	10196	14484	20538	29085	41168	58282	82561	117036
42	49	60	69	85	98	120	140	169	200	238	285	336
862	1222	1738	2481	3540	5049	7182	10199	14458	20484	29022	41138	58350
904	1271	1798	2550	3625	5147	7302	10339	14627	20684	29260	41423	58686

k = 7

...8	9	10	11	12	13	14	15	16	17	18	19	20	21	22	23	24	25	26
...1	2	4	7	11	16	22	29	37	47	61	82	114	162	232	331	467	650	894
...1	2	2	3	3	4	4	5	5	7	7	10	10	14	14	19	19	26	26
...0	0	1	2	4	6	9	12	16	20	27	36	52	74	109	156	224	312	434
...1	2	3	5	7	10	13	17	21	27	34	46	62	88	123	175	243	338	460

27	28	29	30	31	32	33	34	35	36	37	38	39
1220	1660	2262	3096	4261	5893	8175	11351	15747	21803	30121	41535	57210
36	36	50	50	69	69	95	95	131	131	181	181	250
592	812	1106	1523	2096	2912	4040	5628	7808	10836	14970	20677	28480
628	848	1156	1573	2165	2981	4135	5723	7939	10967	15151	20858	28780

(continued)

	...9	10	11	12	13	14	15	16	17	18	19	20	21	22	23	24	25	26	27
	...1	2	4	7	11	16	22	29	37	46	57	72	94	127	176	247	347	484	667
$k=8$	...1	2	2	3	3	4	4	5	5	6	7	8	10	11	14	15	19	20	25
	...0	0	1	2	4	6	9	12	16	20	25	32	42	58	81	116	164	232	321
	...1	2	3	5	7	10	13	17	21	26	32	40	52	69	95	131	183	252	346

28	29	30	31	32	33	34	35	36	37	38	39	40
907	1219	1625	2158	2867	3823	5126	6913	9367	12728	17308	23513	31876
27	33	37	44	51	59	70	79	95	106	128	143	172
440	593	794	1057	1408	1882	2528	3417	4636	6311	8590	11685	15852
467	626	831	1101	1459	1941	2598	3496	4731	6417	8718	11828	16024

	...10	11	12	13	14	15	16	17	18	19	20	21	22	23	24	25	26	27	28
	...1	2	4	7	11	16	22	29	37	46	56	68	84	107	141	191	263	364	502
$k=9$	...1	2	2	3	3	4	4	5	5	6	6	8	8	11	11	15	15	20	20
	...0	0	1	2	4	6	9	12	16	20	25	30	38	48	65	88	124	172	241
	...1	2	3	5	7	10	13	17	21	26	31	38	46	59	76	103	139	192	261

29	30	31	32	33	34	35	36	37	38	39	40	41
686	926	1234	1626	2125	2765	3596	4690	6148	8108	10754	14326	19132
26	26	34	34	45	45	60	60	80	80	106	106	140
330	450	600	796	1040	1360	1768	2315	3034	4014	5324	7110	9496
356	476	634	830	1085	1405	1828	2375	3114	4094	5430	7216	9636

TABLE 2

(continued)

Even without knowing the nim-values, you can see that the (only) good moves are to take out column *d* or column *a*.

Figure 5 shows a Twopins-wheel which enables us to read off the nim-value of any Twopins position of eight or fewer columns of pins, provided that we know the nim-values for a row of *n* pins in Kayles or Dawson's Kayles (for Dawson's Chess, slide the nim-values one place to the left);

n	0	1	2	3	4	5	6	7	8	9	10	11	12
Kayles	0	1	2	3	1	4	3	2	1	4	2	6	4
Dawson's Kayles	0	0*	1	1	2	0	3	1	1	0	3	3	2

Suppose for example, you want the nim-value of

$$***000**.$$

FIGURE 5

A Twopins-wheel for finding the nim-values of small Twopins positions.

Find this arrangement in the outer ring (running from 12 o'clock to 3 o'clock), spiral in from the first and last stars, and meet in a cell containing the value 4. Thus is the nim-value.

What is the best move in the Dawson's Chess game in Figure 2? Our advice is to allow your opponent the privilege of the first move. It is a *P*-position (previous-player-winning) and has nim-value 0.

---

* Note that a single pin must remain standing in Dawson's Kayles.

References

1　Austin, Richard, and Guy, Richard. 1978. Binary sequences without isolated ones. *Fibonacci Quart.* 16.

2　Ball, W. W. Rouse, and Coxeter, H. S. M. 1974. *Mathematical Recreations and Essays.* 12th ed. Toronto: Univ. of Toronto Press. pp. 36–39.

3　Berlekamp, E. R.; Conway, J. H.; and Guy, R. K. 1980. *Winning Ways.* New York: Harcourt, Brace Jovanovich.

4　Bouton, Charles L. 1901–2. Nim, a game with a complete mathematical theory. *Ann. Math.* Princeton (2) 3: 35–39.

5　Conway, J. H. 1976. *On Numbers and Games.* New York: Academic Press.

6　Dawson, T. R. 1934. Problem 1603. *Fairy Chess Review.* p. 94.

7　———. 1935. Caissa's Wild Roses. *Fairy Chess Review.* p. 13.

8　Dudeney, H. E. 1958. *Canterbury Puzzles.* N.Y.: Dover. pp. 118–119, 220.

9　Gardner, Martin. 1960. *More Mathematical Puzzles of Sam Loyd.* N.Y.: Dover. pp. 5, 122.

10　———. 1974. Mathematical Games: Cram, crosscram and quadraphage: new games having elusive winning strategies, *Sci Amer.* 230 2: 106.

11　Green, Thomas M. 1968. Recurrent sequences and Pascal's triangle. *Math. Mag.* 41: 13–21.

12　Grundy, P. M. 1964. Mathematics and games. *Eureka.* 27: 9–11.

13　Grundy, P. M., and Smith, C. A. B. 1956. Disjunctive games with the last player losing. *Proc. Cambridge Philos. Soc.* 52: 527–533; M.R. 18: 546.

14　Guy, Richard K., and Smith, Cedric A. B. 1956. The *G*-values for various games. *Proc. Cambridge Philos. Soc.* 52: 514–526; M.R. 18: 546.

15　Hardy, G. H., and Wright, E. M. 1960. *An Introduction to the Theory of Numbers.* 4th ed. Oxford: Oxford Univ. Press. pp. 117–120.

16　Jarden, Dov. 1966. Recurring Sequences. *Riveon Lematematika* 2nd ed. 86–91.

17　———. 1946–47. Third order recurring sequences. *Riveon Lematematika* 1: 74; 1952–53 6: 41–42.

18　——— and Katz, A. 1947–48. Table of binary linear third order recurring sequences. *Riveon Lematematika* 2: pp. 54–55.

19　Loyd, Sam. 1914. *Cyclopedia of Tricks and Puzzles.* New York: Dover. p. 232.

20　Sloane, N. J. A. 1973. *A Handbook of Integer Sequences.* New York: Academic Press.

21　Sprague, R. P. 1935–36. Über mathematische Kampfspiele. *Tôhoku Math. J.* 41: 438–444; Zbl. 13: 290.

Pretzel Solitaire as a Pastime for the Lonely Mathematician

N. G. de Bruijn

EINDHOVEN UNIVERSITY OF TECHNOLOGY

In general, playing solitaire card games is a rather sad form of amusing oneself. Starting from a random position of cards on the table, arranged in some predefined format, we have to try to find a sequence of legitimate moves that turn the position into some predescribed goal. The moves are usually irreversible, and therefore *dead* positions can result where no further move is possible. If we are fortunate enough to reach the goal, we get some feeling of satisfaction. If we run into a dead end, however, we usually feel sad. The reason for distress is not that luck was against us, but that we wonder whether the failure had been unavoidable. We are sad because we feel as though we may have spoiled the game. More often than not, we have forgotten the original position, and now it is impossible to retrace our steps. Failing to solve a solitaire problem is particularly humiliating if it is a game with complete information; that is, all of the cards are facing up.

1.   It is not in a mathematician's nature to be proud of having been lucky enough to solve a problem more or less by accident. We want to have some method or other by which we can find a solution if there is any. And if there is no solution, we want to be able to prove that there is no solution.

What the mathematician aims for is to be able to study a given position of the cards, and to produce without touching a card, either a solution or a proof that no solution exists. It is clear that this can only be done with solitaire games that are not too difficult. The game we suggest for this purpose of study has two integer parameters, which the player may fix according to his taste and ability, in order to get something that is neither too difficult nor too simple. We shall call it the $k \times n$ *pretzel*. The game is old, certainly for the case $k = 4$, $n = 13$ (see section 5 for a description of how the 4×13 is usually played), and it is difficult to trace its origin. The name is new, and was chosen because a pretzel is a kind of biscuit in the shape of a knot.

2.   For the $k \times n$ pretzel, we take a deck of cards consisting of k suits of n cards each. To make things easier, let's say $k = 4$, and we'll call the suits S, H, D, C (for spades, hearts, diamonds and clubs). The cards in each suit will be numbered from 1 to n (the 1 is usually called "ace"), and our notation will be $S1, \ldots, Sn, \ldots, C1, \ldots, Cn$.

After shuffling the deck, we put the cards face up on the table in a rectangle with 4 rows and n columns. Next we take away the aces and put them in a new column in front of the rectangle, in the order S1, H1, D1, C1. We get a $k \times (n + 1)$ rectangle with 4 empty spaces which we call *holes*; each hole is the size of a card. We still call them holes even when 4 of them are in the last column. Let us imagine a $k \times (n + 1)$ frame around the cards in order to distinguish between the holes and the infinite outside world. In the course of the game we are never allowed to put cards outside the frame. As an example of a position in the 4×4 pretzel, we display

1

♠1	—	♥4	♥2	—
♥1	—	♥3	♣2	♣4
♦1	—	♦4	♠3	♠2
♣1	♦2	♦3	♣3	♠4

Now we start playing. A move consists of putting a card in a hole provided that the next lower card (in the same suit) is lying directly to the left of the hole. So in the position **1** any one of the cards $S2, H3, H2, D2$ may be used for a first move. With the first move we have filled one hole and obtained a new one. This new hole allows us to make our second move. We want to reach this final position;

2

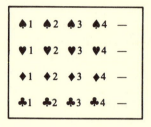

♠1	♠2	♠3	♠4	—
♥1	♥2	♥3	♥4	—
♦1	♦2	♦3	♦4	—
♣1	♣2	♣3	♣4	—

We shall call this position *the goal*.

A position is called *solvable* if a sequence of moves exists that leads to the goal (the goal itself is also called solvable). If there is no such sequence, we will term it unsolvable. A position is called *dead* if it is different from the goal and does not allow a single move. Dead positions are unsolvable, of course.

The rules of the game were explained here for the case $k = 4$, $n = 4$, but for all other cases we have exactly the same rules.

3.   The position **1** is solvable; for example, by the following sequence of moves (each move is indicated by the name of the moving card)

3   S2,S4,C4,C3,D4,D2,D3,C2,H4,H2,S3,S4,D4,C3,C4.

Let us now first change the notation. The columns in **1** do not play any role in our game, so we may just as well print the position in a single line, with separation marks:

S1—H4H2—*H1—H3C2C4*D1—D4S3S2*C1D2D3C3S4*

4.   The 4×2 and the 4×3 pretzel are very dull games, but the 4×4 pretzel can be quite interesting. The 4×5, 4×6 are usually still playable (in the sense explained in section **2**).

For larger values of n, the author does not have enough experience to say whether an average player will be able to settle the majority of cases without touching the cards, and without using pencil and paper and/or computer.

5.   Let $p(k,n)$ be the probability that a random $k \times n$ pretzel position is solvable. It seems to be difficult to determine $p(k,n)$ exactly, unless $n \leqslant 2$. If $n = 2$, the only unsolvable positions are those with a *cyclic blockade*. In a cyclic blockade we have a cyclic arrangement of a subset of the suits, for instance, $H \rightarrow D \rightarrow S \rightarrow H$, such that (D1 H2), (S1 D2) and (H1 S2) are pairs of neighbors. If $k = 4$ and $n = 2$, there are 40 such cases with a 3-cycle, 168 with one 2-cycle, and 6 with 2-cycles—214 unsolvable positions altogether. The total number of positions is 1680.

It can be estimated that $p(4,4)$ is about 0·45. The author has played several hundred 4×4 pretzels at one time or another. In all cases he was able to decide whether or not the position was solvable (although he often deviated from the no-touching rule by allowing "safe moves"—see section 14 below). Moreover, a computer search dealt with 2473 random positions, of which 1123 were solved and the remaining 1350 were established to be unsolvable.

The author has less experience with $n = 5$ and $n = 6$. A very rough guess would be that $p(4,5)$ is of the order of $0 \cdot 1$.

The 4×13 pretzel still seems to have a reasonable number of solvable cases. Usually it is not played with the expectation of reaching the goal but rather to get as far as possible. The player tries to get a long sequence $S1, S2, \ldots, Sp$ in the first row, and similarly $H1, \ldots, Hq$, $D1, \ldots, Dr$, $C1, \ldots, Cs$. When the board is dead, he takes away all the cards that do not belong to these sequences, shuffles the deck, and fills the rectangle again, leaving a hole directly to the right of Sp, Hq, Dr, Cs. Now the game starts over. After this second round, a third round may follow, and the player is considered to be successful if he reaches the goal position in the third round.

Yet it happens now and then that the goal position is reached in the first round. Let us say that it happens in approximately one percent of all cases. This only gives a lower estimate for $p(4,13)$, since nobody knows how many solvable positions have been spoiled. From this kind of experience one may guess that $p(4,n)$ is not exponentially small if n tends to infinity.

6.   In section 5 we mentioned that the 4×4 pretzel game was run on a computer. The program was based on studying the graph of the game. The points of that graph are all possible positions, and there is an oriented edge from P to Q if we can get from position P to position Q by a single move. If P is a position, then by $S(P)$ we denote the set of all points that can be reached by a finite sequence of moves if we start at P. The question is whether the goal position P_0 belongs to $S(P)$.

The set $S(P)$ is, for any given P, easily built up in a computer's memory by trying to find new points obtainable from old points in a single move. The execution is stopped if no further points of $S(P)$ are found (whence P is unsolvable) and also if P_0 is reached. In that case P is solvable, and then we need not know the full $S(P)$.

If P is unsolvable, then the number of elements in $S(P)$ is rarely more than 60; however, the largest number ever found was 380. With solvable P, the number of points inspected before P_0 was reached, was of the order of 150, and the largest number ever found was 802. The size of $S(P)$ is not very interesting, however. If it is very big or very small, it is usually quite easy to calculate P by mere thought.

In the pretzel there can be many different sequences of moves leading from one position to another, because of the possibility of interchanging moves or sequences of moves that do not interfere with each other. This means that it would be inappropriate to run a computer program based on the backtracking technique applied to sequences of moves.

7.   With the 4×4 pretzel and arbitrary solvable starting position P, it is usually not too difficult to find a solution mentally although one may occasionally hit on tricky cases. But if the starting position is unsolvable, how do we show it? Determining the full $S(A)$ (of Section 6) is usually hopeless for a non-computerized brain. But there are many other ways to prove unsolvability.

One possibility is cyclic blockade. In its simplest form, it was mentioned in Section 5. Often it is more complicated, like the following example. We fix our attention on a certain card c_1. Then we show that, if it ever moves, a card c_2 must have moved before c_1 moves for the first time. Then we note that a card c_3 cannot move before c_2 moves for the first time. So in any solution there is a moment where c_2 moves for the first time, and c_1 and c_3 have not moved yet. Next we find card c_4 that has to be moved before c_3 moves for the first time, etc. In the course of this argument we have a growing set of cards that have not moved yet at the moment we have in mind, and that strengthens further argumentation. If sooner or later we run into a loop, observing that c_1 must have moved before all the others, we have proved that the original position was unsolvable.

The mathematician who plays the game will no doubt discover a bag of additional tricks which augment his proving ability.

Such tricks are not always easy to put into words. Often they are a mixture of forward and backward analysis. Backward analysis is what we just explained when discussing cyclic blockade. Forward analysis involves seeing what happens to all possible first moves in the starting position: if from P we can get in a single move to either P_1, P_2, P_3 or P_4, and if P_1, P_2, P_3, P_4 are all unsolvable, then P is unsolvable. We have to be careful not to mix forward and backward arguments on a large scale, since one can easily get to an impossibility proof based on wishful thinking, which is worse than having no proof at all. Usually such a proof can be safeguarded if we replace it by something based on the truncation theorem of section 8.

Failure to establish a solution may help to construct a proof of unsolvability, and failure to construct such a proof may help to find a solution. In short; it is in the spirit of the working mathematician!

8.   If p is an integer, $1 \leqslant p \leqslant n$, then $E_S(p)$ is the set of spades $\leqslant p$ (that is, $\{S1, S2, \ldots, S_p\}$). Similarly we define E_H, E_D, E_C and we put

$$F(p,q,r,s) = E_S(p) \cup E_H(q) \cup E_D(r) \cup E_C(s).$$

Such an $F(p,q,r,s)$ will be called a *trunk*; if $k \neq 4$ the definition is similar.

If P is a position in the $k \times n$ pretzel and if F is a trunk, then we shall call "P *truncated* by F" the position we get by taking away from P every card not belonging to F. This is no longer a position in the sense of Section 2 (unless $F = F(n,n,n,n)$) yet we can play the game on truncated positions, keeping the definition of a move the same as in Section 2. We stipulate that whatever F is, the truncation of the $k \times n$ pretzel still has to be played in the original $(k + 1) \times n$ frame (therefore the $k \times n$ pretzel is not the same game as the $k \times (n + 1)$ pretzel truncated by $F(n,n,n,n)$). We take P_O truncated by F as the goal in the truncated pretzel, where P_O is the goal in the original pretzel. Also, a truncated position is called solvable if there is a sequence of moves leading to the new goal.

Let P_1 and P_2 be positions (or truncated positions) such that a single move leads from P_1 to P_2. Let F be a trunk, and let P_i^* be P_i truncated by F. If the moving card belongs to F then the same move can be used to transform P_1^* into P_2^*. If it does not belong to F we have $P_1^* = P_2^*$. Applying this to a sequence of moves, with a fixed F, we get a theorem that can serve as a necessary criteria for solvability:

THEOREM   If P is solvable, then every truncation of P is solvable.

9.   The simplest applications of the truncation theorem of Section 8 are those where some truncation of a position is dead. For example,

S1D4S3D2H2∗H1D3—H4—∗D1H3C2S2—∗C1C4S4C3—∗

is unsolvable since truncation by $F(1,3,3,1)$ leads to

S1——D2H2∗H1D3———∗D1H3———∗C1———∗

which does not admit a single move.

In more complicated cases we show that the truncated position is unsolvable but not dead. Or we show that every move possible from our starting position leads to a position for which we can select an unsolvable truncation, etc.

10.   A move is called *safe* unless it turns a solvable position into an unsolvable one. In an unsolvable position all moves are safe, but in many solvable positions we can make unsafe moves.

Quite often we can prove a move to be safe. Consider a position P where a card c lies at a place p. Moving c from p to a place q is certainly safe if we have both **1** and **2**:

1   The place directly to the right of p is either outside the frame, that place is occupied by the next higher card in the same suit, or there is no card higher in that suit.

2   By no sequence of moves leading from P to the goal P_0, can place q ever be occupied by a card other than c.

The purpose of condition **1** is to guarantee that there is no further use for card c at place q; condition **2** guarantees that c cannot do any harm at place q.

As to condition **2** we note that it is often easy to say that certain cards can never get to certain places. For example, if we have a row S1H3C2-D4S3D2D3 then we can say, by inspecting the places on the left, that the only possible candidates for the place now occupied by D2 are S6,H7,C5,D6 and S4.

11. Another source of safe moves is the following. If in a sequence of moves the only moving cards are diamonds, and if at the end of the sequence all diamonds are in their proper places (that is, the places they have in the goal position), then the sequence consists of safe moves. This statement remains true if both occurrences of the word "diamond" are replaced by "diamonds and spades," or by "diamonds, spades and clubs."

These statements are easily deduced from the truncation theorem.

12. If in a position P there is only one possible move, then that move is obviously safe.

Now assume that we are in a position P with just two possible moves, with cards c_1 and c_2, and assume that c_2 is neither the next higher nor the next lower to c_1 in the same suit. (This guarantees that the moves do not interfere: after moving c_1 the old move for c_2 is still possible, and vice versa). Moreover we assume that c_1 transforms P into a position in which c_2 is the only available move. Then we can say that c_2 is a safe move for P.

This simple idea for proving safety can be generalized to more than two noninterfering moves and to noninterfering sequences of moves. The lonely mathematician will no doubt discover many arguments of this kind.

We finally mention that in any position, a move can be proved to be safe by showing that all other moves are bad (that is, lead to unsolvable positions).

13. In Section 1 we said that the game should be analyzed without touching a card. This may sometimes be frustrating, and therefore it is not unreasonable to modify the rules by allowing the player to carry out moves as long as he can prove that they are safe.

A further mnemonic aid the player might allow himself is connected with the kind of blockade analysis mentioned in Section 7. Analyzing what happens when a certain card moves for the first time, the player might put a penny on top of each card that he knows he's been unable to move until that moment.

14. Thus far our pretzel game satisfied these conditions:

1 Every suit has the same number of cards.

2 In the frame (see (1)), all rows have the same length.

3 The number of holes equals the number of suits.

We can ignore the above three statements and still have a game for which almost everything holds true that was discussed in the previous sections. As an example we present

$$S1D4D3—S3S5*H1H2S4D2*D1—S2H3*$$

with solution

$$D2,S5,S4,H3,D4,S2,D3,S3,D4,S4,S5.$$

15.   We present a number of exercises, taken from a collection of random positions. The exercises marked † are quite difficult: the author's bag of tricks does not seem to contain nice arguments. Cases **1–9** are 4 × 4's, **10–16** are 4 × 5's. The solutions are listed below.

1   S1D2—D3S3*H1—C2H2D4*D1—S4C3H4*C1S2H3
—C4*

2   S1—H2H3S2*H1H4S4—D2*D1—C3S3C2*C1D4D3
C4—*

3   S1C3—C4H4*H1—D4H2C2*D1D3S4S2H3*C1—D2
—S3*

4   S1S2D4C3D3*H1C4—H3H4*D1S4C2H2—*C1—D2
—S3*

5   S1C4C2D3—*H1H2H3H4—*D1S4S3C3D4*C1D2
—S2—*

6   S1C4H4—H2*H1D3D4C3S4*D1S3—H3—*C1S2
—C2D2*

7†   S1D3C2—H2*H1S3D4D2—*D1H4C3H3C4*C1
—S4S2—*

8   S1C2C3S2—*H1——D2H4*D1D3H3S4H2*C1C4S3
—D4*

9   S1D4S4S2H4*H1C4D3H2C2*D1—C3—H3*C1S3
—D2—*

10   S1D4C4D3—C5*H1—D5—H5H2*D1C2S3H4H3S4*
C1S2S5—C3D2*

11   S1S5D2H2—D4*H1H4S4S3—D5*D1H5C5C2D3
—*C1S2—C3H3C4*

12   S1C4D5D3S5C2*H1—D4—C3S2*D1C5H5H3—H4*
C1D2H2S4—S3*

13   S1—C4H4C5D4*H1D3C2S2H2H5*D1D5—D2
—H3*C1C3S3S5—S4*

14†   S1S5S2D5H3D4*H1—C2D3S3H4*D1
—H2C4H5C3*C1C5—S4D2—*

15 S1C4—D2C5S4∗H1S5D5H5D4D3∗D1H2S2H4—C3∗
C1S3H3—C2—∗

16 S1C4H2D4—D2∗H1C3C5D3S5H5∗D1—H4S4—C2∗
C1—D5H3S3S2∗

Solutions

1 Solvable: D3, D4, H3, S3, S4, D2, D3, H2, S2, S3, C2, C3, D4, S4, C4, H3, H4.

2 Unsolvable. Truncation by $F(1,1,3,3)$, then dead in two moves.

3 Unsolvable. Truncate by $F(4,1,3,4)$. Then D3 is a safe move. Next D2, C2, C3, S2, S3 is a sequence of safe moves leading to a dead position.

4 Unsolvable. Truncate to $F(4,1,4,1)$, then dead in one move.

5 Solvable: D4, C4, D3, C3, S4, S3, D2, D3, C2, C3, S2, S3, S4, C4, D4.

6 Unsolvable: H3 and D4 are blocking each other.

7 Unsolvable.

8 Unsolvable. Truncate by $F(2,1,1,4)$.

9 Solvable: C4, H2, D4, S2, S4, S3, S4, D2, C2, C3, C4, D3, D4, H3, H4.

10 Unsolvable. Move H2 is safe, the following 14 moves are unique and lead to a dead board.

11 Solvable: D4, H3, H4, H2, C4, C5, D3, C3, S3, S4, H3, H4, H5, D4, C4, D2, D3, S5, S2, S3, S4, C2, D4, S5, C3, C4, C5, D5.

12 Unsolvable. After truncation by $F(4,1,1,4)$ a dead board results in 2 moves.

13 Unsolvable. Move S2 is safe because move D3 leads to an unsolvable position (truncation by $F(1,1,5,1)$). For the same reason C3, C2, D4 are safe. The following 13 moves are unique (apart from interchanging the 10th and 11th) and lead to a dead board.

14 Unsolvable.

15 Solvable: H4, C3, S3, S4, S5, C5, C2, H2, D3, D5, H3, C3, H5, H4, D4, H5, C4, C5, S2, S3, D2, D3, D4, S4, S5, D5.

16 Solvable: D2, C2, S5, D4, H3, H4, H5, D3, D5, C3, C4, H2, S2, S3, C5, H3, H4, S4, S5, D4, D5, H5.

Some Remarks about a Hex Problem

Claude Berge
University of Paris

The readers of Gardner's rubric (see *Scientific American*: July 1957; October 1957; July 1975; December 1975) know the depth and the beauty of the game of Hex, launched almost simultaneously by Piet Hein in Denmark and by John Nash in America. This game is especially appealing to mathematicians. Two players play alternately by putting a peg in an empty hole of the lozenge board shown in Figure 1. The first player uses black pegs, and wins if he manages to construct a chain of black pegs connecting East to West; his opponent tries to construct a chain of white pegs connecting North to South. The best size for a board seems to be 14 × 14, and because a proof by contradiction shows that the first player has a winning strategy, his opponent can require him to put the first peg in a restricted area of the board. Of course, all kinds of handicaps can be used if the two players are not of equal strength, and the aesthetic interest of the game is nearly unchanged.

I will pose a Hex problem that I wish to present in dedication to Martin Gardner. Desirable attributes of a "problem" may be, for example: the first move is unique; or the first move is paradoxical, or the underlying idea is unexpected, etc. (We recall Sam Loyd, the greatest chess problemist of all time, who claimed that his main goal was to compose a problem where the first move is the opposite of what 999 players out of 1000 would propose.)

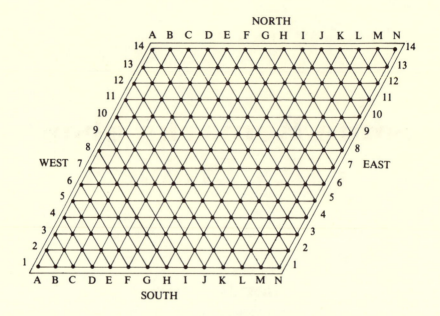

FIGURE 1

Empty lozenge board.

Here, to avoid unnecessary complications and to permit a few intermediate moves, we shall not adopt the criterion of uniqueness of the key move, and we shall not require that the diagram have the same number of black pegs and white pegs, because a feasible position may come from a play with handicaps. The problem to be posed is not meant to be difficult by the complexity or the number of variations, because White can easily complete a connecting chain between North and South playing either J9 or M8: since Black cannot play simultaneously on these two holes, it appears that he is about to lose the game. The problem is for *Black to play and win*!

A simple analysis of the position shows that the black pegs can be split into two separate groups which are connected to the West side, but not to the East side; in fact, these two groups are completely surrounded by two *disjoint* walls of white pegs. The first is:

A11, B10, B11–C10, C11, D11–D10, E10, E11–F10, F11, G11–G10, H10, H11–I10, I11, J11–J10, K10, K11–L10, L11, L12–M11, M12, N12, N13, N14.

and the second is:

A8, B8, C8, C9–D8, D9, E8–E9, F8, F9–G8, G9, H9–H8, I8, I9–J8, J9, K9–K8, L8, L7, L6–M6, M5, N4, N3, N2, N1.

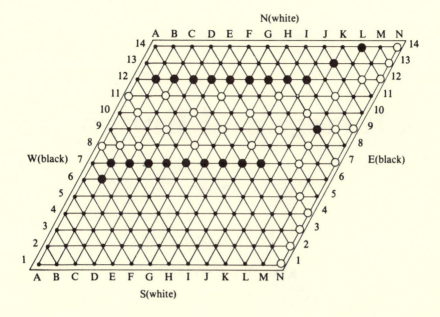

FIGURE 2

Black to play and win.

As you can see, none of the critical holes (like C9, D8, etc....) belong both to the first wall and to the second one. Therefore no combination can be expected to break through "at least one" of the white walls. So who would guess that only the black peg L9 can initiate a new path to connect West and East? After the intermediate black moves M6, M11 and L10, the black lines of pegs will be: K9, J10, I10, I9, H9, G10, F10, F9, E9, etc....

It would be nice to solve some Hex problem by using nontrivial theorems about combinatorial properties of sets (the sets considered are groups of critical holes). It is not possible to forget that a famous chess problem of Sam Loyd (the "comet"), involving parity, is easy to solve for a mathematician aware of the König theorem about bipartite graphs; also in chess, the theory of conjugate squares of Marcel Duchamp and Alberstadt is a beautiful application of the algebraic theory of graph isomorphism (the two graphs are defined by the moves of the kings).

The use of a mathematical tool may be unexpected and therefore adds some new interest to a game; but Hex exists as a most enjoyable game in its own right, for mathematician and layman alike.

A Kriegspiel Endgame

Jim Boyce

STANFORD UNIVERSITY

Kriegspiel is one of the most interesting variants of chess: each player tries to mate his opponent, using ordinary chessmen and following the ordinary rules, but neither player knows where the other player's pieces are. Instead, both players have a concealed board on which they can keep track of their own position and guess at the locations of the opponent's pieces. There is also a third participant in the game, namely the kriegspiel referee; he has a third set of chessmen, with which he keeps the actual position. When it is White's move, White suggests a possible move to the referee. If it is legal in the actual game position, it becomes White's official move; otherwise White must try additional moves until one is legal. Then it is Black's turn, and the game continues in this fashion. Any legal move that places the opposing king in check is announced to both players. There are other rules (which do not concern us) that involve captures and pawn moves; further details can be found in [1].

This article analyzes the ending king and rook vs. king in kriegspiel. In normal chess, this is well known to be an elementary mate [2], but the problem is by no means simple under the kriegspiel ground rules. In fact, experienced chess players have been known to spend hours on this problem without solving it. Therefore the reader is encouraged to try his own hand at the task before looking at the solution below.

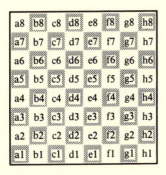

FIGURE 1

Standard algebraic notation.

FIGURE 2

A sample kriegspiel position.

Movement and Notation

In this article, White has a king and a rook; Black has only his king. Moves are written in a version of algebraic notation, in which the files are lettered a–h starting at White's left and the ranks are numbered 1–8 starting at White's end of the board (see Figure 1). All figures show the board as White sees it: the known locations of his pieces are marked, and each possible location for the Black king is marked.

The rules and conventions will be clear after we consider a simple example or two. Suppose that White tries to move his king to d7 in the position of Figure 2. If the referee says that such a move is illegal, White sees the position in Figure 3. Suppose White now tries to play his rook to e2 and the referee announces check. That means that the Black king must be on e8. Black makes a legal move from e8 and it is White's move in the position of Figure 4. White moves his king to c6, a move he knew was legal before he tried it. After Black moves, his king can no longer be on d8; if it was there, it had to move away. He can however, still have a king on f8. A

FIGURE 3

Position after K–d7 is illegal.

FIGURE 4

Position after R–e2 gives check.

FIGURE 5

Position after K–c6.

FIGURE 6

Position after R–e8 gives mate.

king on f8 would also have had to move, but a king on f7 could move to f8. The position after Black's next move appears in Figure 5. If the Black king is on c8, White can mate by moving his rook to e8. Figure 6 shows that result. Of course, if White makes that move, a king on f7 or f8 could capture the rook and assure a draw.

The sequence of moves from Figure 2 to Figure 6 would appear as follows in algebraic notation: 1. K–d7, R–e2+, 2. K–c6, 3. R–e8#. Note, first of all, that there is no number before R–e2+. That is because K–d7 was illegal and White was still looking for his first move. The symbol + denotes a check and the symbol # denotes checkmate. A draw (stalemate or capture of the rook) is denoted by the symbol =, which would appear in the same place as a + or a #.

King and Rook vs. King

We will see that White can almost always force a win in this endgame, even under kriegspiel conditions. Black can draw only in certain starting positions where White cannot unite his king and rook before Black captures the rook. Even if Black might be able to capture the rook (given what White knows about the position), in general, he will not. For if White does put his rook next to the Black king, Black must guess where to move to make the capture. However, this article considers only those positions where White can force mate against any (even the most clairvoyant) defense.

White's plan consists of several stages. First, he must make sure that his rook is safe from capture. Next he plays to a position where all of the possible squares for the Black king are in a rectangle where one corner of that rectangle is a corner of the board and the opposite corner is at the rook. White wants his king in that rectangle also to keep the Black king away from his rook. He then forces the Black king back until it can occupy only those squares on a single edge. It is then a fairly simple task to mate the Black king.

This article demonstrates a simple (albeit slow) way to force checkmate by considering these types of positions:

1 Both kings are in the same quadrant of the board as seen from the rook.

2 The Black king is restricted to one or two quadrants of the board.

3 The White rook is (can be) safe from capture.

Then it examines the positions of Type **1** in more detail:

1a Black king confined to one rank (or file).

1b Black king confined to two ranks.

1c Black king confined to three ranks.

1d Black king confined to four ranks.

1e Black king confined to more than four ranks.

1 Both kings in the same quadrant. In Figure 7, both kings are above and to left of the rook. White wants to confine the Black king to smaller and smaller rectangles of the board. With this in mind, White tries to move his king to e4 and his rook to f3. (He would be even happier to move his rook to the fourth rank.) The game might continue. 1. K–e4, R–f3 (or if K–e4 is legal, continue as in A, below), 2. K–e4, K–e3, K–f5 (or B), 3. K–e4, R–f4. This particular sequence of moves reduces the smaller dimension of the quadrant and results in a position like that after White's first move. If K–e4 was legal on move 2 or 3, then White reduced the larger dimension and obtained a position similar to Figure 7. There are other possible responses (by the referee) to some of White's attempts. If 1. K–e4 is legal, we have line A: 2. R–f3 or 2. R–f3+, 3. R–f4. The second case results in a position where the smaller dimension of the relevant quadrants is decreased, but the Black king can be in either of two quadrants. The other

FIGURE 7

Both kings in one quadrant (as seen by rook); White to move.

FIGURE 8

White to move.

possibility is that 2. … K–e3 is legal. Line B continues 3. K–e4, R–f4. The first move is familiar. The second reduces the smaller dimension, but results in a position in which the Black king is restricted to one quadrant, while the White king is not in that quadrant. A similar sequence of moves can be played from any analogous position (except if the smaller dimension is one). Subsequently it will be shown that the White king can end up in the same quadrant with the Black king in all cases without increasing the smaller dimension. Therefore, White will be able to restrict the king to the edge of the board.

2 Black king in one or two (adjacent) quadrants. Figure 8 shows a position where the Black king might be in either of two quadrants. White wants to reach a position similar to Figure 7; for example, White king on b4, rook on a3, Black king somewhere in ranks 4–8. The first step is to move the rook (and the king if necessary) so that one of the quadrants that might contain the Black king has a side of length two. From Figure 8, play could proceed 1. K–b2, 2. R–c3. Then White plays his king into that quadrant, 3. K–b3, 4. K–b4. If the black king interferes, White simply retreats his rook. Then both kings will be in the same quadrant, one with a dimension of two. Finally, White moves his rook to the edge, 5. R–a3. White can ignore any checks that occur during that sequence of moves. White can use the same plan if the Black king is restricted to one quadrant. Sometimes, White doesn't need to move his king and rook over to the second and third lines from the edge. In Figure 9 White plays his king around to c6 and moves his rook to c5 to get a position that arises after the first move in Section **1**. 1. R–d4 would also result in a position with both kings in the same quadrant, but the smaller dimension of the quadrant would be larger. The game in Figure 9 could continue 1. K–d4, 2. K–c5, 3. K–c6, 4. R–b5.

3 The rook is (or can be) safe. If the king and rook are united in the center of the board and the Black king is not restricted as in the previous

FIGURE 9

White to move.

FIGURE 10

Problem 1. How many moves save the rook?

FIGURE 11

White to move.

FIGURE 12

White to move.

sections, then White wants to limit the Black king in some way. One position which resembles that in Section **1** has the rook on a corner square; for example, h1, and his king nearby, say, g2. So White moves his pieces toward the corner. If the Black king interferes, then White knows that its position is restricted and proceeds as in Section **1** or **2**. Otherwise, White's pieces get to the corner and White can follow the line in Section **1**.

If the rook is subject to capture, White's first task is to protect the rook. This is usually straightforward.Sometimes, there are several ways to defend the rook. Figure 10 is a position with an undefended rook: can the reader discover all of the ways to save it? (The answer appears at the close of this article.)

1a Black king confined to one rank. The previous analysis does not apply to positions with both kings in the same quadrant if the quadrant is a single line on the edge of the board. Figure 11 shows such a position. White can mate quickly by forcing the king back into the corner and mating him with the rook. But White must be a little careful to avoid stalemate. The game could end 1. K–d8, R–g7, 2. K–d8, 3. K–c7, (not 3. K–c8=), 4. R–g6, 5. R–a6#. When the Black king prevents a king move, the rook makes a move and the Black king must then retreat.

1b Black king confined to two ranks. White can improve on the line given in **1** if the black king is already restricted to two lines. It is unnecessary to advance the rook along with the king when chasing the Black king into the corner. From Figure 12, play could continue 1. K–e7, 2. K–d7, K–d6, 3. K–d7, 4. K–c7, K–c6, K–d8, 5. K–c7, 6. K–b6, K–c8, 7. R–a6#. If 5. K–c7 is illegal White can play 5. ... R–h7. Two moves later, 7. R–h7 = would be a serious mistake, but one easily avoided. (The line in section **1** shows that White can shrink the area available to the Black king. One point not mentioned then is that this can stalemate Black. It is easy to determine that, as in the last example, White will have a different move which checkmates. That, of course, is what White should play.)

FIGURE 13

White to move.

FIGURE 14

Problem 2. How quickly can
White force mate?

1c Black king confined to three ranks. There are a few ways to improve
the line in Sections **1** and **2** when the Black king is limited to three ranks or
files. Figure 13 shows a position after White has played R–f5 +. The Black
king is in one of two quadrants. White can avoid the bother of going over to
the edge of the board with his pieces. He plays 1. R–f6,   2. K–e7. If K–e7
is legal, the black king is in the quadrant on the right; if it is illegal, the king
is on the left.

Figure 14 shows another way White can save time in some positions.
On 1. R–d5+ Black may escape to the other side of the board and last
until White's twelfth move. How can White do better? (See the answer
below.)

1d Black king confined to four ranks. Figure 15 is similar to Figure 13.
White wants to play R–f4. If he plays it now and it is check, Black's king
will be in one of two quadrants, and White is too far from the edge for the
idea in Figure 13 to work. White can still avoid the time and effort
involved in the line of Section **2** by playing 1. K–e6. If it is illegal, he plays

FIGURE 15

White to move.

FIGURE 16

White to move.

FIGURE 17

Problem 3. White to move.

FIGURE 18

Problem 4. White to move.

... R–g5 and has reduced the smaller dimension. If it is legal he plays 2. R–f4. If that results in a check, he is in the line that follows Figure 13. **1e** Black king is confined to more than four ranks. When the Black king is *limited* to a large section of the board, White wants to shrink the region as quickly as he can. Figure 16 shows the Black king limited to the largest area possible. If the region is large enough, White can try to shrink the smaller dimension (instead of the larger one, as in Section **1**) of the quadrant the Black king occupies. From Figure 16, play could continue 1. R–g1, 2. K–e3, 3. R–f1. If either rook move is a check, White next moves his rook two squares to the left to produce a position in which both kings are in the same quadrant and the smaller side is at most four squares. Figure 17 is more difficult. Again, the goal is to limit the Black king to at most, four lines, without playing king and rook to the edge of the board as in **2**.

Figure 18 arises from Figure 10 if 1. K–f2 is illegal. How does White play to restrict the Black king to a region of no more than four lines? (Note that a 3 × 4 rectangle contains all possible squares for the Black king, but the edges are not defined by the edges of the board and the rook.)

This article has shown that White, with a king and a rook, can checkmate Black, who has only his king, in a game of kriegspiel. Sections **1a** and **1b** show that White can checkmate if he forces the Black king to the edge of the board. Section **3** suggests that White can easily reach a position with all possible positions of the Black king in a rectangle bounded by a rank and file controlled by his rook and two edges of the board. Sections **1** and **2** show that White can proceed from such a position to one where the rectangle is smaller, where we say that rectangle A is *smaller* than rectangle B if the smaller dimension of A is less than the smaller dimension of B or if the smaller dimensions are equal and the larger dimension of A is less than that of B. The arguments suffice to show that White can mate from a wide variety of positions, and on a rectangular board of arbitrarily large size.

By carefully studying the strategy detailed here, it can be shown that White can force mate from the starting position in Figure 10 in at most 39 moves. Therefore there is no need to worry about games that are drawn because of the "50 moves without a capture" rule.

Acknowledgements

This article is the result of one of the assignments in the class CS204 taught at Stanford University in the fall of 1978. I would like to thank the professor of the course, Donald E. Knuth, who suggested the problem of the kriegspiel ending after he had learned of it from some friends in Germany; and the T.A. of the course, Chris Van Wyk, who discussed this problem at length and helped prepare the article; and the rest of the CS204 students who made several valuable suggestions.

Solutions

1   Three moves save the rook, R–f1, K–f1, and K–f2. After each of the last two, Black cannot take the rook because then White's move would have been illegal. If the move is illegal, White can simply move the rook away as he pleases.

2   White mates with 1. K–c7, R–e6, (If K–c7 is legal, White wins with a similar but shorter line.) 2. K–d7, K–d6,   3. K–e7 and mates in nine more moves as in **1b**.

3   White limits Black to a "small" rectangle with 1. R–e4, R–c4. If the first move is a check, White replies 2. R–e2. If the second move is a check, White continues 3. R–e4.

4   White first moves the rook to safety, 1. R–h5,   2. R–a5. White then moves his king near the rook 3. K–d2,   4. K–c3. If the second king move is illegal, he moves his rook to safety again R–g5, and keeps trying to move his king near to it with

5   K–e3, K–c3,   6. R–a5. If these two king moves are illegal, then the Black king must be on d4. White plays a tempo move, 6. R–h5, to make the king move and then tries again. This time, he must succeed. If 3. K–d2 is illegal, White plays R–a4,   4. R–h4 to produce a position similar to what arises if it is legal.

Bibliography

1   Compayne, Charles. 1976. Kriegspiel. *Games and Puzzles* 50: 12–15.

2   Fine, Reuben. 1941. *Basic Chess Endings*. David McKay.

Mental Poker

Adi Shamir, Ronald L. Rivest and Leonard M. Adleman

MASSACHUSETTS INSTITUTE OF TECHNOLOGY

ABSTRACT

Can two potentially dishonest players play a fair game of poker without using any cards—for example, over the phone? This paper provides the following answers:

1 No. (Rigorous mathematical proof supplied.)

2 Yes. (Correct and complete protocol given.)

Once there were two "mental chess" experts who had become tired of their pastime. "Let's play 'Mental Poker,' for variety" suggested one. "Sure" said the other. "Just let me deal!"

Our anecdote suggests the following question (proposed by Robert W. Floyd): "Is it possible to play a fair game of 'Mental Poker'?" We will give a complete (but paradoxical) answer to this question. First we will prove that the problem is intrinsically insoluble, and then describe a fair method of playing "Mental Poker".

What does it mean to play Mental Poker?

The game of Mental Poker is played just like ordinary poker (see Hoyle [2]) except that there are no cards; *all* communications between the players must be accomplished using messages. Perhaps it will make the ground rules clearer if we imagine two players, Bob and Alice, who want to play poker over the telephone. Since it is impossible to send playing cards over a phone line, the entire game (including the deal) must be realized using only spoken (or digitally transmitted) messages between the two players.

We assume that neither player is beyond cheating. "Having an ace up one's sleeve" might be easy if the aces don't really exist! A fair method of playing Mental Poker should preclude any sort of cheating.

A fair game must begin with a "fair deal". To accomplish this, the players exchange a sequence of messages according to some agreed-upon procedure. (The procedure may require each player to use dice or other randomizing devices to compute his hand or the messages he transmits.) Each player must know which cards are in his hand, but must have no information about which cards are in the other player's hand. The dealing method should ensure that the hands are disjoint, and that all possible hands are equally likely for each player.

During the game the players may want to draw new cards from the "remaining deck", or to reveal certain cards in their hand to the opposing player. They must be able to do so without compromising the security of the cards remaining in their hand.

At the end of the game, each player must be able to check that the game was played fairly and that the other player has not cheated. If one player claimed that he was dealt four aces, the other player must now be able to confirm this.

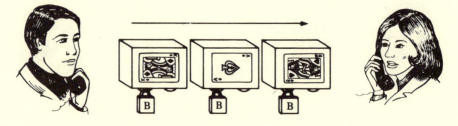

FIGURE 1

Bob encrypts the cards and sends them to Alice in scrambled order.

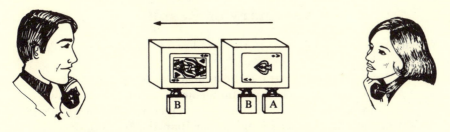

Alice chooses one for Bob, and encrypts another for herself, and sends
them both to Bob.

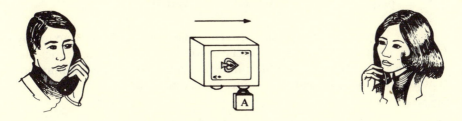

Bob decrypts both cards, and returns Alice's encrypted card to her.

The above set of requirements makes a "fair game" of Mental Poker
look rather difficult to achieve. To make things easier, we'll assume that
both players own computers. This enables the use of complicated protocols
(say, involving encryption). We do not assume that either player will trust
the other's computer. (The players could program their computers to
cheat!)

We suggest that you might find it an interesting challenge to attempt
to find on your own a method for playing Mental Poker, before reading
further.

I win!

FIGURE 2
Alice decrypts her card, and then they compare cards to see who has
won.

Summary of Results

We will present two solutions to the problem of playing Mental Poker:

1 A rigorous proof that it is theoretically impossible to deal the cards in a way which simultaneously ensures that the two hands are disjoint and that neither player has any knowledge of the other player's hand (other than that the opponent's hand is disjoint from his).

2 An elegant protocol for dealing the cards that permits one to play a fair game of Mental Poker as desired.

The blatant contradiction between our two results is not due to any tricks or faults in either result. In fact, we will leave to the reader the enjoyable task of puzzling out the differences in the underlying assumptions that account for our seemingly contradictory results.

The Impossibility Proof

For the sake of simplicity, we consider the minimal non-trivial case of dealing two different cards (one to each player) from a deck of three cards $\{X, Y, Z\}$. The impossibility proof for this case can be easily generalized to any combination of cards and hand sizes.

If a legal protocol for this case exists, then after exchanging a finite number of messages, Alice and Bob each know their card but not their opponent's card. These messages must coordinate the two players' choices of cards to prevent them from getting the same card.

Suppose that for a particular deal

the messages exchanged are $M_1, ..., M_n$,
the card Alice actually gets is X, and
the card Bob actually gets is Y.

We define S_A to be the set of cards that Alice could have gotten in any deals where exactly the same messages are exchanged. (Since each player may want to make some random choices in order to get a card which is unpredictable to the other player, different deals could arise with the same sequence of messages having been exchanged.) Obviously, the card X is in S_A.

If S_A were to contain just the card X, then the deal would violate our requirement that Bob should have no information about Alice's card. Clearly the sequence of messages uniquely determines Alice's card in this case, so in an information-theoretic sense he has (total) information about her card. Furthermore, in any physically-realizable (and terminating) protocol for the deal, Alice has only a finite number of random computations possible, so that Bob can actually determine Alice's card by examining those which are consistent with the given message sequence.

On the other hand if S_A contains all three cards, then Bob cannot get any card—regardless of which card he gets, the message sequence is consistent with the possibility that Alice's card is the same. Consequently, S_A must contain exactly two cards.

The set S_B of cards Bob can get without altering his external behavior is similarly defined, and it must also contain exactly two cards. However, the total number of cards in the deck is three, so that S_A and S_B cannot be disjoint. (In our example, Z belongs to both sets.) Thus it could happen that both Bob and Alice get the card Z in the case that the message sequence is $M_1, ..., M_n$. Thus the protocol cannot guarantee that Bob and Alice will choose distinct cards. We conclude that a fair deal is impossible.

A Protocol for the Deal

The following solution meets all the requirements for the problem. First of all, Bob and Alice agree on a pair of encryption and decryption functions E and D which have the following properties:

1 $E_K(X)$ is the encrypted version of a message X under key K,

2 $D_K(E_K(X)) = X$ for all messages X and keys K,

3 $E_K(E_J(X)) = E_J(E_K(X))$ for all messages X and keys J and K,

4 Given X and $E_K(X)$ it is computationally impossible for a cryptanalyst to derive K, for all X and K,

5 Given any messages X and Y, it is computationally impossible to find keys J and K such that $E_J(X) = E_K(Y)$.

Property **3**, the commutativity of encryption, is somewhat unusual but not impossible to achieve. Properties **4** and **5**, (especially **4**), essentially state that E is *cryptographically strong* or *unbreakable*.

As an example of a function with the above properties, consider

$$E_K(M) \equiv M^K \pmod{n}$$

where n is a large number (prime or composite with a given factorization) which is known to both Bob and Alice, and where

$$gcd[K, \phi(n)] = 1.$$

[$\phi(n)$] is Euler's totient function, which can be easily computed from the prime factorization of n.)

The corresponding decoding function is

$$D_K(C) \equiv C^L \pmod{n},$$

where

$$L \cdot K \equiv 1[\text{mod } \phi(n)].$$

Since

$$E_K[E_J(M)] \equiv E_J(E_K(M)) \equiv M^{JK} \pmod{n},$$

E satisfies property **3**. (For more details on the cryptographic strength and importance of this function see [1,3,4].) We describe this particular encryption function here only to demonstrate that the kind of encryption functions we desire apparently exist; we will not make use of any particular properties this function has other than **1 ... 5**.

Once Bob and Alice have agreed on the functions E and D (in our example this means agreeing on p), they choose secret encryption keys B and A respectively. These keys remain secret until the end of the game, when they are revealed to verify that no cheating has occurred.

Bob now takes the fifty-two messages:

"TWO OF CLUBS",

"THREE OF CLUBS",

$\vdots$

"ACE OF SPADES"

and encrypts each one (whose bit string is considered as a number) using his key B. (That is, he computes E_B ("TWO OF CLUBS"), etc.) He then shuffles (randomly rearranges) the encrypted deck and transmits it all to Alice.

Alice selects five cards (messages) at random and sends them back to Bob; these messages Bob decodes to find out what his hand is. Alice has no way of knowing anything about Bob's hand since the encryption key B is known only to Bob.

Now Alice selects five other messages, encrypts them with her key A, and sends them to Bob. Each of these five messages is now doubly encrypted as $E_A(E_B(M))$, or equivalently $E_B(E_A(M))$, for each M. Bob decrypts these messages obtaining $E_A(M)$ for these five messages and sends them back to Alice. Alice can decrypt them using her key A to obtain her hand. Since Bob does not know A, he has no knowledge of Alice's hand.

Michael Rabin suggested a nice physical analogy for the above process. We can view encryption as equivalent to placing a padlock on a box containing the card. Bob initially locks all the cards in individual undistinguishable boxes with padlocks all of which have key B. Alice selects five boxes to return to him for his hand, and then sends him back five more boxes to which she has also added her own padlock with key A to the clasp ring. Bob removes his padlock from all ten boxes and returns to Alice those still locked with her padlock, for her hand. Notice the implicit use of commutativity in the order in which the padlocks are locked and unlocked.

Should either player desire additional cards during the game, the above procedure can be repeated for each card.

At the end of the game both players reveal their secret keys. Now either player can check that the other was "actually dealt" the cards he claimed to have during play. By property **5** neither player can cheat by revealing a key other than the one actually used (one which would give him a better hand).

The above procedure can also be easily extended to handle more than two players. (Details left to the reader.) Another obvious generalization is to use commutative encryption functions in secret communications systems to send arbitrary messages (rather than just card names) over a communications channel which is being eavesdropped.

Conclusions

Initially we proved that the card-dealing problem is insoluble, and then we presented a working solution to the problem. We leave it to you, the reader, the puzzle of reconciling these results. (Hint: each player would in fact be able to determine the other player's hand from the available information, if it were not for the enormous computational difficulty of doing so by "breaking" the code.)

Acknowledgements

We would like to thank Robert W. Floyd, Michael Rabin, and Albert Meyer for their motivation and valuable suggestions.

References

1 Diffie, Whitfield and Hellman, Martin E. 1976. New Directions in Cryptography. *IEEE Trans. Info. Theory* IT-22: 644–654.

2 Morehead, A. H., Frey, R. L. and Mott-Smith, G. 1947. *The New Complete Hoyle.* New York: Garden City Books.

3 Pohlig, Stephen C. and Hellman, Martin E. 1978. An Improved Algorithm for Computing Logarithms over $GF(p)$ and its Cryptographic Significance. *IEEE Trans. Info. Theory* IT-24: 106–110.

4 Rivest, Ronald L., Shamir, Adi and Adleman, Leonard M. 1978. A Method for Obtaining Digital Signatures and Public-Key Cryptosystems. *CACM* 21: 120–126.

This research was supported by NSF grants MCS78–05849 and MCS78–04343; and by ONR grant N00014–76–C–0366.

Cheap, Middling or Dear

Vašek Chvátal

McGill University

On page 128 of a book entitled *The Games and Diversions of Argyleshire*, written by R. C. MacLagan and published in 1901 by David Nutt in London, one finds the following description of an old Scottish game.

Cheap, Middling, or Dear

This also is played by two. The letters **C**, **M**, **D**, representing respectively the words from which the game is named, are written on a slate, with some interval between them. Under **C** the figures 1, 2, 3 are placed, under **M** 4, 5, 6, and under **D** 7, 8, 9, thus:

C	**M**	**D**
1, 2, 3	4, 5, 6	7, 8, 9

Player A, who is to play first, marks one of the figures from any of the groups, concealing it from player B, whom he challenges to guess to which group it belongs, saying "My father bought a horse at a fair". B asks, "Cheap, middling, or dear?" A answers him, naming the group from which he has selected his figure. Thus if his figure were 5, the answer would be "middling". B then guesses one of the three numbers, and if he hits upon 5, that is a gain to him of 5, but if he says 4 or 6, then the 5 is scored to A. In any case the 5 is blotted out. B then leads, each playing in turn, till all the figures have been expunged. The total marks credited to each are then ascertained, and he who has the highest number is the winner.

We are going to describe the results of an analysis of Cheap, Middling or Dear. For the benefit of readers not acquainted with game theory, we shall first sketch a few basic facts concerning matrix games.

Let us begin with a game of matching coins, somewhat related to Cheap, Middling or Dear but much simpler. Here, each of the two players hides a nickel, a dime or a quarter. If the two coins match, then they go to the first player; otherwise they go to the second player. Clearly, neither player has a safe winning strategy: given bad enough luck, one may lose in every single play of the game. And yet we claim that the second player is better off at least in a statistical sense: if he plays correctly then he can *expect* steady winnings over long periods of time. Naturally, he must make every choice in some random fashion, so that the opponent will be unable to predict its outcome. At the same time, however, the probabilities of the outcomes must be fixed: in a long run, the second player ought to play a nickel 7/34 of the time, a dime 12/34 of the time and a quarter the remaining 15/34 of the time. Assuming that he does follow these instructions, we now consider a long sequence of plays. Each nickel chosen by the first player is met by the second player's nickel 7/34 of the time, by a dime 12/34 of the time and by a quarter 15/34 of the time: no matter how the first player behaves, he cannot obliterate the element of chance introduced into the game by the second player's random choices. Within this group of plays, the second player loses a nickel 7/34 of the time and wins a nickel 22/34 of the time. Similarly, in those plays in which the first player chooses a dime, the second player loses a dime 12/34 of the time and wins a dime 22/34 of the time. In those plays in which the first player chooses a quarter, the first player loses a quarter 15/34 of the time and wins a quarter 19/34 of the time. Within each of these three groups, the second player's average winnings come to 100/34 cents per play. Of course, these winnings are not *guaranteed*: they are only *expected* in the same sense as an unbiased coin is expected, but not guaranteed, to show heads fifty percent of the time. In this sense, the first player can protect himself from heavier losses than 100/34 cents per game by playing a nickel 10/17 of the time, a dime 5/17 of the time and a quarter the remaining 2/17 of the time. No matter how the second player behaves now, every time he plays a nickel, dime, or quarter, his greatest expected wins are respectively

$$-\frac{10}{17}\cdot 5 + \frac{5}{17}\cdot 10 + \frac{2}{17}\cdot 25 = \frac{50}{17}, \qquad \frac{10}{17}\cdot 5 - \frac{5}{17}\cdot 10 + \frac{2}{17}\cdot 25 = \frac{50}{17},$$

$$\frac{10}{17}\cdot 5 + \frac{5}{17}\cdot 10 - \frac{2}{17}\cdot 25 = \frac{50}{17}.$$

This game may be represented by the matrix

$$\begin{bmatrix} 5 & -5 & -5 \\ -10 & 10 & -10 \\ -25 & -25 & 25 \end{bmatrix}.$$

The first player chooses one of the three rows (corresponding to nickel, dime and quarter, respectively) and the second player, unaware of the first player's choice, chooses one of the three columns; the corresponding entry in the matrix specifies the amount won by the first player. Generally, each matrix specifies a game in which the first player chooses a row, the second player chooses a column, and the corresponding entry a_{ij} indicates the amount won by the first player. (Cheap, Middling or Dear may be viewed as this kind of a game, with each row of the matrix corresponding to a set of unambiguous instructions for player A to be followed throughout the nine rounds, and with each column corresponding to a similar set of instructions for B. The number of rows and columns in this matrix is enormous.) Again, the only sensible strategy for each player is to make his choices at random, so that they cannot be predicted by the opponent. Suppose that the first player chooses each row $i = 1, 2, \ldots, m$ with a relative frequency x_i and that the second player chooses each column $j = 1, 2, \ldots, n$ with a relative frequency y_j. The resulting average winnings of the first player come to

$$\sum_{i=1}^{m} \sum_{j=1}^{n} a_{ij} x_i y_j$$

per play. In matrix notation, this quantity may be recorded as xAy. The row vector x with components $x_1, x_2, \ldots, x_m$ and the column vector y with components $y_1, y_2, \ldots, y_n$ share a characteristic feature: all the components are nonnegative and their sum equals one. Such vectors are called *stochastic*. In the late nineteen twenties, John von Neumann (Zur Theorie der Gesselschaftsspiele. 1928. *Math. Ann.* 100: 195–320) proved the celebrated *minimax theorem* asserting that

$$\max_{x} \min_{y} xAy = \min_{y} \max_{x} xAy$$

for an arbitrary matrix A, with the maxima taken over all stochastic vectors x and the minima taken over all stochastic vectors y. To put it differently, for every matrix A there are stochastic vectors x^* and y^* such that

$$xAy^* \leq x^*Ay^* \leq x^*Ay$$

for all stochastic vectors x and y. Now x^* and y^* may be thought of as optimal strategies in the game represented by A: the first player can expect to win at least

$$\min_{y} x^*Ay = x^*Ay^*$$

per play whereas the second player can expect not to lose more than

$$\max_{x} xAy^* = x^*Ay^*$$

per play. The quantity x^*Ay^* is referred to as the value of the game. (The value of the coin matching game is $-50/17$.)

After these preliminaries, we return to Cheap, Middling or Dear. To break the suspense, let us reveal that the game is biased in favour of B: its value is about -3.8. (That is, if both players follow their optimal strategies then B may expect to score about 24.4 whereas the corresponding figure for A is only about 20.6.) Each conceivable position in the game is specified by the set S of numbers left on the slate. If the size of S is odd, then A is about to choose a number; if the size of S is even, then B is about to choose a number. If the player about to choose a number follows his optimal strategy then he may expect to score at least v points during the rest of the game, no matter what strategy his opponent adopts. Similarly, if the player about to guess a number follows his optimal strategy, then he may expect to score at least w points during the rest of the game, no matter what strategy his opponent adopts. By the minimax theorem, the sum $v + w$ equals the total of the number in S. We shall refer to the difference $v - w$ as the value of S. We are about to describe an easy way of computing the value of every S, the optimal ways of choosing a number from S and the optimal ways of guessing a number in a specified group in S. Our claim may be verified by induction on the size of S; we omit the tedious proofs.

Each round begins when one of the players names the group from which he has selected his number. The optimal strategy is:

1 larger groups are always preferable to smaller ones

2 among the three-number groups, Dear is preferable to Middling and Middling is preferable to Cheap,

3 all the two-number groups are equally attractive,

4 among the one-number groups, Cheap is preferable to Middling and Middling is preferable to Dear.

Naturally, the selection of the number itself has to be randomized. The strategy is:

5 from a group containing two numbers x and y,
    choose x with probability $y/(x + y)$,
    choose y with probability $x/(x + y)$,

6 from a group containing three numbers x, y and z,
    choose x with probability $yz/(xy + xz + yz)$,
    choose y with probability $xz/(xy + xz + yz)$,
    choose z with probability $xy/(xy + xz + yz)$.

These rules provide a complete description of optimal strategies for that player who is about to choose a number. In fact, almost all of the optimal choices have the above form; the only exception is the fact that

7 in positions with one Cheap, two Middling and two Dear, one cannot go wrong by playing Cheap.

Now we turn to evaluating the various positions. First, let us assume that two of the three groups have been erased completely. In that case, the value of $\{x\}$ is $-x$, the value of $\{x, y\}$ is $(x^2 + y^2)/(x + y)$, and the value of $\{x, y, z\}$ is

$$\frac{1}{xy + xz + yz}\left(xyz - \frac{xy(x^2 + y^2)}{x + y} - \frac{xz(x^2 + z^2)}{x + z} - \frac{yz(y^2 + z^2)}{y + z}\right).$$

In explicit terms, the last formula says that

the value of $\{1, 2, 3\}$ is $-\dfrac{613}{330} \doteq -1.85757575,$

the value of $\{4, 5, 6\}$ is $-\dfrac{64913}{18315} \doteq -3.544253344,$

the value of $\{7, 8, 9\}$ is $-\dfrac{2129473}{389640} \doteq -5.465232009.$

Next, let us note that the rules **1–6** described above, combined with the twenty-one values presented so far, point out a simple way of evaluating several different positions. For example, we shall consider $S = \{1, 2, 3, 4, 5, 7, 9\}$. Since S has an odd number of elements, A is about to choose a number. An optimal play may take the following course:

A chooses Cheap, B chooses Cheap, A chooses Middling,
B chooses Dear, A chooses Cheap, B chooses Middling,
A chooses Dear.

The two players might as well agree to proceed in the following order:

A chooses Cheap, B chooses Cheap, A chooses Cheap,
A chooses Middling, B chooses Middling,
B chooses Dear, A chooses Dear.

Thus we conclude that the value of S is

value of $\{1, 2, 3\}$ + value of $\{4, 5\}$ − value of $\{7, 9\}$

$$= -\frac{613}{330} + \frac{41}{9} - \frac{130}{16} = -\frac{21491}{3960} \doteq -5.4270202.$$

In order to provide a similar way of evaluating each of the 512 positions, we shall require a few additional building blocks. Again, let S be a position with two of the three groups completely erased. By the out-of-phase game on $S = \{x, y\}$, we shall mean the game with A choosing in both rounds; by an out-of-phase game on $S = \{x, y, z\}$, we shall mean the game with A choosing in the first round only and guessing in the remaining two rounds. The out-of-phase value of $\{x,y\}$ is $-(x^2+y^2)/(x+y)$ and the out-of-phase value of $\{x,y,z\}$ is

$$\frac{1}{xy+xz+yz}\left(xyz+\frac{xy(x^2+y^2)}{x+y}+\frac{xz(x^2+z^2)}{x+z}+\frac{yz(y^2z^2)}{y+z}\right).$$

In explicit terms, the last formula says that

the out-of-phase value of $\{1,2,3\}$ is $\dfrac{973}{330} \doteq 2.948484848,$

the out-of-phase value of $\{4,5,6\}$ is $\dfrac{124313}{18315} \doteq 6.787496587,$

the out-of-phase value of $\{7,8,9\}$ is $\dfrac{4185793}{389640} \doteq 10.74271892$

Now we know all we need for a fast evaluation of an arbitrary position. For example, we shall consider $S = \{1,2,3,5,6\}$. An optimal play may take the following course:

A chooses Cheap,   B chooses Cheap,   A chooses Middling,
B chooses Cheap,   A chooses Middling,

The two players might as well agree to proceed as follows:

A chooses Cheap,   B chooses Cheap,   B chooses Cheap,
A chooses Middling,   A chooses Middling.

Thus, we conclude that the value of S is

out-of-phase value of $\{1,2,3\}$ + out-of-phase value of $\{5,6\}$

$$=\left(\frac{973}{330}\right)-\left(\frac{61}{11}\right)=-\frac{857}{330}\doteq-2.5969696.$$

Finally, we turn to the optimal guessing strategies. Suppose that one of the players is choosing a number from a three-number group $\{x,y,z\}$. Let a,b,c denote the values of $S-\{x\}$, $S-\{y\}$, $S-\{z\}$, respectively. Clearly, the optimal guessing strategy for the other player is identical with the second player's optimal strategy in the game specified by the matrix

$$\begin{bmatrix} -x-a & x-a & x-a \\ y-b & -y-b & y-b \\ z-c & z-c & -z-c \end{bmatrix}.$$

This player ought to

guess x with probability $\dfrac{1}{2(xy+xz+yx)}\left[y(x+c-a)+z(x+b-a)\right],$

guess y with probability $\dfrac{1}{2(xy+xz+yz)}[x(y+c-b)+z(y+a-b)]$,

guess z with probability $\dfrac{1}{2(xy+xz+yz)}[x(z+b-c)+y(z+a-c)]$.

Similarly, if one of the players chooses a number from a two-number group and if a,b stand for the values of $S-\{x\}$, $S-\{y\}$, respectively, then the other player's optimal guessing strategy amounts to the second player's optimal strategy in the game specified by the matrix

$$\begin{bmatrix} -x-a & x-a \\ y-b & -y-b \end{bmatrix}.$$

This player ought to

guess x with probability $\dfrac{1}{2(x+y)}(x+y-a+b)$,

guess y with probability $\dfrac{1}{2(x+y)}(x+y+a-b)$.

A Random Hopscotch Problem or How to Make Johnny Read More

David Berengut
STATE UNIVERSITY OF NEW YORK AT BINGHAMTON

$\mathbf{M}$any of our most popular games combine the element of randomness—produced either by the roll of a die, the twirl of a spinner, or the deal of a shuffled deck of cards—with a playing surface or board consisting of spaces arranged in a sequence, either in a closed loop (as in Monopoly), or with separate start and finish points (as in cribbage or backgammon). The movement of pieces in these games might be whimsically described as *random hopscotch*. As a mathematical statistician, I have more than a passing interest in the probabilistic aspects of such games.

This article originated from a problem that David Klarner mentioned to me. A teacher of young children had devised a game of random hopscotch with the aim of stimulating her students' interest in acquiring reading skills. The board consisted of a sequence of spaces with separate starting and finishing points. Various spaces contained instructions to perform various reading tasks, while the remaining ones were blank. Each student in turn took a run through the board, making moves according to the throw of a modified die whose faces consisted of two 1's, two 2's, and two 3's (so that only moves of 1, 2, or 3 spaces, each equally likely, were possible). When the student landed on a space containing an instruction, he was required to perform the appropriate reading task.

By introducing the element of chance—as well as the challenge to control the outcomes of the tosses—the teacher was very successful at holding the attention of her students. The question posed by the teacher was the following: In which spaces should she place the instructions so as to maximize the expected number of tasks a student would have to perform? Implicit in this question, of course, is the assumption that the number of spaces on the board and the number of tasks are both fixed.

From Words to Numbers

The key to solving many mathematical problems lies in simply formulating the problem the right way; to a large degree, this is the case here. Since it is often easier to solve mathematical problems in a general way rather than for specific cases, I choose to let the number of spaces on the board and the number of tasks both be arbitrary, and denote these quantities by n and t respectively.

To solve any problem, we must first know exactly what the problem is saying. What, precisely, is meant by the expression *expected number of tasks to be performed*? The expected value of a random variable is a standard notion in the theory of probability—in brief, it is the probability-weighted average of the possible values of the random variable. If T represents the number of tasks performed in a hypothetical run through the board, T is a random variable which can take on any one of the integer values from 0 to t inclusively, each with a certain probability (as yet undetermined). Letting $P(j)$ denote the probability that T takes the value j, where j is any integer between 0 and t, the expected value of T, written $E(T)$, is given by the formula:

$$E(T) = 0 \cdot P(0) + 1 \cdot P(1) + \cdots + (t-1) \cdot P(t-1) + t \cdot P(t).$$

It is convenient to number the spaces on the board sequentially from the starting point so that the first space after the starting point is numbered 1, the next space 2, and so on, with the final space on the board being numbered n. Let $i_1, i_2, \ldots, i_t$ represent the numbers of the spaces, in ascending order, which contain tasks. For any given play of the game then, the value of T is simply the number of spaces among the set $\{i_1, i_2, \ldots, i_t\}$ on which the player lands. A convenient mathematical device here is the use of *indicator variables*, which can only take on the values 1 or 0, according to whether or not a particular event occurs. In this case, let I_1 be the indicator variable for the event that the player lands on space i_1; that is, I_1 equals 1 if the player does land on space i_1 during the course of the game, and it equals 0 if not. In a similar fashion, let I_2 be the indicator variable for the event that the player lands on space i_2, and so on, up to I_t. Then, clearly, $I_1 + I_2 + \cdots + I_t$ simply counts the number of times during the game that the player lands on a space containing a task. We will call this number, T.

Calculating the expected value of T is thus equivalent to calculating the expected value of $I_1 + I_2 + \cdots + I_t$. At this point, we make use of a fundamental property of expectations: the expected value of a sum is the sum of the expected values. (Thus, for example, the expected number of rainy days in a year for a given location equals the expected number of rainy days in January plus the expected number of rainy days in February plus, etc.) In symbols, then, $E(T)$ equals $E(I_1) + E(I_2) + \cdots + E(I_t)$.

How do we calculate $E(I_1)$, for example? From the definition of expectation, it follows that $E(I_1)$ is given by $0 \cdot P(0) + 1 \cdot P(1)$, or simply $P(1)$. $P(1)$ is the probability that I_1 takes the value 1; the probability that the player will land on space i_1. Similarly, $E(I_2)$ is simply the probability that the player will land on space i_2, and so on. If we let p_i denote the probability that the player will land on space i, where i ranges from 1 to n, then we can write $E(T) = p_{i_1} + p_{i_2} + \cdots + p_{i_t}$.

The problem, then, is to evaluate the numbers $p_1, p_2, \ldots, p_n$. Once we have these numbers, we are in a position to solve the teacher's problem: choosing the spaces $i_1, i_2, \ldots, i_t$ to place the tasks so as to maximize $E(T)$. Obviously, these are the t spaces whose corresponding p-values are the t largest among all the values $p_1, p_2, \ldots, p_n$. In effect, the teacher ought to choose those t spaces which have the greatest chance of being landed on!

Determining which spaces these are, however, is a nontrivial problem. To illustrate the method of solution, I want to consider first a slightly simpler version of the game.

A Simpler Problem

Suppose the rules of the game are altered slightly, so that each move along the board can be of only one or two spaces, determined by the flip of a coin. In this case, can we calculate the probabilities of landing on each of the spaces, $p_1, p_2, \ldots, p_n$?

Well, p_1 is easy enough—it's just the probability of moving one space on the first move, namely $\frac{1}{2}$. Now p_2 is the probability of landing on space 2. This can happen in either of two distinct ways: by moving two spaces on the first move, or by moving one space on each of the first two moves. The first way has probability $\frac{1}{2}$ of happening; the probability of the second way is just the probability of getting 2 tails in 2 consecutive coin tosses. Since the coin tosses are independent of each other, the multiplicative principle tells us that this probability is simply the product of the probability of a tail on the first toss and the probability of a tail on the second toss, namely $\frac{1}{2} \cdot \frac{1}{2}$ or $\frac{1}{4}$. Since the two ways of arriving at space 2 are exhaustive and distinct, the overall probability of landing on space 2, p_2, is the sum of their probabilities $\frac{1}{2} + \frac{1}{4}$ or $\frac{3}{4}$.

The same type of argument, which requires listing all the possible paths, could be used to calculate p_3 and all the remaining p's; obviously, the task will become increasingly tedious as one moves further up the board. Fortunately, this difficulty can be finessed by using a more ingenious argument to obtain the p's in a recursive manner.

The Recursive Approach

Suppose we want to calculate p_i, the probability that the player will land on space i, where i exceeds 2. Any move landing on space i must have originated from either space $i - 2$ or $i - 1$. Thus, there are two distinct ways of landing on space i that can be described:

1 the player eventually lands on space $i - 2$ and then moves 2 spaces on the subsequent move (thus passing over space $i - 1$); or

2 the player eventually lands on space $i - 1$ and on the subsequent move advances 1 space (see Figure 1).

Because successive moves are independent, the probability of landing on space i via path **1** is the probability of eventually landing on space $i - 2$ (namely p_{i-2}), multiplied by the probability of moving 2 spaces on the next move, namely $\frac{1}{2}$, which gives $\frac{1}{2}p_{i-2}$; similarly the probability of arriving via path **2** is $\frac{1}{2}p_{i-1}$. Thus the overall probability of landing on space i (p_i) equals $\frac{1}{2}p_{i-1} + \frac{1}{2}p_{i-2}$. Therefore, each term in the sequence p_1, p_2, ..., p_n beyond the second is simply the average of the two preceding terms!

Knowing $p_1 = \frac{1}{2}, p_2 = \frac{3}{4}$ enables us to conclude that $p_3 = \frac{1}{2}(\frac{1}{2} + \frac{3}{4})$ or $\frac{5}{8}$. In this manner, we can recursively calculate $p_5 (= \frac{11}{16})$, $p_4 (= \frac{21}{32})$, etc. Of course, if we were interested in calculating the value of p_{20}, say, by this method, we would first have to calculate the previous 19 p-values. This would certainly be a tedious computation by hand. Can we obtain an explicit formula for p_i, for any i? The answer is an emphatic *yes*. Indeed, the formula for p_i turns out to be disarmingly simple: $p_i = \frac{2}{3} + \frac{1}{3}(-\frac{1}{2})^i$ (which incidentally tells us that p_{20} equals $\frac{2,097,152}{3,145,728}$).

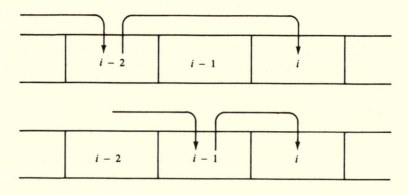

FIGURE 1

The two distinct ways of arriving at the ith space.

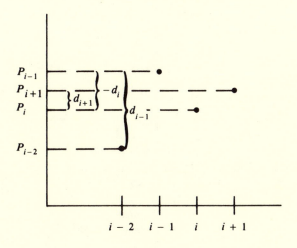

FIGURE 2

Relations between successive p-values.

Obtaining this formula from the recursive formula is a fairly straightforward matter, and follows from the simple observation that the average of two numbers lies midway between the numbers. Thus, p_i lies midway between p_{i-2} and p_{i-1}. The difference between p_i and p_{i-1} is therefore in magnitude, half the difference between p_{i-1} and p_{i-2}, and it is easy to see that the differences are opposite in sign. Letting d_i denote the difference $p_i - p_{i-1}$, for any i, means that $d_i = -\frac{1}{2}d_{i-1}$. This is illustrated in Figure 2.

By repeated application of the relation $d_i = -\frac{1}{2}d_{i-1}$, we get $d_i = -\frac{1}{2}d_{i-1} = -\frac{1}{2}(-\frac{1}{2}d_{i-2}) = \ldots = (-\frac{1}{2})^{i-2}d_2$. But d_2 equals $p_2 - p_1$, which is $\frac{3}{4} - \frac{1}{2}$ or $\frac{1}{4}$. Thus d_i equals $\frac{1}{4}(-\frac{1}{2})^{i-2}$, or $(-\frac{1}{2})^i$. Finally, making use of the fact that p_i can be written as the sum $(p_i - p_{i-1}) + (p_{i-1} - p_{i-2}) + \cdots + (p_2 - p_1) + p_1$, or equivalently $d_i + d_{i-1} + \cdots + d_2 + p_1$, we get by direct substitution that p_i equals $(-\frac{1}{2})^i + (-\frac{1}{2})^{i-1} + \cdots + (-\frac{1}{2})^2 + \frac{1}{2}$. Using the well-known formula for the sum of a finite-length geometric series yields the answer $p_i = \frac{2}{3} + \frac{1}{3}(-\frac{1}{2})^i$. Table 1 lists the first 12 p-values.

An interesting observation at this point is to note that the sequence of p-values oscillates in a regular fashion around the value $\frac{2}{3}$, with the oscillation decreasing in magnitude the farther one proceeds along the sequence. Moreover, the subsequence with odd indices $p_1, p_3, p_5, \ldots$ increases monotonically toward the limiting value of $\frac{2}{3}$, while the complementary subsequence $p_2, p_4, p_6, \ldots$ decreases monotonically to the same limiting value. Clearly, then, the largest p-value is p_2, the second largest is p_4, and so on, while the smallest p-value is p_1, the second smallest is p_3, etc.

We are finally in a position to answer the teacher's question (at least for the modified, simpler version of the game). To maximize the expected

i	p_i
1	.5000
2	.7500
3	.6250
4	.6875
5	.6563
6	.6719
7	.6641
8	.6680
9	.6660
10	.6670
11	.6665
12	.6667

TABLE 1

Values of p_i for i up to 12. Simpler version of game.

number of tasks performed, the tasks should be placed on the even-numbered spaces, beginning with space 2 and proceeding sequentially down the board until either (a) all the tasks are placed or (b) the end of the board is reached; in case (b), any remaining tasks are placed on the unoccupied (that is, odd-numbered) spaces, beginning at the end of the board and working backwards.

What is it Worth to be Smart?

Being a mathematician of somewhat applied bent, I wondered how much is gained by placing the tasks optimally on the board. To make the question more precise, what is the difference in expected number of tasks performed between the optimal placement and the worst possible placement, which is obtained by simply reversing the roles of odd and even in the optimal scheme? The answer involves only straightforward algebra, and turns out to depend on whether or not n (the number of spaces on the board) is at least twice as large as t (the number of tasks); if this is the case, the answer is $(1 - 4^{-t})/3$, or $(1 - 4^{t-n})/3$. This number can never exceed $\frac{1}{3}$, no matter how large n or t may be. In most cases, however, it will be quite close to $\frac{1}{3}$. For example, if there are 5 tasks in all, and the board contains 10 or more spaces, the expected number of tasks performed under the optimal placement is 3.4443, as compared with a value of 3.1113 under the worst placement; the difference is 0.3330 (341/1024 to be exact). Although this number may seem an insignificant difference, it becomes increasingly important the more often the game is played. For example, if each student in a class of 30 plays the game once, the difference between the optimal and the worst placement of the tasks translates into approximately 10 extra reading tasks being performed.

The Original Problem

Having solved the simpler problem, let us try to apply the same methods to solving the teacher's original problem, where moves of one, two, or three spaces are equally likely. Certainly an analogous argument works for producing a recursive relation; in this case, the relation becomes $p_i = (p_{i-1} + p_{i-2} + p_{i-3})/3$, for values of i exceeding 3. In order to get the ball rolling, we need to calculate p_1, p_2, and p_3. Now p_1 is obviously $\frac{1}{3}$. Space 2 can be landed on either by an initial move of two spaces or by two consecutive moves of one space each, hence p_2 equals $\frac{1}{3} + (\frac{1}{3})^2$, or $\frac{4}{9}$. There are four ways of landing on space 3: (1) an initial move of three spaces; (2) a move of two spaces followed by a move of one space; (3) a move of one space followed by a move of two spaces; (4) three consecutive moves of one space each. Thus p_3 equals $\frac{1}{3} + (\frac{1}{3})^2 + (\frac{1}{3})^2 + (\frac{1}{3})^3 = \frac{16}{27}$. The recursive relation, in conjunction with these three values, enables us to calculate in principle the value of p_i for any i. Table 2 gives the first twelve p-values.

i	p_i
1	.3333
2	.4444
3	.5926
4	.4568
5	.4979
6	.5158
7	.4902
8	.5013
9	.5024
10	.4980
11	.5006
12	.5003

TABLE 2

Values of p_i for i up to 12. Teacher's version of game.

Unlike the case of the simpler game, however, there is no easy way of deriving on explicit formula for p_i from the recursive relation.* Nevertheless, some interesting observations can be made. Since the recursive relation states that any p-value beyond the third is simply the average of the preceding three values, it follows that it must exceed the smallest of those three values but be smaller than the largest of them. By repeated application of this argument, one is able to conclude that for any three consecutive values in the sequence $\{p_1, p_2, \ldots\}$, the largest of the three values exceeds all subsequent terms in the sequence, while the smallest of the three values is in turn exceeded by all subsequent terms in the sequence. An immediate consequence of this remark is the fact that the largest of p_1, p_2, p_3, namely p_3, is the largest of all the p-values, whereas the smallest of the three, namely p_1, is the smallest of all the p-values.

Is there a regular pattern to the way in which the p-values are ordered, as there was in the simpler model? You might be tempted to speculate that the largest p-values, in descending order, would be given by the subsequence $p_3, p_6, p_9, \ldots$. Alas, this is not the case, as an examination of Table 2 reveals: the pattern first breaks down with $p_{12}(=0.5003)$ which is not the fourth-largest, but only the sixth-largest p-value, being exceeded by both $p_8(=0.5013)$ and $p_{11}(=0.5006)$. Similarly, a listing of the smallest p-values in ascending order reveals no regular pattern: $p_1, p_2, p_4,$ p_7, p_5, p_{10}, etc.

Table 1 does suggest that the sequence of p-values converges to $\frac{1}{2}$ in the limit; in fact, this can be proved, although the proof requires some rather sophisticated mathematics. An easy consequence of this fact, however, is that no three consecutive p-values can lie on the same side of $\frac{1}{2}$; for if there did exist three such p-values, then by earlier remarks, all subsequent p-values must lie between the largest and smallest of the three; hence they must be further from $\frac{1}{2}$ than the one of the three which is closest to $\frac{1}{2}$. This contradicts the convergence of the p-values to $\frac{1}{2}$.

The p-values are forever oscillating about the value $\frac{1}{2}$, with no more than two consecutive p-values on the same side of $\frac{1}{2}$. There is no apparent pattern, however, to the oscillation. Table 3 gives the pattern of oscillation for the first 25 p-values.

Suppose we apply what we've learned to the specific example considered earlier. If there are five tasks to be placed on the board, and the board contains at least eleven spaces, then the optimal placement is on spaces 3, 6, 8, 9, and 11; this gives a value of 2.6127 for the expected number of tasks performed. By contrast, the worst possible placement would be on spaces 1, 2, 4, 5, and 7, giving a corresponding value of 2.2226. The difference between the best and the worst placement is 0.3901

---

* The more advanced reader may be interested to know that p_i can be expressed in the form $(1/2) + [(-1)^i/4] [(1 + \sqrt{-2})^i + (1 - \sqrt{-2})^i]$.

i	1	2	3	4	5	6	7	8	9	10	11	12
p_i vs. 0.5	−	−	+	−	−	+	−	+	+	−	+	+
i	13	14	15	16	17	18	19	20	21	22	23	24	25
p_i vs. 0.5	−	+	+	−	+	−	−	+	−	−	+	−	−

TABLE 3

Position of first 25 p-values relative to 0.5. + = above, − = below.

expected tasks; it is interesting that this difference is larger than in the case of the simpler game.

Some Final Words

This problem illustrates how certain aspects of a solution may be generalized for all cases, whereas other aspects may be specific to the particular case considered. The argument leading to the recursive relation for the p_i's is valid for all versions of the game, but the pattern of the resulting solution depends very strongly on how many possible moves can be made. Of course, the two most common versions of random hopscotch—based on moves of 1 to 6 spaces or of 2 to 12 spaces—have not been considered here, but the reader should now be in a position to analyze these versions with the aid of a calculator.

After working his way through this article, the reader can have the fun of generalizing the results. After all, all work and no play makes Johnny a dull boy.

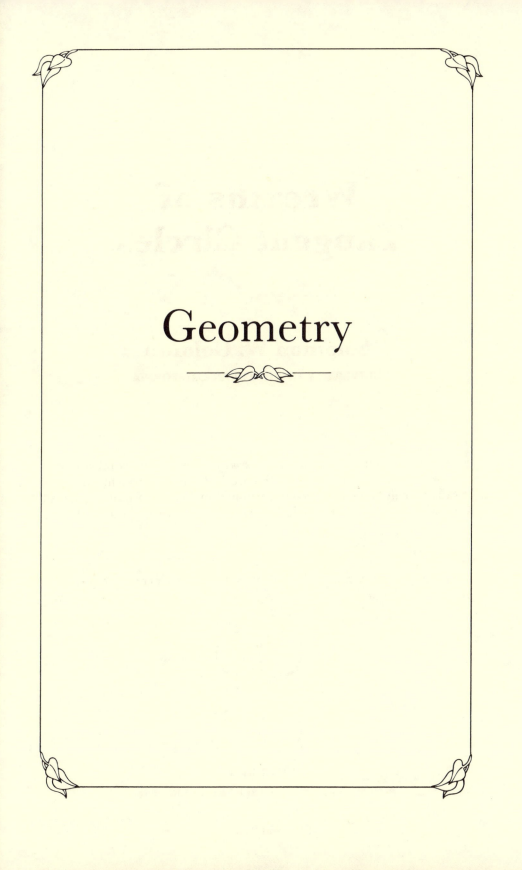

Geometry

Wreaths of Tangent Circles

Solomon W. Golomb

UNIVERSITY OF SOUTHERN CALIFORNIA

$\mathbf{I}$t is well known that a circle can be *exactly surrounded* by six other circles the same size as the original circle (Figure 1). Generally, n identical circles of radius s can be used to exactly surround one circle of radius r, for all $n \geq 3$ where, by elementary trigonometry we have (see Figure 2)

$$\sin \frac{\pi}{n} = \frac{s}{r + s},$$

from which $r = s \, (\csc \pi/n - 1)$. If we fix $r = 1$, then as n increases, s decreases, as shown in Table 1.

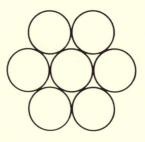

FIGURE 1

A circle exactly surrounded by six identical circles.

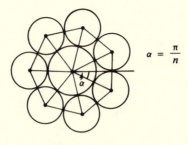

$$\alpha = \frac{\pi}{n}$$

FIGURE 2

The case $n = 7$ of n identical circles surrounding a given circle.

We next consider the case where the inner circle and the n surrounding circles may all have different radii (Figure 3). This problem is encountered when one has circular coins in a variety of sizes, and wishes to surround one of them exactly with several of the others. First, we need a precise definition of the notion that n circles exactly surround a given circle:

DEFINITION   A circle C^* is said to be *exactly surrounded* by the n circles $C_1, C_2, \ldots, C_n$ if each circle C_i is (exteriorly) tangent to C^*, to C_{i-1}, and to C_{i+1}, where the subscripts are taken modulo n.

Given the radii $r_1, r_2, \ldots, r_n$ of the surrounding circles, is there a simple formula for the radius r of the surrounded circle C^*?

The first surprise is that for $n > 3$, the size of the surrounded circle depends not only on the magnitudes of the radii of the circles surrounding it, but on the sequential order in which these circles occur! In general, with n given surrounding circles, there may be as many as $\frac{1}{2}(n-1)!$ different sizes for the exactly-surrounded circle in the middle, depending on the

n	s
3	$6.46410 = 3 + 2\sqrt{3}$
4	$2.41421 = 1 + \sqrt{2}$
5	1.42592
6	1.00000
7	$.76642$
8	$.61991$
9	$.51980$
10	$.44721$

TABLE 1

The radius s of each of n circles exactly surrounding a unit circle.

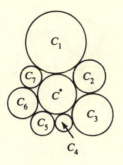

FIGURE 3

A circle exactly surrounded by seven unequal circles.

permuted order of the surrounding circles. (In fact, if the n radii of the surrounding circles are algebraically independent real numbers, there will indeed be $\frac{1}{2}(n-1)!$ different values for the radius of the exactly-surrounded circle.) The importance of the order of surrounding circles is exemplified in Figure 4.

In Figure 4, the circles C_6 and C_6' are the same size. Yet C_6 placed between C_4 and C_5, accomplishes something in surrounding C^*, while C_6', between C_1 and C_2, does not. If C_6' were enlarged a tiny bit, it would force C_1 and C_2 to separate, but would still not accomplish as much as it would between two circles more nearly its own size!

Roughly speaking, the radius of the exactly-surrounded circle is maximized if the surrounding circles are arranged so that each circle is as close as possible to others of the same size. Conversely, this radius is minimized if adjacent circles are as disparate as possible in size. If it were merely a matter of stringing the n circles out along a line, the maximizing and minimizing strategies would be clear (Figure 5). (The straight line can be thought of as a central circle C^* of infinite radius, which of course

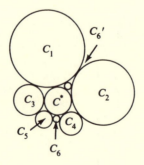

FIGURE 4

Circles C_6 and C_6' are the same size, but play very different roles in surrounding C^*.

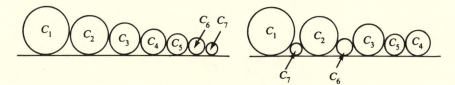

FIGURE 5

Maximizing and minimizing strategies for placing n unequal circles
along a line.

cannot be "surrounded" by a finite number of finite circles.) We number
the circles $C_1, C_2, C_3, \ldots, C_n$ so that their corresponding radii satisfy $r_1 \geq r_2 \geq r_3 \geq \ldots \geq r_n$.

NOTE   The notion of "minimizing" is not well-defined if C_n is smaller
than the circle simultaneously tangent to C_1, C_2, and C^*. For present
purposes, we will not consider the minimization problem for this extreme
case.

As to the problem of placing the n surrounding circles so as to
maximize the radius of the surrounded circle, the empirical algorithm
illustrated in Figure 6 is suggested, continuing from C_1 with $C_2, C_4, C_6, \ldots$ in
one direction, and with $C_3, C_5, C_7, \ldots$ in the other. An empirical algorithm
for minimizing the radius of C^* is illustrated in Figure 7, where, starting
with C_1, we continue in one direction with $C_n, C_2, C_{n-2}, C_4, C_{n-4}, C_6, \ldots$,
and in the other direction with $C_{n-1}, C_3, C_{n-3}, C_5, C_{n-5}, \ldots$. For neither of
these algorithms has the optimality been proved or disproved, for all
possible choices of radii of the circles $C_1, C_2, C_3, \ldots, C_n$.

When $n = 3$, there is, except for rigid Euclidean motions (rotation,
reflection, and translation), only one way to arrange the three circles
C_1, C_2, and C_3 to be mutually tangent, and this uniquely specifies the
radius of the exactly surrounded circle C^* (see Figure 8). If C_1, C_2, and C_3
have radii a, b, and c, respectively, then the radius r of C^* can be expressed
as

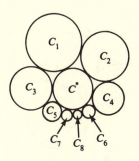

FIGURE 6

Heuristic algorithm for maximizing the radius of C^*.

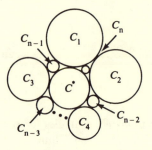

FIGURE 7

Heuristic algorithm for minimizing the radius of C^*.

$$r = \frac{abc}{ab + bc + ca + 2\sqrt{abc(a + b + c)}}$$

When $n = 4$, if the circles C_1, C_2, C_3, and C_4 all have different radii, then there are $\frac{1}{2}(4 - 1)! = 3$ essentially different ways in which they can be arranged so as to surround a central circle C^*, as shown in Figure 9.

We will consider the special case when only two distinct radii occur among the four surrounding circles. Suppose that C_1 and C_2 have radius a, while C_3 and C_4 have radius b. The only two distinguishable cases that occur are shown in Figure 10.

In Case I, by Pythagoras' Theorem we have $(a + b)^2 = (a + r)^2 + (b + r)^2$, from which $2ab = 2ar + 2br + 2r^2$, or $r^2 + (a + b)r - ab = 0$. Thus, by the quadratic formula,

$$r = \frac{\sqrt{a^2 + 6ab + b^2} - (a + b)}{2}$$

Integer values of r occur for many integer choices of a and b. Thus $r(3,2) = 1$, $r(10,3) = 2$, $r(12,5) = 3$, etc. (They are related to Pythagorean Triplets by the rule that if $r(a,b) = r$, then $(b + r, a + r, a + b)$ is the Pythagorean Triplet, which is exemplified in Figure 10, Case I.

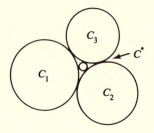

FIGURE 8

Three surrounding circles uniquely specify C^*.

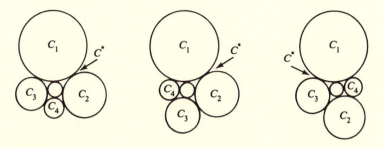

FIGURE 9

The three essentially different sequential orderings where $n = 4$.

Conversely, if (A,B,C) is a Pythagorean Triplet, with $A^2 + B^2 = C^2$, then we have a diophantine solution of Case I with $a = (A - B + C)/2$, $b = (-A + B + C)/2$, $r = (A + B - C)/2$.) It is interesting that a coin of radius 1 can be exactly surrounded by two coins of radius 2 and two coins of radius 3, *provided* that they are arranged as in Case I. If on the other hand, they are arranged as in Case II, they may *seem* to fit perfectly, but we shall see that they do not.

In Case II, let $h = h_1 + h_2$. By Pythagoras' Theorem, we see that $h_1^2 = (a + R)^2 - a^2 = 2aR + R^2$, $h_2^2 = (b + R)^2 - b^2 = 2bR + R^2$, and (if we observe that the dotted line has length h), $h^2 = (a + b)^2 - (a - b)^2 = 4ab$. Thus,

$$4ab = h^2 = (h_1 + h_2)^2 = h_1^2 + h_2^2 + 2h_1 h_2 = 2(aR + bR + R^2 + h_1 h_2),$$

from which

$$h_1 h_2 = 2ab - (aR + bR + R^2),$$

CASE I CASE II

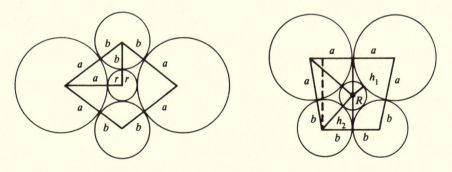

FIGURE 10

The two distinguishable cases of exactly surrounding a central circle with two circles of radius a and two circles of radius b.

and

$$(2aR + R^2)(2bR + R^2) = h_1^2 h_2^2 = [2ab - (aR + bR + R^2)]^2,$$

$$(2a + R)(2b + R) = \left(\frac{2ab}{R} - (a + b) - R\right)^2,$$

$$4ab + 2(a + b)R + R^2 = \frac{4a^2b^2}{R^2} + (a + b)^2 + R^2 - 4ab + 2(a + b)R$$
$$- \frac{4ab(a + b)}{R},$$

$$R^2(a^2 - 6ab + b^2) - 4ab(a + b)R + 4a^2b^2 = 0.$$

By the quadratic formula,

$$R = \frac{4ab(a + b) \pm \sqrt{16a^2b^2\{(a + b)^2 - (a^2 - 6ab + b^2)\}}}{2(a^2 - 6ab + b^2)}$$

and with the requirement $R > 0$,

$$R = \frac{2ab\{2\sqrt{2ab} - (a + b)\}}{8ab - (a + b)^2} = \frac{2ab}{(a + b) + 2\sqrt{2ab}}.$$

We see that for integral a and b, R is rational if and only if $2ab$ is a perfect square. In particular, $R(3,2) = 12/(5 + 4\sqrt{3}) = (12/23)\,(4\sqrt{3} - 5) = 1.006019^-$, an increase of about 0.6% over $r(3,2) = 1$. This difference is too small to have caused much concern "in practice." However, as the ratio a/b increases, so too does the ratio R/r. Thus, while $r(10,3) = 2$, we have $R(10,3) = 60/(13 + 4\sqrt{15}) = 60/71\,(4\sqrt{15} - 13) = 2.10586^-$; an increase in the radius R of C^* of some 5% over r.

When the circle C^* is exactly surrounded by one circle of radius b and n circles of radius a, the order of the surrounding circles does not affect the radius r of C^*. When $n = 2$, this is a special case of four mutually (externally) tangent circles, where the circle C^* of radius r is exactly surrounded by three circles, of respective radii a,b,c. As previously mentioned, and as derived in [1], the formula for this more general case is

$$r = \frac{abc}{ab + bc + ca + 2\sqrt{abc(a + b + c)}} = \frac{S_3}{S_2 + 2\sqrt{S_1 S_3}}$$

where $S_1 = a + b + c$, $S_2 = ab + bc + ca$, $S_3 = abc$, are the three elementary symmetric functions of a,b,c.

Our special case, when $a = c$, is more elementary, and satisfies

$$a = \frac{4br(b + r)}{(b - r)^2}, \qquad b > r.$$

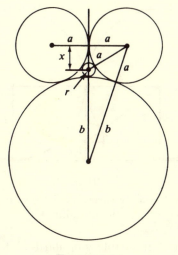

FIGURE 11

A circle exactly surrounded by two circles of radius a and one of radius b.

Figure 11 shows us that:

$$(a + r)^2 = a^2 + x^2, \ (a + b)^2 = a^2 + (x + r + b)^2.$$

Thus,

$$x + r + b = \sqrt{2ab + b^2}$$

with

$$x = \sqrt{2ar + r^2}.$$

Hence,

$$r + b = \sqrt{2ab + b^2} - \sqrt{2ar + r^2},$$

$$r^2 + 2rb + b^2 = (2ab + b^2) + (2ar + r^2) - 2\sqrt{(2ab + b^2)(2ar + r^2)},$$

$$(ab + ar - rb)^2 = (2ab + b^2)(2ar + r^2),$$

$$a^2b^2 + a^2r^2 = 2a^2br + 4abr^2 + 4arb^2,$$

$$(b - r)^2 = (1/a)(4br)(b + r),$$

$$a = 4br(b + r)/(b - r)^2.$$

The apparent symmetry between b and r is not realized geometrically, since clearly $b > r$. If a and b are given, we find from the quadratic formula that

$$r = b\frac{(a + 2b) - 2\sqrt{b^2 + 2ab}}{a - 4b}$$

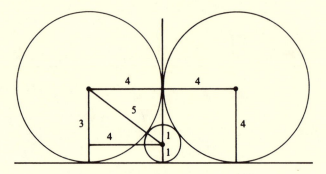

FIGURE 12

A circle of radius 1 exactly surrounded by a straight line and two
circles of radius 4.

The case $a = 4b$ is a "removable singularity," for in this case $r = \frac{1}{3}b$.
However, in the related equation

$$b = r\,\frac{(a + 2r) + 2\sqrt{r^2 + 2ar}}{a - 4r}\,,$$

the case $a = 4r$ is a genuine singularity, corresponding to $b = \infty$.
Specifically, with $r = 1$ and $a = 4$, we see from the 3-4-5 right triangle in
Figure 12 that the three mutually tangent circles are also tangent to a line
(that is, a circle of radius $b = \infty$).

There are many cases of integer triples (r, b, a), including $(1, 2, 24)$
$(1, 3, 12)$, $(5, 7, 420)$, $(6, 14, 105)$, etc. These may be obtained from such
formulas as:

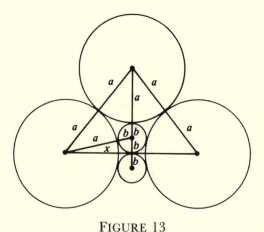

FIGURE 13

A circle of radius b exactly surrounded by three circles of radius a and
one of radius b.

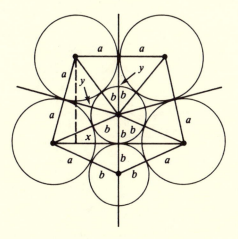

FIGURE 14

A circle of radius b exactly surrounded by four circles of radius a and one of radius b.

1 $r = n, b = n + 1, a = 4n(n + 1)(2n + 1)$

2 $r = n, b = n + 2, a = 2n(n + 1)(n + 2)$

3 $r = 2n, b = 2n + 8, a = n(n + 2)(n + 4)$, etc.

Let us also require the circle C^*, surrounded by one circle of radius b and n circles of radius a, to have radius $r = b$. For each $n \geq 3$, this configuration uniquely determines a ratio $Q_n = a/b$. We will discuss the values of Q_n for $n = 3, 4, 5$, and 6 in some detail.

In Figure 13, we see the case $n = 3$. Here we have $(2a)^2 = x^2 + (a + 2b)^2$ and $(a + b)^2 = x^2 + b^2$. Thus, $(2a)^2 - (a + b)^2 = (a + 2b)^2 - b^2$, from which $a^2 - 3ab - 2b^2 = 0$. By the quadratic formula, $Q_3 = a/b = (3 + \sqrt{17})/2 = 3.5615528\ldots$.

When $n = 5$, of course $Q_5 = a/b = 1$, since this is reduced to the configuration in Figure 1. However, when $n = 4$, the situation is somewhat more complex. From Figure 14, we observe each of the following:

$$(a + b)^2 = b^2 + x^2,$$

$$(a + b)^2 = a^2 + y^2,$$

$$(2a)^2 = (y + b)^2 + (x - a)^2.$$

Thus $x = \sqrt{a^2 + 2ab}$, $y = \sqrt{b^2 + 2ab}$, and $a^2 - 2ab - b^2 = b\sqrt{b^2 + 2ab} - a\sqrt{a^2 + 2ab}$, from which

$$3a^2 - ab - b^2 = \sqrt{(a^2 + 2ab)(b^2 + 2ab)},$$

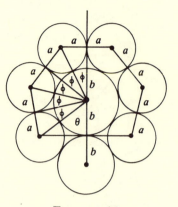

FIGURE 15

When a circle of radius b is exactly surrounded by n circles of radius a
and one of radius b, it is evident that $\pi = \theta + (n-1)\phi$, where
$\theta = \cos^{-1}[b/(a+b)]$ and $\phi = \sin^{-1}[a/(a+b)]$.

and

$$9a^4 - 8a^3b - 10a^2b^2 + b^4 = 0.$$

Thus $Q_4 = a/b$ is a root of $9x^4 - 8x^3 - 10x^2 + 1 = 0$, which has the
numerical value $Q_4 = 1.5684897\ldots$. (In principle, all quartic equations
have solutions which can be expressed in terms of radicals, but the explicit
solution in radicals for this case is too complicated to be worth including
here.)

The case $n = 6$ (see Figure 15) was first posed as a problem in an
unpublished letter from Gary A. Ford to Martin Gardner in 1973. Ford
commented that the problem was inspired by the arrangement of coins
(specifically, dimes and quarters), and that the best he and his colleagues at
the University of Maryland had been able to do was to express Q_6 as a root
of a tenth degree polynomial. (Ford subsequently published this problem
in MIT's *Technology Review* [2].) Can we do as well or perhaps better using
the methods already illustrated? More generally, is it possible to express Q_n
either algebraically or trigonometrically, as a function of n, for all $n \geq 3$?
We will see that a general trigonometric expression exists, from which a
polynomial equation can always be obtained. In particular, we will express
Q_6 as a root of a polynomial of degree eight.

In Figure 15, we see that the straight angle in the central circle of
radius b is the sum of an angle θ and 5 copies of an angle φ where θ
$= \cos^{-1}[b/(a+b)]$ and $\varphi = \sin^{-1}[a/(a+b)]$. For general n, the result is

$$\pi = \cos^{-1}\left(\frac{b}{(a+b)}\right) + (n-1)\sin^{-1}\left(\frac{a}{(a+b)}\right).$$

If we set $a/(a+b) = \alpha$ and $b/(a+b) = \beta$, then $\alpha + \beta = 1$, and the ratio Q_n
$= a/b = \alpha/\beta$, with

1 $$\pi = \cos^{-1}(1-\alpha) + (n-1)\sin^{-1}\alpha.$$

For practical computation, this formula is sufficient to allow α, and therefore Q_n, to be determined to any degree of accuracy. However, it is also possible, for each n, to replace the equation **1** by a polynomial equation having α as a root. Moreover, it is also possible to obtain a polynomial equation having $Q_n = \alpha/(1-\alpha)$ as a root, for if $f(x) = 0$ has $x = \alpha$ as a root, then it is readily verified, by direct substitution, that $g(x) = f[x/(1+x)] = 0$ has $\alpha/(1-\alpha)$ as a root.

The method is to rewrite **1** as $(n-1)\sin^{-1}\alpha = \pi - \cos^{-1}(1-\alpha)$, and then take the cosine of both sides, to obtain

$$\cos[(n-1)\sin^{-1}\alpha] = \cos[\pi - \cos^{-1}(1-\alpha)] = \alpha - 1.$$

Let $\sin^{-1}\alpha = z$. Then it is well known that $\cos(n-1)z$ is a polynomial of degree $n-1$ in $\cos z$, and $\cos z = \cos(\sin^{-1}\alpha) = \sqrt{1-\alpha^2}$. Thus, in the worst case, α may be the root of a polynomial of degree $2(n-1)$. In fact, for n odd, Q_n satisfies a polynomial equation of degree $\leq n-1$ for all $n \geq 3$; and for n even, Q_n satisfies a polynomial equation of degree $\leq 2(n-2)$, for all $n \geq 4$.

We will illustrate the actual computations for the (already solved) cases $n = 3$ and $n = 4$.

When $n = 3$, $\cos 2z = \alpha - 1$, $2\cos^2 z - 1 = \alpha - 1$, $2\cos^2 z = \alpha$, $2(1-\alpha^2) = \alpha$, $2\alpha^2 + \alpha - 2 = 0$, and α satisfies $f(x) = 2x^2 + x - 2 = 0$. Q_3 then satisfies $g(x) = f[x/(1+x)] = 0$, and $(1+x)^2 g(x) = 2x^2 + x(1+x) - 2(1+x)^2 = x^2 - 3x - 2 = 0$. Then Q_3 is the root $(3+\sqrt{17})/2 = 3.56155$ of $g(x) = x^2 - 3x - 2 = 0$.

Similarly, when $n = 4$, $\cos 3z = \alpha - 1$, $4\cos^3 z - 3\cos z = \alpha - 1$, $\sqrt{1-\alpha^2}\ \{4(1-\alpha^2) - 3\} = \alpha - 1$, $(1-\alpha^2)(1-4\alpha^2)^2 = (\alpha-1)^2$, $(1+\alpha)(16\alpha^4 - 8\alpha^2 + 1) = 1 - \alpha$, $16\alpha^5 + 16\alpha^4 - 8\alpha^3 - 8\alpha^2 + 2\alpha = 0$ and since $\alpha = 0$ is not a possible solution, α is a root of $f(x) = 8x^4 + 8x^3 - 4x^2 - 4x + 1 = 0$. Hence $Q_4 = \alpha/(1-\alpha)$ satisfies $g(x) = f[x/(1+x)] = 0$, so that $(1+x)^4 g(x) = 8x^4 + 8x^3(1+x) - 4x^2(1+x)^2 - 4x(1+x)^3 + (1+x)^4 = 9x^4 - 8x^3 - 10x^2 + 1 = 0$ has $Q_4 = 1.56849 \ldots$ as a root.

In Table 2, we give the polynomials for Q_n for $3 \leq n \leq 9$, with the corresponding values of Q_n. Many patterns in the coefficients of these polynomials are readily apparent. The even and the odd values of n clearly correspond to separate populations of polynomials.

Finally, we mention the elegant result that when three circles, of respective radii a, b, and c, are mutually externally tangent to one another and to a common tangent line, with $a \geq b \geq c$, then [3]

$$\frac{1}{\sqrt{a}} + \frac{1}{\sqrt{b}} = \frac{1}{\sqrt{c}}.$$

(This generalizes the situation depicted in Figure 12.)

n	degree	polynomial for Q_n	Q_n
3	2	$x^2 - 3x - 2.$	3.56155
4	4	$9x^4 - 8x^3 - 10x^2 + 1.$	1.56849
5	4	$x^4 - 11x^3 + x^2 + 7x + 2.$	1.00000
6	8	$25x^8 - 188x^7 + 236x^6 + 436x^4 - 2x^4 - 180x^3$ $- 68x^2 - 4x + 1.$	0.73403
7	6	$x^6 - 31x^5 + 40x^4 + 42x^3 - 7x^2 - 11x - 2.$	0.58027
8	12	$49x^{12} - 956x^{11} + 5090x^{10} - 3036x^9 - 11121x^8$ $+ 1800x^7 + 10140x^6 + 4200x^5 - 865x^4 - 972x^3$ $- 222x^2 - 12x + 1.$	0.48015
9	8	$x^8 - 55x^7 + 259x^6 + 77x^5 - 215x^4 - 101x^3$ $- 17x^2 + 15x + 2.$	0.40977

TABLE 2

Polynomials and values for Q_n, $3 \leq n \leq 9$.

NOTES TO TABLE 2

1 It appears that for $n = 4k + 1$, the polynomial of degree $n - 1$ which we obtain always has $x = 1$ as a root. For $n = 5$, this corresponds to $Q_5 = 1$. However, for $n = 9, 13, 17, \ldots$, the number Q_n satisfies a polynomial equation of degree $\leq n - 2$.

2 For $n > 4$, none of these polynomials have been proved irreducible. Note that the polynomial for $n = 5$ is the product of $x - 1$ and the irreducible cubic $x^3 - 10x^2 - 9x - 2$. For $n = 9$, the factorization is

$$(x - 1)(x^7 - 54x^6 + 205x^5 + 282x^4 + 67x^3 - 34x^2 - 17x - 2).$$

References

1 Beecroft, Philip. 1842. Properties of circles in mutual contact. *Lady's and Gentleman's Diary*, pp. 91–96.

2 Ford, Gary A. 1974. *Technology Review*, problem June 5, vol. 76: 57–8. (See also problem NS 13, vol. 81, November 1978, p. 84.)

3 Trigg, C. W. 1940. Problem E432, *American Math. Monthly*, 47: 487.

4 ———. 1941. Solution to Problem E432 *American Math. Monthly* 48: 267–68.

Bicycle Tubes Inside Out

Herbert Taylor

University of Southern California

The old rubber sheet geometry discussed surfaces which could be bent, stretched, or twisted, while they were kept smooth and whole. One popular topological pastime is to try to visualize what a bicycle tube would look like turned inside out. As far as I know, these curiosities have no serious implications for mathematics, but they can be used to cultivate flexibility in visual thinking.

Let us start by moving the surface of Figure 1A to that of Figure 1B as an example. The reader is asked to visualize, or draw, a sequence of

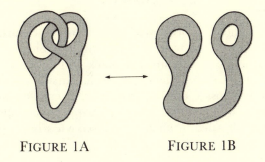

FIGURE 1A   FIGURE 1B

pictures moving Figure 1A to Figure 1B, without cutting, and without letting one part of the surface touch another part. Figure 1C is a possible sequence.

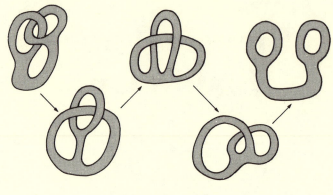

FIGURE 1C

The next exercise involves cutting a temporary hole in the surface. Instead of merely turning a bicycle tube inside out, how about turning a more complicated surface inside out? It will soon be apparent that the mildly complicated Figure 2A could become very complicated, so, to simplify things we paint the inside black, and the outside grey. We are going to cut a small hole in the surface temporarily and put a rim on the hole to keep track of it.

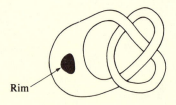

Rim

FIGURE 2A

Figure 2B shows a sequence for turning the surface inside out; shrinking the complicated part, and passing it through the hole.

After the hole is closed up, the black will cover the whole outside, whereas it formerly covered the whole inside. An advantage of the method of turning surfaces inside out, just pictured, is evident in Figure 3. This method makes it just as easy to see what happens to a sphere with many handles when it is turned inside out as to see what happens to the bicycle tube.

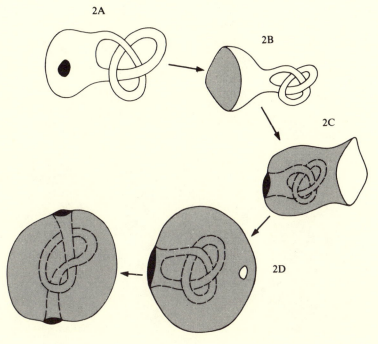

2A

2B

2C

2D

2E

FIGURE 2B

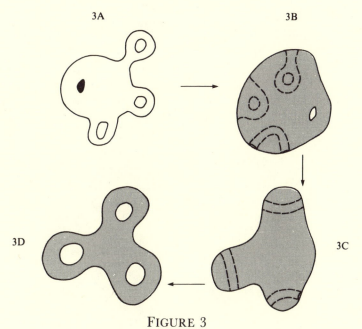

3A

3B

3C

3D

FIGURE 3

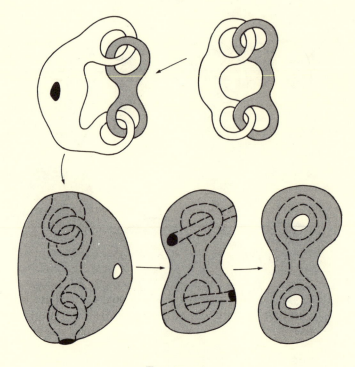

FIGURE 4

The last sequence will aim for a simple picture of what Figure 4 will look like, after one of the two linked surfaces is turned inside out.

These pictorial ideas are not new. They occurred to me 25 years ago, and perhaps to several people before that. The twister of Figure 5 was posed recently by Dennis L. Johnson, who is well-versed in the theory of knots. Now the reader is invited to finish up with a little light exercise, moving from Figure 5A to Figure 5B in the same fashion as 1A → 1B was done.

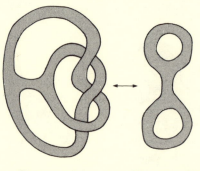

FIGURE 5A   FIGURE 5B

Flexing Surfaces

Robert Connelly

CORNELL UNIVERSITY

Suppose a closed polyhedral surface is built from flat pieces of stiff cardboard taped together along their edges. Will the surface flex? That is; will it change its shape continuously without ripping the tape or bending the cardboard? As an example, let us consider the octahedron shown in Figure 1. If one builds this out of cardboard it turns out to be very rigid and does not flex. However, if the top is slightly smaller than the bottom, the top will pop down as shown in Figure 2. To do this, one must bend the cardboard. Figure 1 will not continuously move into Figure 2 without distortion.

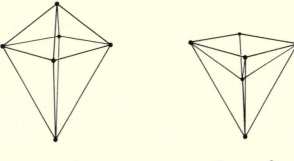

FIGURE 1   FIGURE 2

In 1813 Cauchy, the well-known french mathematician, proved that any *convex* polyhedral surface is rigid. (It is understood that the flat natural faces are the pieces of cardboard and are held rigid). So it seems natural to conjecture that all surfaces, convex or not, are rigid. Unfortunately, this "rigidity conjecture" is false. There is an embedded polyhedral surface, without self-intersections, that flexes. In what follows I will describe some of the examples that I have found, and subsequent modifications by others which refute this conjecture.

The Construction

To understand why the forthcoming surfaces flex, we describe some of the flexible octahedra of R. Bricard, a French engineer who discovered them in 1897. These surfaces do have self-intersections so we regard them as a collection of incompressible, inextendible rods connected by flexible rubber nodes at their endpoints. To build these octahedral frameworks we start with a skew quadrilateral $aba'b'$ as in Figure 3, with opposite sides of the same length. It turns out that there is then a line L in 3-space such that the quadrilateral is symmetric about L. That is, if the quadrilateral is rotated 180° about L, it is rotated into itself. We think of $aba'b'$ as the

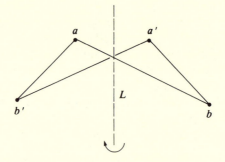

FIGURE 3

equator of the octahedral framework. Choose a point c not on L; the line of symmetry. Join c to each of the nodes a, b, a', b' by a rod. It is not hard to check that this framework flexes as it is. So flex it and join it at each instant to the congruent framework $c'(a'b'ab)$ obtained from the first by rotation by 180° about L. Note that c is rotated into c'. The union is one of the flexible octahedra of Bricard and is easy to build. See Figure 4 for the completed framework. Note that if all the triangles are "filled in" the resulting surface will have many self-intersections. Our goal is to reduce and simplify these self-intersections as much as possible.

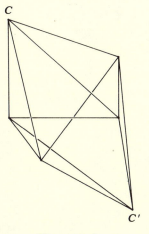

FIGURE 4

In Figure 5 we see another version where all the rods start out lying flat in a plane. The line of symmetry L is perpendicular to this plane and as the framework flexes, the vertices do not remain in a plane, but this is a very convenient position for starting.

Another slight variation on this framework is to start with the points a and a' in a horizontal plane H as in Figure 5. Then choose points b, b' at a height $\varepsilon > 0$ above H and c, c' at height $\delta > \varepsilon$ above H so that all the points project orthogonally onto the picture of Figure 5. Line L is again perpendicular to H, the octahedral framework is still flexible, and the boundaries of the triangles $ab'c$ and $a'bc'$ link; that is, they cannot be pulled apart without breaking.

To construct the embedded flexible surface, we start with the surface that is used for Figure 5. Instead of filling in all of the triangles with flat planar pieces, we change the surface somewhat, still keeping the rods of the old framework as edges in our surface. We regard the octahedral surface as being made of two pieces—a bottom and a top. Let us say that the bottom

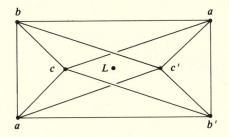

FIGURE 5

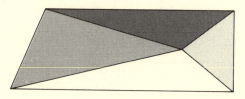

FIGURE 6

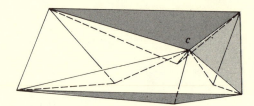

FIGURE 7

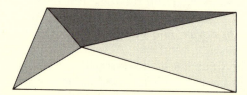

FIGURE 8

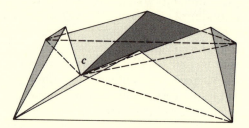

FIGURE 9

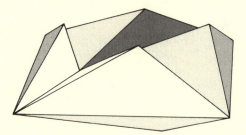

FIGURE 10

is exemplified in Figure 6. We push down on each of the triangular faces to get a new surface that looks like Figure 7, in which each triangle is replaced with an upsidedown bottomless tetrahedron—or, a pit.

Similarly we replace the top surface (Figure 8) with the surface in Figure 9, where each triangle is replaced by a bottomless pyramid. Just as the surfaces of Figures 6 and 8 flex, so do the surfaces of Figures 7 and 9, with the extra vertices; the apexes of the pyramids move rigidly with respect to their bases.

We next glue the surfaces of Figures 7 and 9 together along their common boundary to get the surface of Figure 10. This surface is flexible, just like the surface of Figure 5, but unfortunately, it has a couple of self-interactions—s and s'. Figure 11 shows the parts of the surface of Figure 10 that intersect: points s and s' correspond to the crossing points of Figure 5.

In order to get rid of s and s' we build what I call a crinkle, which is again, based on the Bricard flexible octahedra. Choose a planar quadrilateral $defg$ with opposite sides equal, $de = fg$, $ef = gd$ as in Figure 12, with

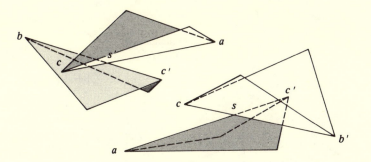

FIGURE 11

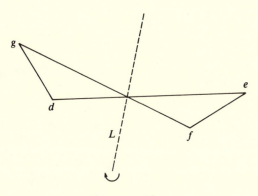

FIGURE 12

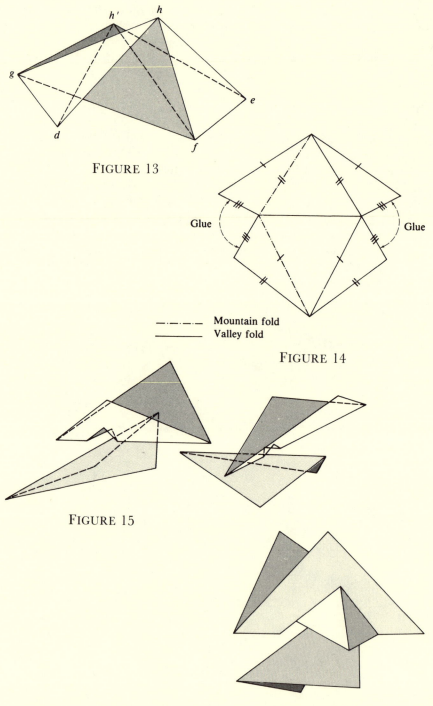

FIGURE 13

Glue   Glue

----·----· Mountain fold
———— Valley fold

FIGURE 14

FIGURE 15

FIGURE 16

the segment *de* intersecting *fg*. Choose a point *h* directly over the center of the circle through *defg* and *h'*—the same distance under the center. Thus $hd = he = hf = hg = h'd = h'e = h'f = h'g$. Then the frameworks $h(defg)$ and $h'(defg)$ flex in conjunction. (The quadrilateral *defg* actually remains coplanar.) The union of the triangular faces, *hef, hfg, hgd, h'ef, h'fg, h'gd* is the crinkle; an octahedron with two triangular faces removed (see Figure 13), with boundary *hdh'e*. The distance from *d* to *e* remains fixed during the flex. Figure 14 shows how to build a crinkle.

To construct the final embedded flexible surface, take the surface of Figure 10 and cut out, as shown in Figure 15, one small quadrilateral hole around each of the self-intersection points. Then insert a crinkle of the appropriate size into each of the holes. If the crinkle is positioned properly, there will be no self-intersections in the resulting surface (Figure 16). Since *de* remains at a fixed position in the crinkle, the surface with the crinkle and the two holes removed flex in conjunction. Thus, the whole crinkled surface flexes. It looks something like Figure 17.

This surface was one of the first flexible surfaces I found, and I paid no attention to the simplicity of construction or how few vertices would be needed. Subsequently N. H. Kuiper and Pierre Deligne modified my construction to get a surface with 11 vertices and 18 faces. They started with the framework described in the text following Figure 5. Instead of adding four pits to the bottom surface, they added only one (as in Figure 18). The other three triangles they kept flat. For the upper surface

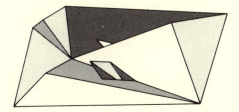

FIGURE 17

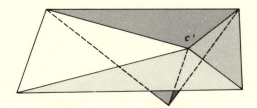

FIGURE 18

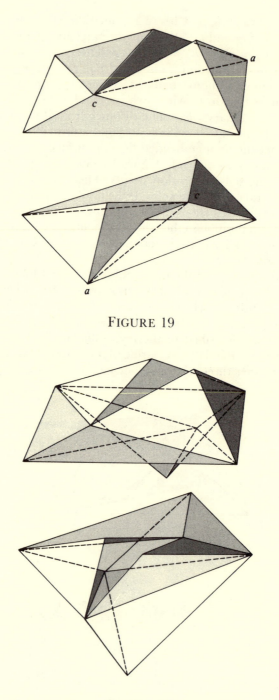

FIGURE 19

FIGURE 20

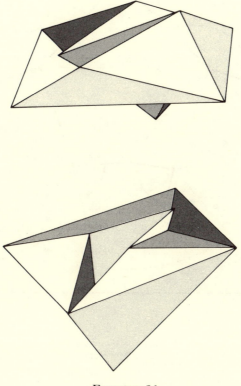

FIGURE 21

they only added two mountains (as in Figure 19—shown with two views). When these new upper and lower surfaces were glued together along their common boundary, the line segment $c'b$ intersected the sides of the two mountains above ca. Due to the slight raising of c,c',b,b' this is the only place where the surface intersected itself. They then removed ca and the inside of the two triangles with ca as an edge and placed a carefully proportioned crinkle, as in Figure 13, in the hole that was created. Points d and e in Figure 13 fit into c and a respectively, and the apexes of the two mountains are h and h'. Figure 20 shows two views of the upper and lower surfaces glued together with ca removed from the upper surface. Figure 21 shows two views of the final flexible surface with the crinkle added.

To top this, Klaus Steffen found a flexible surface with only 9 vertices. He started with two identical crinkles like those in Figure 14. They were joined with two other triangles as in Figure 22, where the figure has a symmetry about a vertical line so that the corresponding lengths are equal. The result was a flexible surface, something like Figure 23.

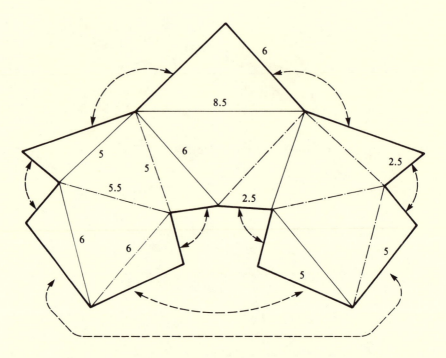

———— Valley folds
—·—·— Mountain folds

FIGURE 22

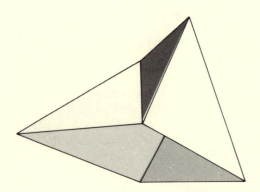

FIGURE 23

Some Conjectures

An interesting property of the previous examples is that as they flex, the volume enclosed by these surfaces remains constant. I do not see however, how to prove that the volume is constant for *every* possible flexible surface.

> **CONJECTURE 1**   If a triangulated polyhedral surface flexes, its volume remains constant during the flex.

Even more surprising things seem to be true. Let P and P' be two 3-dimensional polyhedra in 3-space. We say P is equivalent to P' by dissection, and we write $P \sim P'$, if we can dissect P into a finite number of polyhedral pieces, $P_1, P_2, \ldots, P_k$, and then reassemble them to get P'. (So $P = P_1 \cup \ldots \cup P_k$, $P_i \cap P_j \subset$ (boundary P_i) $\cap$ (boundary P_j), for $i \neq j$, $P' = P'_1 \cup \ldots \cup P'_k$, $P'_i \cap P'_j \subset$ (boundary P'_i) $\cap$ (boundary P'_j), for $i \neq j$, and P_i is congruent to P'_i for $i = l, \ldots, k$.) M. Dehn's solution to the problem posed by Hilbert resulted in the finding that the regular cube and tetrahedron of the same volume are *not* equivalent by dissection. (See page 35 of Martin Gardner's *Second Book of Mathematical Puzzles and Diversions*.) However, suppose that P_t is the 3-dimensional solid enclosed by one of the above flexible surfaces at time t. A result of J. P. Sydler implies that $P_0 \sim P_t$ for all t in the flexing interval. A good discussion of Hilbert's third problem and this (non-trivial) result of Sydler can be found in the recently translated book of Boltianskii (Boltianskii, V. 1978. *Hilbert's Third Problem*. New York: John Wiley and Sons.) Still the general question remains.

> **CONJECTURE 2**   If P_t is the polyhedral solid enclosed at time t by any flexing polyhedral surface, then $P_0 \sim P_t$, for all t.

Even to see the specific dissections for the surfaces described above would be interesting.

This material is based upon work partially supported by the National Science Foundation under Grant No. MCS-7902521.

Planting Trees

Stefan Burr

CITY UNIVERSITY OF NEW YORK

> Your aid I want, nine trees to plant
> In rows just half a score;
> And let there be in each row three.
> Solve this: I ask no more.

In 1821, John Jackson published this mathematical conundrum in a book of problems called *Rational Amusement for Winter Evenings* [4]. These days, verse is not as popular, and a modern-day puzzle poser might even dispense with the trees, saying: Arrange nine points on a plane so that there occur ten rows of three points. When a mathematician encounters such a problem, he feels a natural urge to generalize it and then wants to make it more precise. This leads to the following version: Given a positive integer p, how can p points ($p \geq 3$) be arranged on a plane, no four in a straight line, so that the number of straight lines with three points on them is maximized? We will call this maximal number of lines $l(p)$.

The formidable mathematician J. J. Sylvester pursued the elusive $l(p)$ in the nineteenth century, and it has attracted sporadic attention since, from amateurs as well as professionals. In the past, amateurs have often made valuable contributions to mathematical problems like this. It is a shame that the amateur mathematician seems to be a dying breed today. This is partly due to the inaccessibility of a lot of mathematics. There are however many areas, especially those related to combinatorics, where an amateur can work. Unfortunately, the public doesn't hear much about these accessible questions. Generally, they hear about the glamorous problems like Fermat's Last Theorem, where even a professional has little

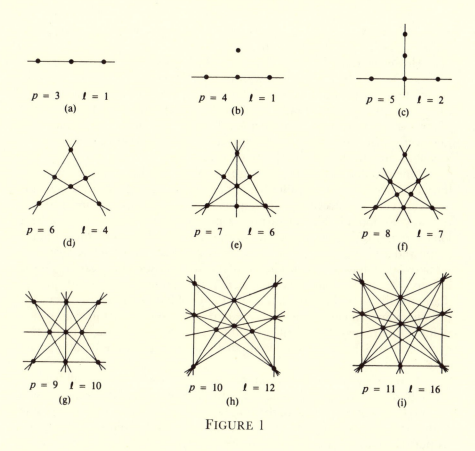

FIGURE 1

chance of making major progress. One of the charms of combinatorial geometry then, (of which this orchard problem is an example) is the fact that amateurs can often make substantial contributions to it.

Figure 1 illustrates some orchards corresponding to $p = 3, 4, \ldots, 11$. All of these are optimal solutions, achieving $l(p)$ lines. Only two other values of $l(p)$ are known exactly—we will get to these shortly. But for now, note that four points are no better than three, and that sometimes one orchard is contained in another, such as for $p = 10$ and 11.

A natural question to ask is whether the above arrangements are unique solutions. The answer is no; Figure 2 shows a different arrangement

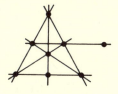

FIGURE 2

for $p = 8$, derived from the one for $p = 7$; of course the additional point can go anywhere on the new line.

However, there is another way in which these arrangements are not unique: they can be projectively transformed. We won't attempt to explain completely what a projective transformation (also called a projection) is, but one type can be literally visualized. Tilt the page and look at the diagrams on the slant: the distances and angles change with the change in perspective, but straight lines still appear straight, so that the transformed arrangement still has the properties we want.

One problem arises from doing this: As with railroad tracks, parallel lines can become non-parallel, changing the nature of an arrangement, and vice versa. However, mathematicians were perhaps the inventors of the now-fashionable idea of turning a problem into an opportunity, and long ago they did so for projective transformations. They took the ordinary plane and added an imaginary *line at infinity* consisting of *points at infinity*. A set of parallel lines were considered to intersect at some point on the line at infinity. Such a point was considered the same if one went in the diametrically opposite direction, so that a set of parallel lines intersected in just one point. In fact it was now true that any two lines met in exactly one point. This created the so-called Euclidean projective plane and projective geometry, which has proved itself to be a fruitful source of interesting mathematics.

In our case, putting points at infinity is very useful for simplifying and making more symmetrical some of the more complicated diagrams. It was also useful for finding some of them in the first place, and for proving some of the results we will refer to. As an example of putting points at infinity, let us put the points of the top line in the arrangement for $p = 9$ at infinity; this leads to Figure 3.

The arrows labeled a, b, and c indicate the directions of the three points at infinity in the arrangement; any of the arrows could just as well have pointed in the opposite direction. Of course, the line at infinity is to be considered one line of the arrangement.

Figure 4 shows the only other orchards that are known to be optimal, namely $p = 12$ and $p = 16$. Note that for $p = 16$ the orchard contains an optimal 7-point orchard. Each arrangement contains three points at

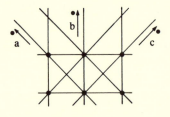

FIGURE 3

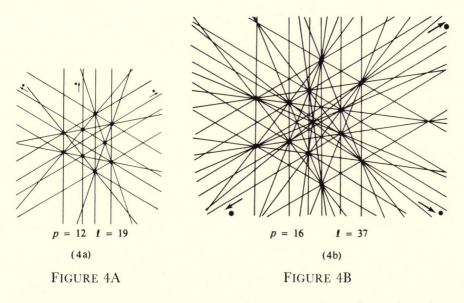

$p = 12 \quad l = 19$

(4a)

FIGURE 4A

$p = 16 \qquad l = 37$

(4b)

FIGURE 4B

infinity and the line at infinity. These diagrams can be projected in such a way that the points and line at infinity in each become real points, but the symmetry would be lost, and they would be hard to draw on a small piece of paper.

What about other values of p? Table 1 tells what has been learned about $l(p)$ for $p = 3, 4, \ldots, 25$, giving the best-known lower and upper bounds. The twelve cases in which $l(p)$ is known exactly are indicated with an asterisk, and for them the upper bound is dispensed with. Table 1, and in fact almost all of the results here, are taken from a paper by myself, B. Grünbaum and N. J. A. Sloane, entitled "The Orchard Problem" [1], although many of those were taken, in turn, from previous work.

From $p = 20$ on, all the bounds in Table 1 are the result of two general theorems, given below.

THEOREM 1 $l(p) \geq \lfloor p(p-3)/6 \rfloor + 1$, where $\lfloor x \rfloor$ denotes the largest integer $\leq x$.

THEOREM 2   If $p \geq 4$, then $l(p) \leq \lfloor (p(p-1)/2 - \lceil 3p/7 \rceil)/3 \rfloor$, where $\lceil x \rceil$ denotes the smallest integer $\geq x$. (Of course, $l(3) = 1$.)

We will not prove Theorem 1, but we will give some indication how it comes out of the theory of cubic curves, that is, curves satisfying an algebraic equation of degree 3. Figure 5 shows the symmetric cubic curve defined by the equation $(x - 1)((x + 2)^2 - 3y^2) = 8$, together with twelve points on it, including three points at infinity. These twelve points are arranged in the same way in the twelve-point orchard in Figure 4a.

p	Lower bound for $l(p)$	Upper bound for $l(p)$
3	1*	
4	1*	
5	2*	
6	4*	
7	6*	
8	7*	
9	10*	
10	12*	
11	16*	
12	19*	
13	22	24
14	26	27
15	31	32
16	37*	
17	40	42
18	46	48
19	52	54
20	57	60
21	64	67
22	70	73
23	77	81
24	85	88
25	92	96

TABLE 1

Why does the use of a cubic curve produce good arrangements of points? The secret lies in the fact that certain cubic curves can be given a parametric representation, based on the so-called *Weierstrass elliptic functions*. This representation gives every point on the curve a real number of at least 0 and less than 360, in such a way that three points are in a line if, and only if, the three corresponding numbers add up to a multiple of 360. (We could make the magic number anything we liked, but 360 is convenient, and lets us think of the numbers as being something like angles.) Figure 6 shows the curve of Figure 5, but with most lines removed and the numbers of the points given by the parametric representation inserted. It is easy to check that each straight line in Figure 5 satisfies the above criterion (including the line at infinity).

It is not hard to work out, for any p, how to choose p numbers of at least 0 and less than 360 in such a way that three of them add up to a

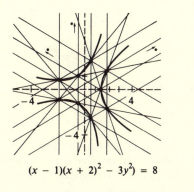

$(x - 1)(x + 2)^2 - 3y^2 = 8$

FIGURE 5

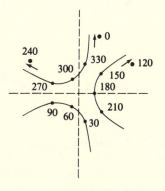

FIGURE 6

multiple of 360 as often as possible. Such a choice leads to an orchard, and therefore to a lower bound on $l(p)$. The resulting lower bound is that of Theorem 1. In 1868, Sylvester [6] proved a lower bound that is the same as in Theorem 1 except when p is a multiple of 3. When this is the case, Theorem 1 is better by one. (It is always very satisfying to improve on the work of someone who is as brilliant as Sylvester—even by a bit!)

Theorem 1 accounts for every lower bound in Table 1, except for p = 7, 11, 16, and 19. In each of these cases, it turns out to be possible to use an arrangement of $p - 1$ points on a specially-chosen cubic curve, and then to add one point not on the curve to get a better orchard than that associated with Theorem 1. As an example, Figure 7 shows the construction of the 16-point orchard of Figure 4b. Note that the extra point $\emptyset$ is not 0—since it is not on the cubic, it has no number. To make the figure easier to see, only the lines through $\emptyset$ have been given.

Now let's turn to upper bounds for $l(p)$, including Theorem 2. Here the approach is quite different. A useful tool is what we will call the *graph* of an orchard. Draw the points of an orchard, and then connect two points with a line segment (called an edge) if they are not in line with a third

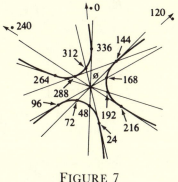

FIGURE 7

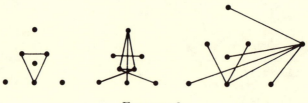

FIGURE 8

point. For example, Figure 8 shows the graphs corresponding to the 7-point and 8-point orchards of Figure 1 and the alternative 8-point orchard of Figure 2. The fundamental difference between the two 8-point orchards shows up very clearly in their graphs; for instance, one contains a triangle and the other doesn't.

Suppose that a point in an orchard has k lines of three going through it. Then, in the corresponding graph, the point is on $p - 1 - 2k$ edges, since the k lines eliminate $2k$ points that would have otherwise been adjacent to the point in the graph. (We call the number of such edges the *degree* of the point.) Because of this fact, it must be that if p is even, every point in the graph must have odd degree, and if p is odd, every point in the graph has even degree (including zero), since p and $p - 1 - 2k$ have opposite parity.

Furthermore, consider the number of edges in the graph of an orchard. If an orchard has no three points in line, then every pair of points gives an edge in the resulting graph, and it is not hard to see that such a graph (called a complete graph) has $p(p - 1)/2$ edges. However, any time three points are on a line in the orchard, three edges are eliminated from the graph.

Therefore, if the orchard has l lines, the corresponding graph has $p(p - 1)/2 - 3l$ edges. Hence, if e is the number of edges, we get

$$e = \frac{p(p - 1)}{2} - 3l$$

so

$$l = \left(\frac{p(p - 1)}{2} - e \right) \Big/ 3.$$

Since $e \geq 0$,

$$l \leq \left\lfloor \frac{p(p - 1)}{6} \right\rfloor.$$

But if p is even, the degree of every point of the graph is odd, which means the degree is at least one. For this to happen, it must be true that $e \geq p/2$, so

$$l \leq \left\lfloor \frac{p(p - 2)}{6} \right\rfloor, \text{ when } p \text{ is even.}$$

With a little fiddling, the above inequalities for l can be combined into the following:

$$l(p) \leq \left\lfloor \frac{p}{3} \left\lfloor \frac{p-1}{2} \right\rfloor \right\rfloor.$$

This result is not as strong as Theorem 2. For that theorem, we need a theorem of Kelly and Moser [5]. This says that, in any arrangement of p points such that not all p are on one line, there must be at least $\lceil 3p/7 \rceil$ pairs of points which are not on a line with any third point. For an orchard with at least four points, this means that in the corresponding graph, $e \geq \lceil 3p/7 \rceil$, so that by the previous paragraph,

$$l \leq \left\lfloor \left(\frac{p(p-1)}{2} - \left\lceil \frac{3p}{7} \right\rceil \right) \middle/ 3 \right\rfloor,$$

which is essentially Theorem 2.

It will be noticed that this proof of Theorem 2 doesn't really use the clever construction of a graph from an orchard very heavily. However, the idea of the graph of an orchard is very useful in dealing with special cases. In fact, in the only cases in which a better upper bound is known than that given by Theorem 2, orchard-graphs have proved useful. These cases are $p = 8$, 10, 12, and 14; each time the upper bound (given in Table 1) is one better than that given by Theorem 2. We will indicate a proof for $p = 8$— the others are similar but rather long.

To prove that $l(8) \leq 7$ (which is the same as saying $l(8) = 7$, since a 7-line orchard exists) we must show that no 8-point, 8-line orchard exists. Assume, to the contrary, that such an orchard did exist. Consider its graph; the number of edges e is given by

$$e = \frac{p(p-1)}{2} - 3l = \frac{8 \cdot 7}{2} - 3 \cdot 8 = 4.$$

Furthermore, because 8 is even, every point in the graph has odd degree. The only possible 4-edge graph satisfying this requirement must consist of four separated edges, as shown in Figure 9. (In Figure 9 the graph is drawn abstractly, showing the connections while ignoring the actual positions of the points.)

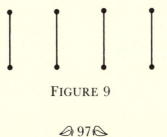

FIGURE 9

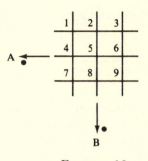

FIGURE 10

Consider the points A and B in the graph. Turning to the corresponding orchard, we can use a projective transformation to put them on the line
at infinity. By studying the graph, we see that the other six points in the
orchard must fall on three lines emanating from A, and also on three lines
emanating from B. The situation is shown somewhat schematically in
Figure 10. (The only way in which Figure 10 is schematic is the even
spacing of the lines.)

The six points (other than A and B) in the orchard must fall among
the nine intersection points shown in Figure 10. Moreover, we have
accounted for only six of the lines in the (hypothetical) orchard; the other
two lines must use the nine intersection points. The only way to add two
new lines is if they go through the points 1,5,9 and 3,5,7—but this
accounts for only seven points, namely A, B, 1, 3, 5, 7, and 9. (Furthermore, the lines A5 and B5 still have only two points on them.) Therefore,
the sought-for orchard is impossible, and $l(8)$ is indeed 7.

Although more could be said about this problem, and even more
about related problems, this discussion should convey the flavor of the
subject. We will close with some comments on what more might be done.
Obviously, the best thing of all would be to determine $l(p)$ exactly for all p.
This seems difficult, but may be possible. In [1], we conjectured that

$$l(p) = \lfloor p(p-3)/6 \rfloor + 1$$

for all $p \neq 7$, 11, 16, 19; in other words, that Theorem 1 tells almost the
whole story.

One could hope at least, to narrow the gap between Theorems 1 and
2. By ignoring the brackets in these theorems and subtracting, we see that
the gap is approximately $4/21\ p - 1$ for each p. The gap therefore grows
fairly slowly, as Table 1 confirms.

In any case, it would be very interesting to determine $l(p)$ or to
narrow the gap for some small p. Clearly the place to start is $p = 13$, 14, or
(possibly) 15. In view of the above conjecture, it would be best to try to
lower the upper bounds in Table 1. Besides, attacking the upper bounds
should not involve knowing a lot about cubic curves or the like, whereas

attacking the lower bounds might. Such an attack probably would require the pursuit of many lines of argument in great detail. One would start out with the approach used in the case $p = 8$ given here, but many alternative subcases would have to be examined separately. Perhaps an artful computer program would be useful in this project.

It has been the purpose of this paper to shed some light on an interesting corner of combinatorial mathematics. I hope also to inspire interested amateurs to try their hands at this or similar problems. (For reading on some other questions in the field, see Branko Grünbaum's fine book, *Arrangements and Spreads* [3].) It seems particularly appropriate to include such problems in this book because the person who has done the most to make mathematics popularly accessible today is Martin Gardner. He has, in fact, discussed some aspects of this tree-planting problem in his column [2]. Surely, planting trees is a fruitful activity for a mathematical gard(e)ner.

References

1 Burr, S. A.; Grünbaum, B.; and Sloane, N. J. A. 1974. The Orchard Problem. *Geometriae Dedicata* 2: 397–424.

2 Gardner, M. 1976. Mathematical Games. *Scientific American*, 102–109.

3 Grünbaum, B. 1972. *Arrangements and Spreads*. Providence, R.I.: Amer. Math. Soc.

4 Jackson, J. 1821. *Rational Amusement for Winter Evenings*. London: Longman, Hurst, Rees, Orme, and Brown.

5 Kelley, L. M., and Moser, W. O. J. 1958. On the Number of Ordinary Lines Determined by *n* Points, *Canad. J. Math.* 10: 210–219.

6 Sylvester, J. J. 1886. Problem 2572. *Math Questions from the Educational Times* 45: 127–128.

Slicing it Thin

Howard Eves
UNIVERSITY OF MAINE

Over six hundred years ago, the Blessed John Colombini of Siena founded a new religious order, originally devoted to nursing and burying the victims of the rampant bubonic plague that swept away more than a third of the population of Europe. The group, known as the *Jesuats* (in no way related to the *Jesuits*—which at the time had not yet been founded), was officially approved by Pope Urban V in 1367. As time went on, the order declined, and in 1606 a partially successful attempt was made to revive the group. But certain abuses, apparently involving the manufacture and sale of distilled liquors in a manner not sanctioned by Canon Law, crept in. This, along with a difficulty in maintaining a reasonable membership quota, led to the order's abolishment by Pope Clement IX in 1668. The order had existed for just over three hundred years.

It was in 1613, shortly following the attempt to revive the Jesuat order, that a young Italian boy of fifteen, by the name of Bonaventura Cavalieri, was confirmed as a new member. Cavalieri spent the rest of his life in the ministry of the Jesuat order. Because of this dedication, and because of the subsequent dissolution of the order and the close similarity between the title *Jesuat* and the much more familiar title *Jesuit*, many major encyclopedias, biographical dictionaries, histories, and source books of today erroneously state that Cavalieri was a Jesuit instead of a Jesuat. We have here an excellent example of a perpetuated error hidden in our written histories. Errors of this sort are not uncommon, some having endured over long periods of time.

Bonaventura Cavalieri was born in Milan, Italy, in 1598. As a young man he studied under Galileo. Later, in 1619, he received an appointment as a professor of mathematics at the University of Bologna, serving in this capacity until his death in 1647, at the relatively young age of forty-nine. He was one of the most influential mathematicians of his time amd he wrote a number of works on geometry, trigonometry, astronomy, astrology, and optics. Among the first to recognize the great value of Napier's invention of logarithms, he played a leading role in their early introduction into Italy. His most important contribution to mathematics, however, is the treatise *Geometria indivisibilibus*, published in its initial version in 1635. This work is devoted to the precalculus *method of indivisibles*, a method that, like so many mathematical achievements of more modern times, can be traced back to the early Greeks—in this case to Democritus (*ca.* 410 B.C.) and Archimedes (*ca.* 287–212 B.C.)—though the direct stimulus probably lay in attempts at integration that had been made by Johann Kepler (1571–1630). In any event, the publication of Cavalieri's *Geometria indivisibilus* in 1635 marks an important milepost in the history of mathematics and in the history of the calculus in particular.

Cavalieri's great treatise is voluble and unclear, and it is difficult to learn from it just what Cavalieri meant by an "indivisible." It seems that an indivisible of a given planar piece is a chord of that piece, and the planar piece can be regarded as made up of an infinite parallel set of such indivisibles, or atomic parts. Similarly, it appears that an indivisible of a given solid is a planar section of that solid, and the solid can be regarded as made up of an infinite parallel set of these kinds of indivisibles, or atomic parts. Cavalieri argued that if we slide each member of a parallel set of indivisibles of some given planar piece along its own axis so that the end points of the indivisibles still trace a continuous boundary, then the new planar piece will have the same area as the original planar piece, since the two planar pieces are made up of the same indivisibles. Similarly, if we slide each member of a parallel set of indivisibles of some given solid along its own plane, so that the boundaries of the indivisibles still form a continuous surface, then the new solid will have the same volume as the original solid, since the two solids are made up of the same indivisibles. This last result can be strikingly illustrated by taking a vertical stack of cards and then pushing the sides of the stack into curved surfaces; the volume of the disarranged stack is the same as the volume of the original vertical stack.

The above remarks, slightly generalized, yield the so-called *Cavalieri principles*:

1 If two planar pieces are included between a pair of parallel lines, and if the lengths of the two sections intercepted by them on each line parallel to the including lines are always in a given ratio, then the areas of the two planar pieces are also in this ratio.

2   If two solids are included between a pair of parallel planes, and if the areas of the two sections intercepted by them on each plane parallel to the including planes are always in a given ratio, then the volumes of the two solids are also in this ratio.

Cavalieri's hazy conception of indivisibles as atomic parts making up a figure spurred considerable discussion, and serious criticisms were leveled by some students of the subject, particularly by the Swiss goldsmith and mathematician Paul Guldin (1577–1642). Cavalieri accordingly recast his treatment in the hope of meeting these objections, but his new attempt was scarcely any more successful than his first. The French geometer and physicist, Gilles Persone de Roberval (1602–1675), claimed to have invented the method before Cavalieri, but this matter of precedence is difficult to settle, for Roberval was consistently tardy in disclosing his discoveries and so became embroiled in several arguments of priority.

Roberval's tardiness has been explained by the fact that for forty years, beginning in 1634, he held a professorial chair at the Collège Royale. This chair automatically became vacant every three years, to be filled by open competition in mathematical contests in which the questions were set by the outgoing incumbent. Naturally, to perpetuate his position, Roberval saved his discoveries to be used for questions in the competition. In any case, Roberval ably employed the method of indivisibles to the discovery of a number of areas, volumes, and centroids. The method, or some procedure very like it, was also effectively used by Evangelista Torricelli (1608–1647), Blaise Pascal (1623–1662), Pierre de Fermat (1601?–1665), Grégoire Saint-Vincent (1584–1667), Isaac Barrow (1630–1677), and others.

Cavalieri's two principles constitute valuable tools in the computation of areas and volumes, and they can easily be made rigorous with the use of modern integral calculus. Accepting the principles as intuitively apparent, many problems in mensuration can be solved that normally require the more advanced technique of the calculus. Indeed, many authors of beginning textbooks on solid geometry have, on pedagogical grounds, advocated the assumption and then the consistent use of Cavalieri's second principle, as this greatly simplifies the subject for the student. For example, in deriving the familiar formula for the volume of a tetrahedron ($V = Bh/3$), the sticky part is first to show that any two tetrahedra having equivalent bases and equal altitudes on those bases have equal volumes. The inherent difficulty here is reflected in all treatments of solid geometry from Euclid's *Elements* on. With Cavalieri's second principle, however, the difficulty simply melts away.

We now turn to some illustrations of the use of Cavalieri's principles. Some of the examples constitute teasing puzzles for those who enjoy challenging mathematical diversions. We first formulate a convenient definition: Two planar pieces that can be placed so that they intercept sections of equal length on each member of a family of parallel lines, or two solids that can be placed so that they intercept sections of equal area on each member of a family of parallel planes, are said to be *Cavalieri congruent*. It follows from Cavalieri's principles that two figures that are Cavalieri congruent have equal areas (in the one case) or equal volumes (in the other case).

Illustrations

1   Let us find the area of an ellipse of semiaxes a and b. Consider the ellipse

$$\frac{x^2}{a^2} + \frac{y^2}{b^2} = 1, \quad a > b,$$

and the circle

$$x^2 + y^2 = a^2,$$

plotted on the same rectangular coordinate frame of reference, as shown in Figure 1. Solving each of the above equations for y, we find, respectively,

$$y = \left(\frac{b}{a}\right)(a^2 - x^2)^{1/2}, \quad y = (a^2 - x^2)^{1/2}.$$

It follows that corresponding ordinates of the ellipse and the circle are in the ratio b/a. The corresponding vertical chords of the ellipse and the circle are also in this ratio, and thus, by Cavalieri's first principle, so are the areas of the ellipse and the circle. We conclude that

$$\text{area of ellipse} = \left(\frac{b}{a}\right)(\text{area of circle})$$

$$= \left(\frac{b}{a}\right)(\pi a^2) = \pi ab.$$

This is essentially the procedure Kepler employed in finding the area of an ellipse of semiaxes a and b.

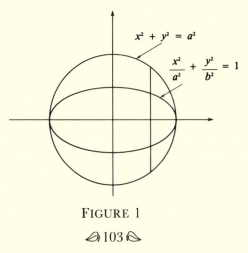

$$x^2 + y^2 = a^2$$

$$\frac{x^2}{a^2} + \frac{y^2}{b^2} = 1$$

FIGURE 1

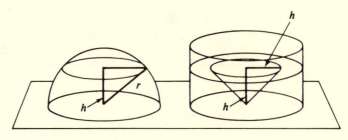

FIGURE 2

2   Now let us find the familiar formula for the volume of a sphere of radius r. In Figure 2 we have a hemisphere of radius r on the left, and on the right, a circular cylinder of radius r and altitude r with a cone removed whose base is the upper base of the cylinder and whose vertex is the center of the lower base of the cylinder. The hemisphere and the gouged-out cylinder are resting on a common horizontal plane. We now cut both solids by a plane parallel to the base plane and at a distance h above it. This plane cuts the one solid in a circular section and the other in an annular, or ring-shaped, section. By elementary geometry one can easily show that each of the two sections has an area equal to $\pi(r^2 - h^2)$. It follows, by Cavalieri's second principle, that the two solids have equal volumes. Therefore the volume of the sphere is given by

$$V = 2 \,(\text{volume of cylinder}-\text{volume of cone})$$

$$= 2\left(\pi r^3 - \frac{1}{3}\pi r^3\right) = \frac{4}{3}\pi r^3.$$

The whole trick here, of course, was to find a "comparison solid" (in this case, a gouged-out cylinder) that is Cavalieri congruent to a hemisphere.

3   As a second illustration of the planar case of Cavalieri's principles, consider the planar piece, bounded by a straight line and two curved arcs, pictured in the left of Figure 3, wherein the two distances marked m are equal to one another. One readily obtains, for a comparison area, the

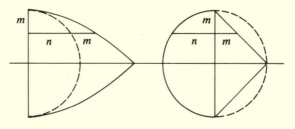

FIGURE 3

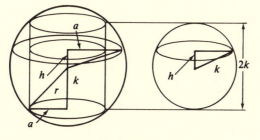

FIGURE 4

planar piece made up of a semicircle and an isosceles right triangle pictured in the right of Figure 3. It follows that the required area is

$$A = \frac{1}{2}\pi r^2 + r^2 = \left(\frac{\pi}{2} + 1\right)r^2.$$

4   For another illustration of the solid case of Cavalieri's principles, let us find the volume of the *spherical ring* obtained by removing from a solid sphere of radius r a cylindrical boring of radius a coaxal with the polar axis of the sphere (pictured in the left of Figure 4). Consider a sphere of diameter equal to the altitude of the spherical ring and placed with its center on the same horizontal plane that contains the center of the spherical ring (pictured in the right of Figure 4). Now cut the two solids by a horizontal plane at distance h from the centers of the two solids. In the spherical ring we get an annular section of area

$$\pi(r^2 - h^2) - \pi a^2 = \pi(r^2 - a^2 - h^2).$$

In the sphere we get a circular section of area

$$\pi(k^2 - h^2) = \pi(r^2 - a^2 - h^2).$$

We conclude, by Cavalieri's second principle, that the volume V of the spherical ring is the same as the volume of a sphere of radius k. That is

$$V = \frac{4}{3}\pi k^3.$$

It is interesting that *all spherical rings of the same altitude have the same volume, irrespective of the radii of the rings*.

5   In Problem E 465, *The American Mathematical Monthly*, March 1941, one is asked to find the area enclosed by the curve

$$b^2 y^2 = (b + x)^2 (a^2 - x^2),$$

where $b \geq a > 0$. The real graph of the curve (see Figure 5) consists of a noncircular ring passing through $(0, \pm a)$ and $(\pm a, 0)$, along with an

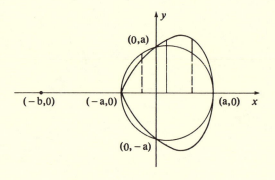

FIGURE 5

isolated point at $(-b, 0)$. The curve is symmetrical in the x-axis. Consider the circle $x^2 + y^2 = a^2$. We shall show that the required area is equal to the area of this circle. Because of the symmetry of the figure in the x-axis, all that we have to do to accomplish our aim is to show that the crescent-shaped area between the curve and the circle in the second quadrant is equal to the crescent-shaped area in the first quadrant. This we accomplish by showing that the intercepted ordinates in the two crescents at equal distances from the origin are equal to one another. To this end, let y_1 and y_2 be the corresponding positive ordinates of the given curve and the circle for the same x. Then we have

$$y_1 - y_2 = \frac{1}{b}(b + x)\sqrt{a^2 - x^2} - \sqrt{a^2 - x^2} = \frac{x}{b}\sqrt{a^2 - x^2}.$$

We see that, except for sign, $y_1 - y_2$ has the same value for $+x$ and $-x$. It now follows that the two crescent-shaped areas are Cavalieri congruent, and the area enclosed by the given curve is equal to πa^2.

Puzzles

Designing a proper comparison solid to assist in finding a required volume by Cavalieri's second principle can sometimes constitute a very enjoyable puzzle. Perhaps there is a reader who might like to try his hand at some puzzles of this sort. Here are a few; hints and suggestions appear at the end of the paper.

6 Find, by Cavalieri's second principle, the volume of a *torus*, or *anchor ring*, formed by revolving a circle of radius r about a line in the plane of the circle at distance $c \geq r$ from the center of the circle.

7 It is not difficult to show that there is no polygon to which a given circle is Cavalieri congruent (because equally spaced chords between two sides of a polygon change length uniformly, whereas equally spaced chords

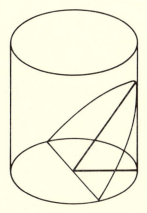

FIGURE 6

in a circle do not). One might also think that there is no polyhedron to which a given sphere is Cavalieri congruent. Show that this is not so by obtaining a *tetrahedron* that may serve as a comparison solid for a sphere.

8   An oblique plane through the center of the base of a right circular cylinder cuts off from the cylinder a cylindrical wedge, called a *hoof* (see Figure 6). Find, by Cavalieri's second principle, the volume of a hoof in terms of the radius r of the associated cylinder and the altitude h of the hoof.

9   A *generalized prismoid* is any solid having two parallel base planes and having the areas of its sections parallel to the base planes given by a quadratic function of their distances from one base.

    a Show that the volumes of a *prism*, a *wedge* (a right triangular prism turned so as to rest on one of its lateral faces as a base), and a *pyramid* are given by the *prismoidal formula*

$$V = \frac{h(U + 4M + L)}{6},$$

where h is the altitude, and U, L, and M are the areas of the upper and lower bases and midsection, respectively.

    b Show, by Cavalieri's second principle, that the volume of a generalized prismoid is given by the prismoidal formula.

    c Show that (1)   a *sphere*, (2)   an *ellipsoid*, (3)   a *hoof* (see Puzzle 8 above), and (4)   a *Steinmetz solid* (the solid common to two right circular cylinders of equal radii and having their axes intersecting perpendicularly) are all examples of a generalized prismoid, and thus find expressions for their volumes.

*   *   *   *   *   *

We conclude with a warning and with a surprising property of Cavalieri congruence. First, the warning.

In elementary textbooks the Cavalieri principles are accepted as intuitively evident. Now consider two solids included between a pair of parallel planes and having the perimeters of the two sections cut by them on any plane parallel to the including planes always equal in length. One may feel, on intuitive grounds, as in the case of the Cavalieri principles, that the lateral areas of the two solids must then be equal. One would be led to this conclusion by thinking of the lateral surfaces of the two solids as each being made up of a number of atomic elements in the form of loops of very thin string (forming perimeters of section of the solids). Since corresponding loops are of the same length and there is a one-to-one relation between the loops on one solid with those on the other, it would seem that the lateral areas of the two solids must be equal. To show that this is not necessarily so, consider two square prisms P and P' of equal bases and altitudes, where P is a right prism and P' is an oblique prism having a pair of opposite lateral faces perpendicular to the base (see Figure 7). The lateral area of the oblique prism is greater than the lateral area of the right prism. This points out the danger of relying, in mathematics, upon one's intuition.

Now let us consider the surprising property of Cavalieri congruence. We first formulate a definition. Two triangles ABC and $A'B'C'$ lying in the same plane are said to be *affine reflections* of one another if AA', BB', CC' are parallel with their midpoints on some line m (see Figure 8). The two following related theorems can be established.

THEOREM Any two coplanar triangles of equal area can be carried one onto the other by not more than three affine reflections.

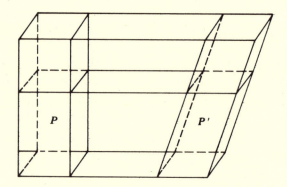

FIGURE 7

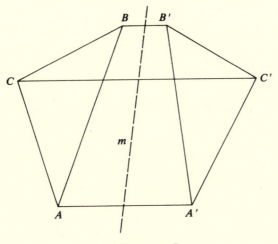

FIGURE 8

THEOREM Any two triangles of equal area are Cavalieri congruent.

The above two surprising theorems have only recently been discovered. The analogue in three-space of the second theorem is not true. That is; two tetrahedra of equal volume are not necessarily Cavalieri congruent. We shall not undertake here the task of establishing these results.

Hints and Suggestions for the Puzzles

6 Place the torus on a plane p perpendicular to the axis of the torus. Take for a comparison solid a right circular cylinder of radius r and altitude $2\pi c$, and place it lengthwise on the plane p. Cut the torus and the cylinder by a plane parallel to p. The section A in the torus is an annular region of outer and inner radii a and b, say, and the section A' in the cylinder is a rectangle of length $2\pi c$ and width w. Now

$$A = \pi a^2 - \pi b^2 = \pi(a^2 - b^2) = \pi(a + b)(a - b) = 2\pi c(a - b)$$

and

$$A' = 2\pi c w = 2\pi c(a - b).$$

Since $A = A'$, it follows that the volume of the torus is equal to the volume of the cylinder. That is

$$V = \pi r^2(2\pi c) = 2\pi^2 r^2 c.$$

7   Let AB and CD be two line segments in space such that (1) $AB = CD$ $= 2r\sqrt{\pi}$, (2) AB and CD are each perpendicular to the line joining their midpoints, and each are distance $2r$ apart, (3) AB is perpendicular to CD. The tetrahedron $ABCD$ may serve as the comparison polyhedron.

8   Divide the hoof into two equal parts by a plane p through the axis of the cylinder and let A be the area of the resulting cross-section of the hoof. Construct a right prism having as its base a square of area A (the base lying in the plane p) and having an altitude equal to the radius r of the cylinder. Cut from this prism a pyramid whose base is the base of the prism not lying in p, and whose vertex is a point in the other base of the prism. This gouged-out prism may serve as a comparison solid for one of the halves of the hoof.

Or one may choose for a comparison solid for half the hoof, a solid whose base is a right triangle of legs r and h, whose altitude is r, and whose upper base is a line parallel and equal to the hypotenuse of the lower base.

The volume of the hoof is $2hr^2/3$.

9b   Any section, which is a quadratic function of the distance from one base, is equal to the algebraic sum of a constant section area of a prism, a section area (proportional to the distance from the base) of a wedge, and a section area (proportional to the square of the distance from the base) of a pyramid. Thus the generalized prismoid is equal to the algebraic sum of the volumes of a prism, a wedge, and a pyramid. Now apply part (a).

9c   The section of the ellipsoid

$$\frac{x^2}{a^2} + \frac{y^2}{b^2} + \frac{z^2}{c^2} = 1$$

formed by the plane at distance z from the xy-plane is the ellipse

$$\frac{x^2}{a^2} + \frac{y^2}{b^2} = 1 - \frac{z^2}{c^2}$$

having semiaxes

$$\left(\frac{a}{c}\right)\sqrt{c^2 - z^2} \quad \text{and} \quad \left(\frac{b}{c}\right)\sqrt{c^2 - z^2}.$$

The area of this ellipse is

$$\frac{\pi ab(c^2 - z^2)}{c^2},$$

showing that the ellipsoid is a generalized prismoid. We find

$$V = \frac{4\pi abc}{3}.$$

For the Steinmetz solid we have $V = 16r^3/3$. There is a story that says the electrical genius, Charles Proteus Steinmetz (1865–1923), was once presented with the problem of finding the volume of the solid now named after him. He astounded everyone by immediately giving the answer, without any use of paper and pencil. Asked how he arrived at the result so quickly, he declined to tell. It is believed that he recognized the solid as an example of a generalized prismoid, and then used the prismoidal formula.

Adapted from one of the lectures in the author's lecture sequence entitled *Great Moments in Mathematics*.

How Did Pappus Do It?

Leon Bankoff

MATHEMATICIAN AT LARGE

One of the chief joys of mathematics is problem-solving and the greatest thrill of problem-solving is the attainment of an elegant result—preferably by elegant means. Pappus was a master of the elegant theorem, as evidenced by the countless "theorems of Pappus" found in the mathematical literature, and it comes as no surprise that he was successful in finding an ingenious way of proving the truth of what he called "an ancient theorem". This is a proposition described in Book IV of his Mathematical Collection and has to do with a sequence of consecutively tangent circles inscribed in the Arbelos, or the Shoemaker's Knife, a geometrical figure first treated by Archimedes in his Book of Lemmas. We may safely presume that this Circle Theorem was probably known empirically in times gone by, but had never before been proved.

If an arbitrary point C is chosen anywhere on a line segment AB and if semicircles are described on the same side of AC, CB and AB as diameters, the space within the curvilinear triangle bounded by the three semicircular arcs is called an Arbelos. One of the three properties of the Shoemaker's Knife considered by Archimedes pertained to the calculation of the diameter of an inscribed circle touching each of the three given arcs. Pappus started with this inscribed circle and extended his arena of operations by following it with a series of successively tangent circles, each touching two of the arcs of the Arbelos, as shown in Figure 1. According to this "ancient theorem" the distance from the center of the first inscribed circle to the line AB is equal to twice its radius; the distance from the center

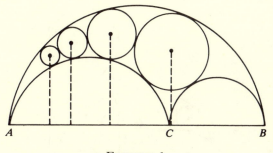

FIGURE 1

of the second inscribed circle to AB is four times its radius—and for the n-th circle, $2n$ times its radius.

In the related configuration shown in Figure 2, the arc CB is omitted and the first inscribed circle touches arcs AC and AB and the line segment CB. Again, a chain of successively tangent circles is bounded by the arcs of the mixtilinear triangle (a triangle which may have some curved sides) and now the distance from the center of the n-th circle of the chain to the baseline AB is equal to $2n - 1$ times the corresponding radius. For example, the distance from the center of the seventh circle of the chain to the line AB is thirteen times its radius.

At first glance the proof of these properties appears to be quite simple. We twentieth century mathematicians, smug in our possession of sophisticated tools, look upon a problem of this sort as rather routine. All we have to do is to apply principles of Inversion to the basic configuration and let the inverted figure stare back at us, stripping the veil of mystery from the original figure and revealing the naked truth of the Pappus Circle Theorem.

For the benefit of those readers not well-acquainted with Inversion, a brief review of the procedure may be in order. Erect a perpendicular to AB at C, cutting the circumference of arc AB in D. Then with A as center and with AD as radius, describe an arc representing a portion of the circle of

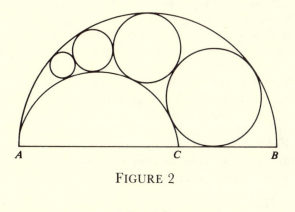

FIGURE 2

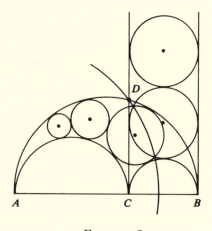

FIGURE 3

inversion. Since $AD^2 = AC \cdot AB$, the arc AB becomes the extended line CD perpendicular to AB at C; the arc AC inverts to the line tangent to the arc AB at B. Since the circle of inversion cuts the arc CB orthogonally, the semicircle on CB is self-inverse and remains unchanged. As for the Pappus chain of circles, they are transformed into the chain of equal tangent circles bounded by the two vertical parallel lines, just as their original counterparts were bounded by the arcs AB and AC. In Figure 3, it is immediately apparent that the distance of the center of the n-th inverted circle from the baseline is $2n$ times its radius. A similar inversion applied to Figure 2 yields the result $y_n = 2n - 1$, where y_n is the height of the center of the n-th circle above the baseline.

Now let us return to the topical question: "How did Pappus do it?" After all, he had to work without the assistance of Inversion (that was to come fifteen centuries later). There was nothing in Euclid, Archimedes or Apollonius to serve as a clue for the attainment of his astonishing result. So Pappus prepared for his attack on the main problem by first proving several pertinent introductory propositions or lemmas.

We commend Pappus for his remarkable achievement and we are grateful to him for bequeathing us his complicated, yet rigorous proof. We appreciate the difficulties he encountered by having to forge his own tools for the solutions of his problems. But if we apply present day standards of evaluation and criticism, we are compelled to conclude that the proof offered by Pappus is not one we would want to foist upon a trained mathematician, let alone a bright, eager high school student. This becomes evident if we pursue our natural curiosity about the Pappus proof to the point of actually locating it and wading through its intricacies. But where are we to find it? Although sixteen centuries have gone by since the days of Pappus, no-one has ever taken the trouble to do an English translation of the Collection in its entirety. However, an abbreviated version of a small

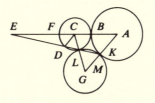

FIGURE 4

part of the works of Pappus was undertaken by Sir Thomas Heath and among the topics covered are the Circle Theorem and its associated lemmas. Heath's scholarly treatment was published in his History of Greek Mathematics by the Oxford University Press in 1921. However, in his condensed Manual of Greek Mathematics (Oxford, 1931) he merely mentions the theorem but deftly bypasses any explanation of the method of solution. Other works on Greek mathematics, such as those by James Gow and Ivor Thomas, assiduously avoid tackling the lemmas and the main proof.

My personal library boasts a copy of the Collection in Latin, translated from the Greek by Commandino in 1659. To do justice to Pappus's proof one would have to plow through eleven legal-sized pages of densely packed small print to try to ferret out what Pappus had to say. On the other hand, there does exist a two-volume edition in French, faithfully translated and meticulously edited by Paul Ver Eecke and published in Paris in 1933 by Desclée de Brouwer et Cie. Nineteen pages of this edition are devoted to the description of how Pappus did it. In his footnotes, Ver Eecke frequently cites a German edition by Frederic Hultsch, published in Berlin in 1876 and consisting of three volumes.

Having travelled thus far, the reader is entitled to at least a summary of the approach undertaken by Pappus. The following outline will consist of brief comments on the four diagrams designated here as Figures 4 to 7 inclusive. In Figure 4, illustrating the first lemma, the self-explanatory diagram serves to show that $KE \cdot EL = EB^2$, a relation to be used later on. The modern equivalent of this result would be the establishment of L and K as anti-homologous points with respect to E.

In Figure 5 the circle FGH centered at A is any circle touching the

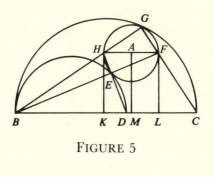

FIGURE 5

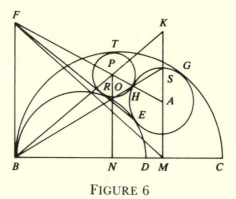

FIGURE 6

arcs of the semicircles erected on the diameters BD and BC, with D lying on BC. Starting with the diameter HF parallel to BC, the rest of the figure is constructed as shown. Then by setting up ratios BC/BG and BF/BL in the similar triangles BGC, BKH and BLF, BED, Pappus arrives at the ratio $2BM/KL = (BC + BD)/(BC - BD)$. Since $KL = 2r$, this relation can be written as $BM/r = (BC + BD)/(BC - BD)$.

To arrive at the main part of his proof, Pappus now establishes a few more essential lemmas. From the similar triangles BKH and FLC he obtains $BK \cdot LC = AM^2$. And, since $BC/BD = BL/BK$, he develops two more relations, $BL \cdot CD = BC \cdot 2r$ and $BK \cdot CD = BD \cdot 2r$.

We now go to Figure 6, in which circles (A) and (P) are any two tangent circles placed as shown. The projections of A and P upon BC are denoted by M and N respectively. Pappus has already established that the ratios BM/AS and BN/PO are constant and are equal to $(BC + BD)/(BC - BD)$. Using the other established lemmas and setting up appropriate relationships in similar triangles, he finds that $FH = FB$. Finally, with a bit more maneuvering, Pappus succeeds in showing that $(AM + d)/d = PN/d'$, where d and d' are respectively the diameters of the circles (A) and (P).

Only now can Pappus attack the main theorem. Referring to Figure 7, we use the lemmas ingeniously contrived by Pappus to find that

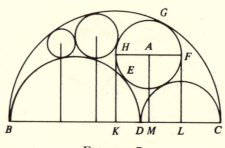

FIGURE 7

$BK \cdot LC = KL^2$ and that $BK \cdot LC = AM^2$, whence $KL = AM$, or $p_1 = d_1$, where p_1 is the distance from A to the line BC. Let p_2 and d_2 denote the corresponding elements of the second circle. Then the last lemma gives us $(p_1 + d_1)/d_1 = p_2/d_2$, so that $p_2 = 2d_2$. Continuing in the same manner, we obtain the same result as that shown by inversion, namely that the distance of the center of any circle of the Pappus Chain to the baseline BC is equal to $2n$ times its radius. A similar argument applied to the case shown in Figure 2 results in the conclusion that the required distances are $2n - 1$ times the corresponding radii.

The foregoing is a condensed version of Heath's synopsis of how Pappus did it; that is, how Pappus developed a framework on which to build his famous Circle Theorem. If we plod through the Pappus proof with painstaking patience we are convinced that Pappus was a genius who knew his way around geometrical configurations.

Like Euclid and Archimedes, Pappus proved his theorems by pure geometry with occasional ventures into a primitive sort of algebra, one involving nothing more complicated than the equating of ratios. We have seen how Pappus might have proved his famous Circle Theorem if he had only had access to the principles of Inversion. Let us imagine how he might have done it with the help of our more flexible algebra.

Adopting a fresh notation for Figure 8, let D and E denote the centers of any two tangent circles (radii x and y respectively), bounded by the arcs AB and AC (centered at O and P), and let F, H denote the projections of D and E on AB. Let $AO = R$, $AP = r$, $DF = mx$ and $EH = ny$.

From the relation $FO^2 - FP^2 = DO^2 - DP^2$, we obtain

$$(R - AF)^2 - (r - AF)^2 = (R - x)^2 - (r + x)^2,$$

or

$$(R + r - 2AF)(R - r) = (R + r)(R - r - 2x),$$

thus, $\qquad \dfrac{x}{AF} = \dfrac{(R - r)}{(R + r)},$ and similarly $\dfrac{y}{AH} = \dfrac{(R - r)}{(R + r)}.$

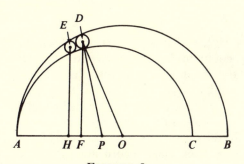

FIGURE 8

Then

$$HF = AF - AH = (x - y)\frac{(R + r)}{(R - r)}.$$

Now

$$DE^2 = HF^2 + (DF - EH)^2. \quad \text{So}$$

1
$$(x + y)^2 = \frac{(R + r)^2(x - y)^2}{(R - r)^2} + (mx - ny)^2.$$

Now express the area of triangle DOP first by the Heronian Formula and then by the traditional "half the base times the altitude". We find that

$$\sqrt{Rrx(R - x - r)} = \frac{mx(R - r)}{2}.$$

Similarly, for triangle EHO, $\sqrt{Rry(R - y - r)} = ny(R - r)/2$.
Solving for x and y, we obtain

2
$$x = \frac{4Rr(R - r)}{4Rr + m^2(R - r)^2}$$

and

3
$$y = \frac{4Rr(R - r)}{4Rr + n^2(R - r)^2}.$$

Equations **2** and **3** yield the relation

4
$$\frac{x}{y} = \frac{n^2(R - r)^2 + 4Rr}{m^2(R - r)^2 + 4Rr}.$$

Now equation **1** can be converted to the form

$$\left(\frac{x}{y}\right)[4Rr + m^2(R - r)^2] + \left(\frac{y}{x}\right)[4Rr + n^2(R - r)^2]$$

$$= 2[2R^2 + 2r^2 + mn(R - r)^2],$$

which together with equation **4**, yields

$$4(R - r)^2 = (R - r)^2(m^2 + n^2 - 2mn).$$

Finally, $m - n = \pm 2$, and since $n > m$, we obtain $n - m = 2$.

Note that D could be chosen as the center of the initial circle or of any other circle of the Pappus chain and that the ratio EH/y and the corresponding ratios for successively tangent circles ad infinitum follow the pattern $n - m = 2$, that is, the successive ratios belong to an arithmetic progression with a common difference of 2.

Whether solved by geometry, algebra or inversion, the Pappus Circle Theorem continues, after many centuries, to arouse our mathematical imagination and to intrigue us with its elegance.

Numbers and
Coding Theory

Fault-free Tilings of Rectangles

R. L. Graham

BELL LABORATORIES

Imagine that we have an unlimited supply of rectangular tiles of size 2 by 1 and we wish to tile the floor of a rectangular room of size p-by-q. Of course, we must cover all pq square units of floor area. Furthermore, two tiles are never allowed to overlap. As an example, we show in Figure 1 a tiling of a 5-by-6 rectangular floor.

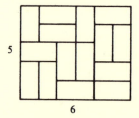

5

6

FIGURE 1

A tiling of 5-by-6.

It is easy to see that if such a tiling is to be possible then pq must be even, since the area of each tile is 2. On the other hand, if pq is even, then at least one of p and q must be even, say $p = 2r$. In this case we can place the tiles as shown in Figure 2 to construct the desired tiling.

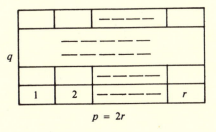

$$p = 2r$$

FIGURE 2

A tiling of $2r$-by-q.

Fault-free Tilings

If we examine the tiling shown in Figure 1 more carefully, we notice that it contains a "fault-line", that is, a straight line completely cutting through the rectangle which doesn't cut through any tile. Let us call a tiling which has *no* such fault-lines *fault-free*. If we think of a tiling as a cross-section of a wall built of bricks then it is clear why we might like to avoid fault-lines. In Figure 3 we show a fault-free tiling of a 5-by-6 rectangle.

A curious phenomenon occurs however, in trying to construct a fault-free tiling of a 6-by-6 rectangle. (The reader is encouraged to find one before proceeding further.) The same difficulty occurs for a 4-by-6 rectangle. In fact, there are no fault-free tilings for either of these cases! This leads to a question* that no mathematician can resist asking:

Exactly which p-by-q rectangles have fault-free tilings?

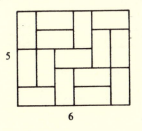

FIGURE 3

A fault-free tiling of 5-by-6.

Answering the Question

To begin with, a rectangle cannot have a fault-free tiling if it has no tiling at all, that is, if its area is an odd number. Stating this another way, a necessary condition for a p-by-q rectangle to have a fault-free tiling is that pq is divisible by 2.

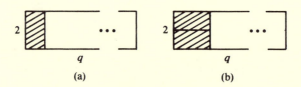

FIGURE 4

A fault-free tiling of 2-by-q?

As we have already seen, however, this is not enough. Suppose for example we try finding a fault-free tiling of a 2-by-q rectangle (where $q \geq 2$). There are just two ways to fill in the left-hand end of the rectangle, which we show in Figure 4. In either case however, (because $q \geq 2$) we must form a (vertical) fault-line. We therefore conclude that such rectangles have no fault-free tilings, even though their area is even.

In a similar spirit, consider what happens in the attempt to avoid fault-lines when tiling a 3-by-q rectangle. There are basically just two ways of tiling the end of the rectangle (see Figure 5).

The start shown in (a) is never any good since if $q = 2$ there is a horizontal fault-line and if $q > 2$ there is a vertical fault-line. What happens in case (b)? We show the various possibilities in Figure 6.

As before, case (a) creates fault-lines and must be discarded. The remaining possibility is to place Tile 1 as shown in (b). However, this forces use to place the additional Tiles 2 and 3 as shown in (c) (nothing else will fit into the gaps created in (b)). But notice that the "profile" of the pattern

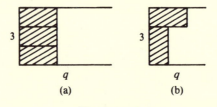

FIGURE 5

Trying to tile 3-by-q.

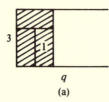 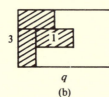 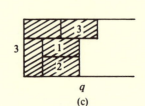

FIGURE 6

Still trying to tile 3-by-q.

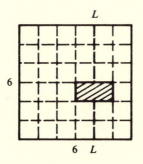

FIGURE 7

Potential fault-lines in 6-by-6.

in Figure 6 (c) is exactly the same as that in Figure 5 (b). Thus, we have only *delayed* the problem of filling the indentation created in Figure 5 (b). Eventually we must face the issue of completing the tiling (*q is* finite, after all) and this can only be done by finally filling this annoying indentation with a vertical tile as in Figure 6 (a). However, as soon as this happens we have formed a fault-line!

So, we conclude that no 3-by-q rectangle has a fault-free tiling.

In a similar way (but with a few more cases) it follows that no 4-by-q rectangle has a fault-free tiling as well. (Naturally, from the symmetry of the situation this means that no p-by-4 rectangle has a fault-free covering either.) Of course, one must resist the temptation to generalize at this point by trying to show that the same holds for 5-by-q rectangles, etc. After all, we do have a fault-free tiling of a 5-by-6 rectangle.

What we have shown at this point is that another necessary condition for a p-by-q rectangle to have a fault-free tiling is that both p and q must be at least 5.

This still leaves the nonexistence of a fault-free tiling of the 6-by-6 square unexplained. This gap can be filled by the following stunning argument of S. W. Golomb and R. I. Jewett.

Suppose for the moment that we have managed to find a hypothetical fault-free tiling of the 6-by-6 square. In the square there are 5 vertical and 5 horizontal fault-lines which, we assume, must all be broken by tiles (see Figure 7). Notice that each tile breaks *exactly one* potential line. Furthermore (and this is the crucial observation), if any fault-line (say L in Figure 7) is broken by just a *single* tile, then the remaining regions on either side of it must have an odd area, since they consist of 6-by-t rectangles with a single unit square removed. However, such regions are impossible to tile by 2-by-1 tiles. Each of the 10 potential fault-lines must be broken by at least two tiles. Since no tile can break more than one fault-line, then at least 20 tiles will be necessary for the tiling. But the area of the 6-by-6 square is only 36 while the area of the 20 tiles is 40! Thus, we have reached a contradiction. No such tiling of a 6-by-6 square can exist.

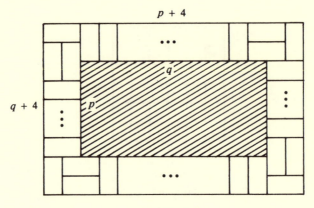

FIGURE 8

Extending fault-free tilings.

We can summarize what we know up to this point as follows: *Necessary* conditions for the existence of a fault-free tiling of a *p*-by-*q* rectangle are:

1 pq is divisible by 2;

2 $p \geq 5$, $q \geq 5$;

3 $(p,q) \neq (6,6)$.

Surprisingly, it turns out that these conditions are also *sufficient*. That is, if p and q satisfy **1**, **2** and **3** then the *p*-by-*q* rectangle will always have a fault-free tiling. One way this can be proved is by starting with small fault-free tilings, such as 5-by-8, 6-by-8 and our earlier 5-by-6, and building up larger fault-free tilings from these. For example, if we have a fault-free tiling of a *p*-by-*q* rectangle then in Figure 8 we show how to form a fault-free tiling of a $(p + 4)$ by $(q + 4)$ rectangle. A similar construction can be used to form a fault-free tiling of a $(p + 2r + 2)$ by $(q + s + 2)$ rectangle from the original *p*-by-*q* for any positive integers r and s.

Other Tiles

It is only natural to wonder about the possibilities of fault-free tilings when other-sized tiles are used; for example, 3-by-1 or 7-by-5. At first, an exact characterization of just which *p*-by-*q* rectangles can be appropriately tiled appears to be a hopelessly difficult problem. However, it turns out that there is a surprisingly beautiful answer to this question. Suppose we are to use tiles of size *a*-by-*b* where we can assume by changing our units if necessary, that *a* and *b* have no common factor greater than 1. (We can also assume that we don't have $a = b = 1$ since otherwise *no* fault-free tilings are possible, except for the trivial tiling consisting of a single tile!) As in the previous result, there are two basic necessary conditions: one dealing with a *divisibility* condition and one dealing with a *size* condition.

Regarding divisibility, the area pq of the rectangle must be divisible by the area ab of the tile if any tiling, fault-free or not, is to be possible. Other requirements are necessary as well. By *coloring* the tile and the rectangle with ab colors in an appropriate cyclic manner, it can be argued that it is necessary that

1′ Both a and b each divide at least one of p and q.(This is stronger than just requiring that ab divides pq).

Regarding size, there should be an analog in the general case to the earlier condition **2** which required $p, q \geq 5$. The corresponding condition is unexpected:

2′ Each of p and q must be able to be expressed as a sum $xa + yb$ with positive integers x and y in at least two ways.

Basically, this guarantees that there is enough freedom in placing tiles across the sides of the rectangle; that is, we don't always have to place the same numbers of tiles horizontally and vertically. For the case $a = 2, b = 1$, **2′** reduces to the earlier condition **2**, since 2, 3 and 4 do not have two representations as $x \cdot 2 + y \cdot 1$, $x, y > 0$, while any integer ≥ 5 does; for example, $5 = 1 \cdot 2 + 3 \cdot 1 = 2 \cdot 2 + 1 \cdot 1$, etc.

Curiously enough, the impossibility of a fault-free tiling of a 6×6 square with 2 by 1 tiles remains as the unique anomalous exception. In all other cases it is possible by construction techniques similar to those mentioned before to produce the required fault-free tilings whenever p and q satisfy the necessary conditions. The general result is summarized in the following statement.

THEOREM A fault-free tiling of a p-by-q rectangle with a-by-b tiles exists (where we assume $pq > ab$ and $(a,b) = 1$) if and only if

1′ Each of a and b divides p or q;

2′ Each of p and q can be expressed as $xa + yb$, $x, y > 0$, in at least two ways;

3′ For $\{a,b\} = \{1,2\}$, $(p,q) \neq (6,6)$.

The actual proof of this result is not difficult and is left to the energetic reader. We remark that m can be expressed as $xa + yb$ in at least two ways if and only if $m - ab$ can be expressed as $xa + yb$ in at least one way (assuming $(a,b) = 1$). In general, such integers do not form an interval as they do for $a = 2$, $b = 1$. For example, by using 3-by-2 tiles, it is possible to find fault-free tilings for 11-by-18 and 14-by-15 rectangles but it is not possible for a 12-by-12 rectangle (in particular, $11 = 3 \cdot 3 + 2 \cdot 1 = 3 \cdot 1 + 2 \cdot 4$ but $12 = 3 \cdot 2 + 2 \cdot 3$ is all).

What Next?

It is typical of this business that one answer leads to n more questions. For example, how many fault-free tilings does a rectangle have? What if we can use two sizes of tiles instead of just one? What about these same questions in 3 (or more) dimensions? What if we require each fault-line to be broken by at least 2 tiles? At least n? We have reached the frontier of our current knowledge on this topic. We encourage the interested reader to explore this fascinating byway of geometry and discover for himself the gems which must surely lie waiting to be discovered.

---

* Another more general question is: How many different fault-free tilings does a p by q rectangle have? We don't confront this question here.

Dissections into Equilateral Triangles

W. T. Tutte
University of Waterloo

I am delighted to have the opportunity to contribute to this Collection honouring Martin Gardner. I once wrote a paper for his column in *Scientific American*, about dissections of rectangles into squares [7]. Perhaps another article on dissections would be appropriate here.

The afore-mentioned paper was concerned with "squared rectangles"; that is, dissections of rectangles into squares. A dissection of a rectangle or square into unequal squares is a *perfect* rectangle or square respectively. The paper was based on earlier work of Brooks, Smith, Stone and Tutte published in the *Duke Mathematical Journal* in 1940 [2]. This earlier paper gives an example of a perfect square and describes a method (exploiting symmetrical subgraphs) for constructing others. There is also a note on possible generalizations. One of these is the study of dissections of equilateral triangles into other equilateral triangles. There are two later papers on this generalization by myself, ([6] and [8]), and one by Brooks, Smith, Stone and Tutte [3].

I have no cause to complain about the reception of papers on squared rectangles by mathematicians and other scientists. They are all pleased that there should be a connection between Kirchoff's Laws of electrical flow and the dissection problems of pure geometry. But there seems to be no corresponding interest in the triangular problem. Yet, in my opinion the theory of triangulated triangles is a simple and elegant generalization of those theories of squared rectangles. In the remainder of this paper I will try to justify the preceding statement, in the hope of awakening the interest that I feel these dissections deserve.

When we are discussing dissections into equilateral triangles it does not really matter whether the original figure is a triangle or a parallelogram (with angles of 60° and 120°). A dissection of a triangle can be changed into that of a parallelogram by deleting one constituent triangle (at a vertex) and adding another. A dissection of a parallelogram can be changed into that of a triangle by adding new constituent triangles on two adjacent sides.

It is the dissected parallelogram that is the natural analogue of the squared rectangle. Suppose we start with a squared rectangle R and shear it so that it becomes a dissection of a parallelogram into rhombuses. Each rhombus can be dissected along an appropriate diagonal to form two congruent equilateral triangles. The squared rectangle R is thus transformed into a triangulated parallelogram P. Because of this observation we can regard the theory of squared rectangles as a special case of the theory of triangulated parallelograms.

Suppose that R is a perfect rectangle. Does it make sense to say that P is perfect also? Given any triangulated triangle or parallelogram we can fix one of its sides as "horizontal". Then any constituent triangle T has one horizontal side. Let us say that T is positively or negatively oriented according to its position above or below its horizontal side. Let us define the "size" of T as plus or minus the length of its side, depending on whether T is positively or negatively oriented. Then we say that a triangulation is "perfect" if no two of its constituent triangles have the same size. With this definition we can say that the perfect rectangle R shears into a perfect parallelogram P. For the two congruent triangles of P, arising from any square of R, have opposite orientations and therefore have different sizes.

Figure 1 shows a perfect parallelogram that cannot be obtained by shearing a perfect rectangle. In [6] I have satisfied myself that this parallelogram has the least number of constituent triangles, namely 13, consistent with perfection. For the sake of comparisons with squared rectangles, I like to say that its order is 13/2. For each square of a rectangle corresponds when sheared to two triangles. The number entered in each triangle of Figure 1 is its size.

At this stage in papers on squared rectangles, there appears a diagram showing the relation between a squared rectangle and its electrical network. Figure 1 fulfills the same purpose in the theory of triangulated parallelograms. But now the term "electrical" must be used in a generalized sense. It refers to a form of "electricity" not yet found in physics, (as far as I know). But we think of it as flowing in directed graphs and obeying analogues of Kirchhoff's Laws.

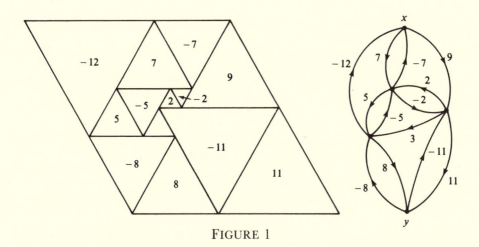

FIGURE 1

The parallelogram P is on the left and the corresponding electrical network is on the right. Each vertex of the network N corresponds to a maximal horizontal segment, made up of sides of constituent triangles in P and is shown on the same level. The network is directed. Each of its darts, or directed edges, corresponds to a constituent triangle of P (the direction of the dart is from the apex to the base of the triangle). The size of each triangle is written against the corresponding dart of N and is called the *current* in that dart. The vertices of N corresponding to the upper and lower horizontal sides of P are called the positive and negative *poles* of N respectively.

We observe that N is a balanced digraph. (*Digraph* is short for *directed graph.*) *Balanced* means that at each vertex the number of incoming darts is equal to the number of outgoing ones. It is not difficult to see that N is planar (for any P), and that it can be drawn in the plane so that the order of the darts around each vertex is the order of the corresponding triangles around the horizontal segment of P. In such a drawing, incoming and outgoing darts must alternate around each vertex. We describe a plane figure with this property as an *alternating map*.

At each vertex v of N, the sum of the currents in the darts directed to v is called the *current leaving N at v.* Its negative is the *current entering N at v.* Thus, the current entering the positive pole is equal to the current leaving the negative pole, and each of these numbers measures the length of a horizontal side of P. But the current entering or leaving N at any non-polar vertex is zero, and the current leaving is the sum of the sizes of the triangles of P having their bases in the corresponding horizontal segment. This rule of currents at non-polar vertices is our analogue of Kirchoff's First Law.

Let us define the potential of a horizontal segment of P as *its distance above the lower horizontal side of P, measured parallel to a slanting side of P.*

Transferring this number to N we call it the *potential* of the corresponding vertex. Now the vertices of N incident with a dart D are called the *head* and *tail* of D, with the direction of the dart being from tail to head. We can assert the usual form of Kirchhoff's Second Law, as follows. The total fall of potential around any circuit C of N, regardless of the directions of the darts, is zero. We note that the current in a dart D is equal to the fall of potential from tail to head. We can thus apply the Second Law by saying that the sum of the currents in any circuit C is zero, provided that we agree that currents in darts going against the chosen direction of the circuit are to be replaced by their negatives.

The modified Kirchhoff Laws can be used here like the conventional ones in the theory of squared rectangles. Having drawn an alternating map M, with at least four darts at each vertex, we can choose positive and negative poles, incident with a common face. We can solve the equations for a set of currents. We can then reverse the construction of Figure 1 to get a corresponding triangulated parallelogram. With luck this will be perfect. The complete theoretical justification of this procedure may be difficult, but the principle is clear.

It is now possible to deduce properties of triangulations from those of alternating maps. Thus, we can infer from the Euler Polyhedron Formula that the alternating map corresponding to a triangulated parallelogram must have either a 2-sided face or a non-polar vertex incident with only four darts. In either case two of the constituent triangles of the parallelogram must be congruent. This result leads to the assertion in [2], couched in a different terminology, that *there is no perfect equilateral triangle*.

Consider a squared rectangle R with its ordinary electrical network G. Let it be sheared to form a triangulated parallelogram P, and let the corresponding N be derived. We find that N can be obtained from G by replacing each edge with two oppositely directed darts. In the corresponding alternating map these two darts bound a digon. The modified Kirchhoff Laws for N are easily seen to be equivalent to the ordinary ones for G, on the assumption of unit resistances. The squared rectangle, then, is still a special case in the triangular theory.

Precedents set by the studies of squared rectangles would lead us to expect that someone, at this stage, would have constructed extensive catalogues of triangulated parallelograms. Alas, it is not so.

A detailed discussion of the Kirchhoff equations for an undirected graph G would be out of place here. Let us note that the theory can be based on the *Kirchhoff matrix* $K(G)$ of the graph concerned. We suppose the vertices of G to be enumerated as $v_1, v_2, \ldots, v_n$. Then $K(G)$ has n rows and n columns. Its jth diagonal entry is the number of edges joining v_j to other vertices. The non-diagonal entry in the ith row and j^{th} column is minus the number of edges joining v_i to v_j. Thus, $K(G)$ is a symmetrical matrix and its elements sum to zero in each row and column.

The Matrix-Tree Theorem asserts that if we strike out from $K(G)$ the jth row and column then the determinant of the resulting matrix $K_j(G)$ is the number $T(G)$ of spanning trees of G. Let us agree that the resistance of each edge is to be unity, and let us choose two vertices as positive and negative *poles*. It is convenient to take $T(G)$ as the magnitude of the current entering G at the positive pole and leaving at the negative, for then the currents in the various edges are integers. These currents constitute the *full flow* in G with respect to the chosen poles. The potential differences in this flow can be expressed as the determinants of certain submatrices of $K(G)$, appropriately multiplied by $+1$ or -1. In particular, the potential difference between the poles is the determinant of the submatrix obtained from $K(G)$ by striking out the two rows and two columns corresponding to the poles.

All this generalizes very nicely to a theory of unsymmetrical electricity in directed graphs. Let G be any directed graph, not necessarily balanced. We define $K(G)$ much as before. The jth diagonal entry is the number of darts directed to v_j from other vertices. The non-diagonal entry in the ith row and j^{th} column is minus the number of darts directed from v_i to v_j. We note that this $K(G)$ is not necessarily symmetrical. Its elements sum to zero in the rows but not necessarily in the columns.

We can define $K_j(G)$ as before, and its determinant is found to be a number of spanning trees of G. But not all the spanning trees of G are to be counted—only those in which each edge is directed away from v_j. We call these structures the *out-arborescences* of G on v_j. A spanning tree of G in which each edge is directed towards v_j is an *in-arborescence* of G on v_j. (In the undirected case, the value of det $K_j(G)$ is independent of j, but in the directed case this is not necessarily true.)

Let us choose a positive pole v_p and a negative pole v_q, and use the modified Kirchhoff Laws to calculate a corresponding distribution of currents. The current entering at v_p is not necessarily equal to the current

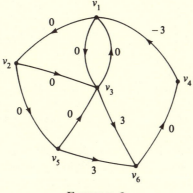

FIGURE 2

leaving at v_q. However, we can find a full flow in which current det $K_q(G)$ enters at v_p and current det $K_p(G)$ leaves v_q. Determinants of appropriate submatrices of $K(G)$ fix the currents and potential differences in this full flow. These currents and potential differences are found to be integers. The rule for the potential difference between the poles is the same as before.

As an example, let us take the graph of Figure 2, with v_1 as positive pole and v_6 as negative. The currents of the corresponding full flow are indicated.

We easily verify the matrix $K(G)$ to be as follows.

$$\begin{bmatrix} 2 & 0 & -1 & -1 & 0 & 0 \\ -1 & 1 & 0 & 0 & 0 & 0 \\ -1 & -1 & 3 & 0 & -1 & 0 \\ 0 & 0 & 0 & 1 & 0 & -1 \\ 0 & -1 & 0 & 0 & 1 & 0 \\ 0 & 0 & -1 & 0 & -1 & 2 \end{bmatrix}$$

Let us write $T_j(G)$ for the number of out-arborescences of G on v_j. Either by inspection of Figure 2 or by evaluating det $K_1(G)$ and det $K_6(G)$ we find that $T_1(G) = 6$ and $T_6(G) = 3$. Hence the flow depicted in Figure 2 is full, for a current of 3 enters at v_1 and a current of 6 leaves at v_6. The fact that the currents in the darts from v_1 to v_2 and from v_2 to v_5 are zero, is an immediate consequence of the modified First Law.

What happens if we reverse the direction of every edge, to get a new digraph G'? As far as the non-diagonal elements are concerned we replace $K(G)$ by its transpose. But diagonal elements may change. If G is the graph of Figure 2 then $K(G')$ is the following matrix.

$$\begin{bmatrix} 2 & -1 & -1 & 0 & 0 & 0 \\ 0 & 2 & -1 & 0 & -1 & 0 \\ -1 & 0 & 2 & 0 & 0 & -1 \\ -1 & 0 & 0 & 1 & 0 & 0 \\ 0 & 0 & -1 & 0 & 2 & -1 \\ 0 & 0 & 0 & -1 & 0 & 1 \end{bmatrix}$$

The digraph G' is shown in Figure 3, again with a full flow.

We have $T_1(G') = 8$ and $T_6(G') = 9$. There is an evident $1-1$ correspondence between the out-arborescences of G' on v_j and the in-arborescences of G on v_j.

When we restrict our theory to balanced digraphs we avoid many complications. For if N is a balanced digraph, the elements of $K(N)$ sum to zero in the columns as well as in the rows. Hence we can show, by elementary determinant theory, that det $K_j(N)$ is independent of j. The number of out-arborescences of N on a given vertex v is the same for each v.

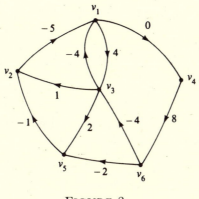

FIGURE 3

We therefore refer to this number simply as the tree-number $T(\mathcal{N})$ of $\mathcal{N}$. As an example we can take the balanced digraph of Figure 1. If the vertices are numbered from above to below in the diagram, then

$$K(\mathcal{N}) = \begin{bmatrix} 2 & -1 & 0 & -1 & 0 \\ -1 & 3 & -1 & -1 & 0 \\ -1 & -1 & 3 & 0 & -1 \\ 0 & -1 & -1 & 3 & -1 \\ 0 & 0 & -1 & -1 & 2 \end{bmatrix}.$$

It follows that in the full flow, the fall of potential between the poles is

$$\begin{vmatrix} 3 & -1 & -1 \\ -1 & 3 & 0 \\ -1 & -1 & 3 \end{vmatrix} = 20.$$

We deduce that the flow shown in Figure 1 is full. Accordingly $T(\mathcal{N})$, being equal to the current entering at the positive pole, is 19.

Given a balanced digraph $\mathcal{N}$ we can reverse all its darts to obtain a digraph $\mathcal{N}'$. Clearly $\mathcal{N}'$ is balanced. Moreover if $\mathcal{N}$ defines an alternating map in the plane, then so does $\mathcal{N}'$. But now $K(\mathcal{N}')$ is simply the transpose of $K(\mathcal{N})$. Hence $T(\mathcal{N}') = T(\mathcal{N})$. Hence, on each vertex of $\mathcal{N}$ the numbers of in-arborescences and out-arborescences are equal.

Let two vertices be chosen to be positive and negative poles in both $\mathcal{N}$ and $\mathcal{N}'$. Consider the two full flows. The current entering at the positive pole is the same in each, by equality of tree-numbers. Moreover the fall of potential between the two poles is the same in each. For it is given in $\mathcal{N}$ by the determinant of a submatrix symmetrically related to the main diagonal, and in $\mathcal{N}'$ by the determinant of the transpose of this matrix.

If $\mathcal{N}$ corresponds to a triangulated parallelogram, then so does $\mathcal{N}'$. The parallelograms, P and P' respectively, have the same size and shape, as a result of the preceding paragraph. Unfortunately this effect is not well-illustrated by the digraph of Figure 1. Reversing its darts merely gives an isomorphic digraph; moreover each pole is invariant under the isomorphism. If N corresponds to a squared rectangle R, then P and P' correspond to two ways of slicing R. In shearings, we can move the upper horizontal side either to the left or to the right with respect to the lower one.

On page 478 of [6] we are shown two perfect triangulated parallelograms which are the P and P' corresponding to an alternating map N. Each parallelogram has horizontal side 3441 and slanting side 2999. Each is dissected into 36 constituent triangles, which are all different sizes. Each has two congruent constituent triangles of side 129. But apart from this, there is no repetition of triangle-size between the two dissections.

The leaky or asymmetrical electricity relevant to the theory of triangulated parallelograms is discussed in more detail in [3], [6] and [9]. In the most general theory, the conductance (reciprocal of resistance) of a dart is not restricted to 1, but is an indeterminate over the integers.

A squared rectangle R has sides in two perpendicular directions, and either of these directions can be taken as horizontal. The two choices lead to two electrical networks G and G_1. These, however, are simply related. If each is completed by the adjunction of a new edge joining the poles then the two completed networks, or *c-nets*, are plane duals. Since the two corresponding systems of electrical equations give rise to the same pattern of squares, it seems reasonable to expect the determinants det $K_i(G)$ and det $K_j(G_1)$ to be equal. This is known to be the case; it can be shown that dual-plane-connected graphs have equal tree-numbers.

Is there anything corresponding to a pair of dual *c*-nets in the theory of triangular dissections? There is indeed, but to exhibit it we must pass from triangulated parallelograms to triangulated triangles. As an example we can adjoin two new constituent triangles to the triangulated parallelogram P of Figure 1 to obtain the triangulated triangle of Figure 4.

The corresponding modification of the electrical network of Figure 1 is obvious. To represent the new triangle of size 20 we need a new dart on

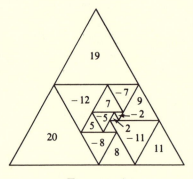

FIGURE 4

the left side, directed from the old positive pole X to the old negative pole Y. For the triangle of size 19 we need a new vertex Z and a new dart from Z to X. We can say that Z represents the apex of the triangulated triangle, even though this apex is not strictly a horizontal segment.

The currents in the new electrical network obey the modified Kirchhoff Laws with Z as positive pole and Y as negative pole. The number of out-arborescences from Z can be shown to be 39; the side-length of the dissected triangle. The number of out-arborescences from Y (or from any vertex but Z) is obviously zero. This corresponds to the fact that the current entering the network at Z is zero.

It is, in some ways, convenient to make Z coincide with Y in the electrical diagram, so that the diagram exhibits a new alternating map M. The dart originally incident with Z can be called the *polar dart*, and it can be distinguished in the diagram by a cross-bar. Figure 5 shows an electrical diagram of this kind corresponding to the triangulation of Figure 4. The vertices correspond to the maximal horizontal segments of the triangulation.

Figures 4 and 5 exhibit one example of a general procedure. Given any alternating map M we can mark in it a polar dart D, and we may then expect to derive a corresponding triangulated triangle. To form the corresponding electrical network we detach D from its tail-vertex Y and give it a new tail Z incident with no other dart. After this, Z is taken as the positive pole, and Y is taken as the negative pole. The calculated currents are interpreted as triangle-sizes. (Sometimes we find a zero current, and then complications arise.)

Returning to Figure 4 we observe that the sides of its constituent triangles make up three families of parallel segments, and that any one of these families could be chosen as horizontal. We might, for example, consider maximal segments parallel to the left slanting side of the dissected triangle, as shown in Figure 4, and make these segments correspond to the vertices of another electrical network. This is shown, with a marked polar

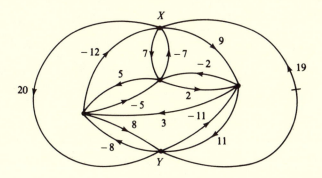

FIGURE 5

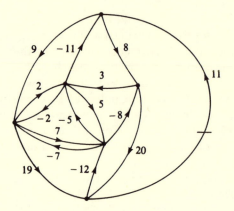

FIGURE 6

edge, in Figure 6. The remaining family of maximal segments, corresponding to the right slanting side, gives rise to the electrical network of Figure 7.

In Figures 5, 6 and 7 we find three different electrical networks, all
carrying the same set of currents. Such triads of alternating maps can be
regarded as analogues of the pairs of dual maps in the theory of squared
rectangles. Our analogue of duality is called *triality* in [6] and *trinity* in [8].
The three maps making up a triad are said in [8] to be *trine* to one another.

Given any alternating map we can define its trine maps directly,
without constructing a related triangulated triangle. It is shown in [8] how
a triad M_1, M_2, M_3 of alternating maps can be derived from a bicubic
map M, and how the triad associated with any triangulated triangle has its
corresponding M.

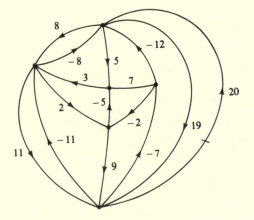

FIGURE 7

In a bicubic map, the vertices can be distinguished as black and white, each edge joining a white vertex to a black one. Moreover, the faces can be 3-coloured (say in red, green and blue) so that each edge separates faces of two different colours. Apart from permutations of the three colours, the 3-colouring is unique. We can form an alternating map M_1 of the corresponding triad as follows. We give it one vertex in each red face of M. We form its darts from the edges of M, separating green and blue faces. Each of these is directed from its black end to its white, and each is extended at each end through the adjacent red face of M to the vertex of M_1 contained in that face. The other members of the triad are defined similarly, each with a permutation of the three face-colours. (In [8] the definition of a triad is given, not directly in terms of the bicubic map M but in terms of its dual map. This dual is an Eulerian triangulation of the sphere.)

Suppose M_1 is derived from an undirected planar map N by replacing each edge by two oppositely directed darts. It is then found that one other member of the triad (say M_2) is similarly derived from the dual map of N. The third member of the triad is an oriented form of the medial map of N. This observation justifies our assertion that trinity is a true generalization of duality.

Since dual plane maps have equal tree-numbers we may expect trine alternating maps to have equal tree-numbers too, and this has been proven to be the case. (Proofs are given in [6] and [8]. A simpler proof was discovered recently by K. A. Berman [1].)

The common tree-number of three trine alternating maps M_1, M_2 and M_3 can be regarded as a property of the associated bicubic map M. In [3] it is interpreted directly in terms of the structure of M, as a count of matchings, not trees.

It seems appropriate to conclude the exposition here. Any reader interested in further study can proceed to the papers on our list of references. However, a warning should be given regarding one minor point. In a triangular dissection it may happen that six constituent triangles meet at a single point. Such a point is called a *cross*. It corresponds to a cross in the theory of squared rectangles; that is, to a point at which four constituent squares meet. In neither theory is it usual to encounter a cross in an interesting example unless it has been forced to occur by some symmetry in the construction. But when crosses do occur, it is necessary to introduce new conventions so that we may still have pairs of dual *c*-nets or triads of trine-alternating maps. In short we subdivide some of the maximal segments at the crosses. It is the resulting subsegments that correspond to the vertices of the associated electrical networks. Each cross, in the triangular case, is the intersection of three maximal segments. The rule is that exactly two of them (the choice is arbitrary) are to be subdivided there.

Addendum

Writing the above revived my interest in triangulated parallelograms, and I set out to calculate one or two more. I calculated two, and they seem to be sufficiently interesting to record here. They are perfect triangulations—each of the order $11\frac{1}{2}$. Each has horizontal side 401 and slanting side 264. There is no case of a constituent triangle in one being congruent to a constituent triangle in the other. This is the only case known to me of two perfect triangulations of the same parallelogram with no constituent size repeated from one dissection to the other.

The electrical networks of the two parallelograms are related by a reversal of darts. These networks, with their currents, are shown in Figures 8 and 9. I have verified that the flows shown are full. In each case X is the

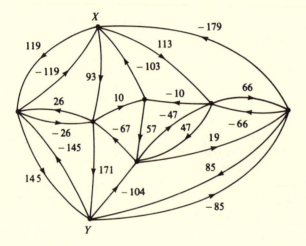

FIGURE 8

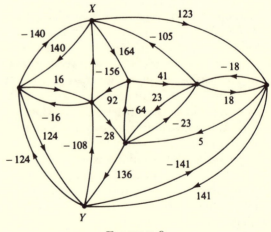

FIGURE 9

positive pole and Y is the negative pole. A current of 401 enters at X and leaves at Y, and the fall of potential from X to Y is 264. The reader should by now have no difficulty in deriving the actual triangulations.

References

1   Berman, K. A. 1978. Spanning trees, arborescences and 4-valent graphs. *Thesis*. Waterloo.

2   Brooks, R. L.; Smith, C. A. B.; Stone, A. H. and Tutte, W. T. 1940. The dissection of rectangles into squares. *Duke Math.*, 7: 312–340.

3   _____. 1975. Leaky electricity and triangulated triangles. *Philips Res. Reports* 30: 205–219.

4   Duijvestijn, A. J. W. 1978. Simple perfect squared square of lowest order. *J. Combinatorial Theory B* 25: 240–243.

5   Sprague, R. 1939. Beispiel einer Zerlegung des Quadrats in lauter verschiedene Quadrate. *Math. Zeitschrift* 45: 607.

6   Tutte, W. T. 1948. The dissection of equilateral triangles into equilateral triangles. *Proc. Cambridge Phil. Soc.*, 44: 463–482.

7   _____. 1961. Squaring the square. In *The 2nd Scientific American Book of Mathematical Puzzles and Diversions*, ed. Martin Gardner. New York: Simon and Schuster.

8   _____. 1973. Duality and trinity. *Colloquia Mathematica Societatis Janos Bolyai*. 10: 1459–1472.

9   _____. 1976. The rotor effect with generalized electrical flows. *Ars Combinatoria* 1: 3–31.

In Praise of Amateurs

Doris Schattschneider

MORAVIAN COLLEGE

One of the most appealing aspects of Martin Gardner's column "Mathematical Games" is its presentation of mathematical problems designed to intrigue amateurs and encourage their personal efforts at solution. By his own insistence, Gardner is an amateur mathematician and gives no special deference to formal mathematical education—his column pays tribute to the efforts of the mathematical "great" and the mathematically unknown, names often appearing side by side with no titles to distinguish one from the other. Amateurs are his most avid followers and enjoy the challenge of matching their wits against others in solving problems. Amazingly, their lack of formal mathematical education is often an advantage rather than a hindrance and their ingenious solutions to problems sometimes top the efforts of the professionals.

A striking case in point occurred as a result of Gardner's July 1975 column "On Tessellating the Plane with Convex Polygon Tiles". Challenged by the column, Richard James decided to try his own hand at solution and his obvious approach (altering the familiar) produced a solution that was overlooked in a formal mathematical scheme. The subsequent report of James's discovery aroused intense curiosity in Marjorie Rice, providing her with staying power for a thorough and methodical search (carried out mostly at her kitchen counter), which ultimately yielded a wealth of new results. The invitation to write this article gives me the chance to relate some details of the events that one Gardner column set in motion—events that still continue to ripple. It is a tribute to the thousands of amateurs who have made Gardner's column such a success.

Tiling problems have been a favorite subject of Gardner's over the years. More than a dozen columns in the last twenty years have been devoted in large measure to this subject. How to fit pieces together snugly to fill a desired space seems to be one of our earliest childhood pastimes. Even as adults we continue to be engaged by these problems, either for pleasure or out of necessity, with tiles, bricks and the like. One of the most basic tiling questions to ask is "What shape tile will fill the plane with its replicas without gaps or overlaps?" Many obvious shapes come to mind— no mathematician is needed to supply a lengthy list of examples. Some of the beautifully shaped tiles found in ancient mosaics around the world attest to the imagination of decorative artists in solving the problem. The most general answer to our question is not known. In order to get partial answers, conditions on the tiles are specified and then these specialized questions become problems whose solution is sought. What if the tiles are composed of stuck-together squares (polyominoes)? Or equilateral triangles (polyiamonds)? Or regular hexagons (polyhexes)?

Gardner's July 1975 article discussed the problem in which the tiles were convex polygons. "What convex polygons will tile? Explicitly describe conditions on a convex polygon to insure that it tiles." Here, Gardner had chosen a topic which had been worked on by many mathematicians over a 50-year period and it had been announced that the problem was completely solved. It is easily discovered that any triangle or any quadrilateral can tile, but that convex polygons of five or more sides do not always tile. For instance, regular pentagons do not tile but any pentagon having a pair of parallel sides will tile. Regular hexagons do tile but many other hexagons do not. Convex polygons having seven or more sides cannot tile. This last statement declared by Gardner as "not hard to show", is best described as a mathematical "folk theorem". Thus, everyone quotes it, but no one can seem to cite a complete and accurate proof. Fortunately, a recent article by Ivan Niven, in the December 1978 *American Mathematical Monthly*, fills this gap in the mathematical literature and presents a thorough as well as convincing proof.

Gardner based his article on a 1968 paper by R. B. Kershner, which surveyed the answers to the question, "What convex polygons will tile?" The complete answer for hexagons (3 types) and partial answer for pentagons (5 types) had been found by K. Reinhardt in 1918, and extensive investigations by Kershner had yielded 3 more types of pentagons which tile. Kershner felt the list was now complete (Figure 1). His fascination with the problem was described in a letter quoted by Gardner: "For reasons that I would have difficulty explaining, I have been intrigued by this problem for some thirty-five years. Every five or ten years I have made some kind of attempt to solve the problem. Some two years ago I finally discovered a method of classifying the possibilities for pentagons in a more

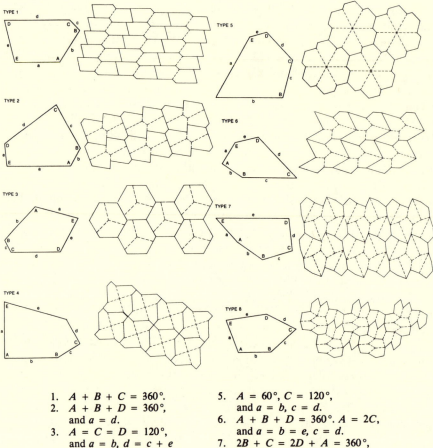

1. $A + B + C = 360°$.
2. $A + B + D = 360°$, and $a = d$.
3. $A = C = D = 120°$, and $a = b$, $d = c + e$
4. $A = C = 90°$, and $a = b$, $c = d$.
5. $A = 60°$, $C = 120°$, and $a = b$, $c = d$.
6. $A + B + D = 360°$. $A = 2C$, and $a = b = e$, $c = d$.
7. $2B + C = 2D + A = 360°$, and $a = b = c = d$.
8. $2A + B = 2D + C = 360°$, and $a = b = c = d$.

FIGURE 1

The eight types of convex pentagons which tile, reported by Martin Gardner in July, 1975.

convenient way than Reinhardt's. ... The result of this investigation was the discovery that there were just three additional types of pentagon ... that can pave the plane. These pavings are totally surprising. The discovery of their existence is a source of considerable gratification." Kershner's fascination with the problem would prove to be contagious. Gardner's exposition removed the subject from dusty mathematical journals (where it had been unquestioned for many years) and placed it in the hands of a wide readership, including many amateur puzzle enthusiasts. It is here that our story begins.

When Richard James III saw Gardner's article, he read only the first part; he decided to test his puzzle-solving skills before reading the remainder of the article. He wrote (in a letter kindly supplied by H. S. M. Coxeter), "Before reading Mr. Gardner's description of Mr. Kershner's eight tessellating pentagons, I set about to try it for myself. The first thing that came to mind was taking ... octagons (with squares filling the holes) and adapting them so that pentagons replace the squares (thus moving the octagons out of a lattice into "parallel" strips). The octagons neatly split into four pentagons. The description was $A = B = E = 90°$, $C = D = 135°$, and $a = b = 2c = 2e$. The tessellation was interesting but the pentagon was dull. Rotating the cross in the octagon (and making other adjustments as needed) produced [the pentagon and tiling shown in Figure 2]." James sent his discovery to Martin Gardner with the inquiry: "Do you agree that Kershner missed this one?" The exciting news was immediately communicated by Gardner to Kershner and to a few other mathematicians. Kershner's good-humored response and the James tessellation were reported to readers in the December 1975 "Mathematical Games" column. An amateur had, with one new example, shown that the list of tessellating pentagons was *not* complete. Were there still others to be discovered?

Perhaps an aside is in order here; a glimpse at the "mathematical grapevine" maintained by Gardner (which I would like to denote MG^2, for *Martin Gardner's mathematical grapevine*—shown in Figure 3). MG^2 is kept humming year round by a steady flow of correspondence, telephone calls, and personal conversations. When researching or writing a column, Gardner contacts experts with knowledge and/or the latest information on the subject; he seeks their comments on the accuracy of his manuscript before submitting it for publication. Conversely, when Gardner receives correspondence on any problem, before filing it away for future reference, he sends copies to those he knows are actively interested. In this way, the latest news is made available to those interested; its accuracy is checked and comments are returned to Gardner; and most importantly, those working on a common problem are put in touch with each other by Gardner. H. S. M. Coxeter was one of those who received a copy of James's discovery from Gardner; it was from Coxeter that I learned of the discovery. I couldn't resist trying my own hand at the problem—it wasn't long before I had a general description of a class of tessellating pentagons which generalized the single example James had sent to Gardner. Following Kershner's method of description, the general pentagon of James's type is one which satisfies the equations: $A = 90°$, $E = 180° - B$, $D = 90° + B/2$, $C = 180° - B/2$, $a = b = c + e$. After my communication of this to Gardner and Coxeter, I found myself the recipient of copies of Gardner's correspondence on "the pentagon problem" and soon was in direct communication with others actively working on the problem. If my bulging files on this problem can be taken as an indicator of the correspondence generated by one

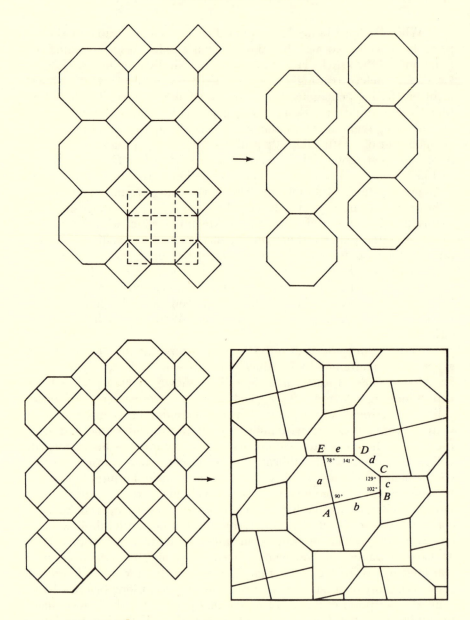

FIGURE 2

HOW RICHARD JAMES DISCOVERED A NEW PENTAGON THAT TILES

The familiar tiling by octagons and squares (with underlying square grid in dotted outline) is shifted to see if pentagons might replace the squares. A successful new tiling is produced. Further alteration of the octagons and pentagons produces an example of a previously undiscovered family of pentagons which tile.

FIGURE 3
MG²

Gardner column, then his files surely must fill several rooms!

When the December 1975 issue of *Scientific American* was delivered, another avid Gardner fan in California turned immediately to the "Mathematical Games" column. Marjorie Rice, a San Diego housewife and mother of five, was usually the first one in the household to read her son's magazine. She had been intrigued by the July article on tiling by pentagons and had "thought how wonderful it must have been [for Kershner] to discover the new types of pentagon tiles." Now, reading of James's newly discovered pentagon tile, her interest was strongly aroused and she set out to see if she might find still other new pentagons which tile. "I thought I would like to understand these fascinating patterns better and see if I could find still another type. It was like a delightful new puzzle to me and I considered how I could best go about this." Her search began quite differently from that of James and soon became a full-scale assault on the problem, extending over a period of two years.

Marjorie Rice had no formal education in mathematics beyond a General Mathematics course required for graduation from high school in 1939. Thus, as she faced the challenge of finding new pentagonal tiles she not only worked out her own method of attack, but also invented her own notation as well. Her first step was to catalogue all of the information available in the two Gardner columns on tiling pentagons (Figure 4). By doing this, she hoped to discover any common relationships satisfied by the pentagons and their tilings. "To begin, I needed to visualize the 9 types to

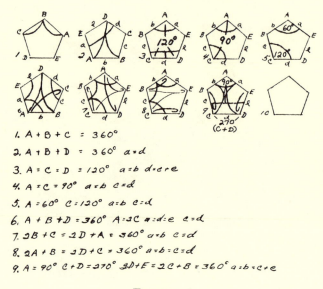

1. $A + B + C = 360°$

2. $A + B + D = 360°$ $a = d$

3. $A = C = D = 120°$ $a = b$ $d = c = e$

4. $A = C = 90°$ $a = b$ $c = d$

5. $A = 60°$ $C = 120°$ $a = b$ $c = d$

6. $A + B + D = 360°$ $A = 2C$ $a = d = e$ $c = d$

7. $2B + C = 2D + A = 360°$ $a = b$ $c = d$

8. $2A + B = 2D + C = 360°$ $a = b = c = d$

9. $A = 90°$ $C + D = 270°$ $2D + E = 2C + B = 360°$ $a = b = c = e$

FIGURE 4

Marjorie Rice's codification of information on the tiling pentagons of types 1 through 8 and James's discovery.

see how they differed from one another. I listed the formulas [the equations on sides and angles] on a 3 × 5 card and drew 10 pentagons on another card. I drew lines in color within the pentagons to show the information in the formulas, red for 360° combinations of 3 angles, blue for edges of the same length, black for 360° combinations of 4 angles, greeen for other information." "Now I could see that in [types] 7 and 8 each vertex had been touched twice by a red or black line (the hooks counting for two times) and that if I had drawn a line on [types] 1 or 2 between angles totaling 180°, each vertex would be touched once by a line. Continuing to [type] 3, I used the symbol ↓ to indicate 3 identical angles would come together, this for 3 of the vertices. Three straight lines were needed then between [angles] B and E. Every corner was touched three times. I saw that for every pattern each vertex of the pentagon must be touched by a line or symbol the same number of times."

This observation that each vertex of a tiling pentagon must be used the same number of times in a tiling was the key to Marjorie's investigation which followed. Her symbolic notation of information was further refined to a form which suppressed all but essential information (Figure 5). "Here I used [symbolic] pentagons in a form easier to draw." Lines connected corners of the pentagon which would come together at a vertex of the tiling by the pentagon. "This notation is the key to all my further work. By labeling the corners (it didn't matter where I started), I could then develop a sort of signature of letter combinations for every diagram and develop these with little sketches."

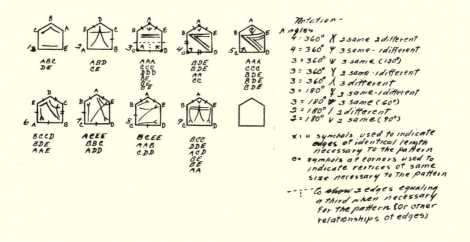

FIGURE 5

The pictorial notation developed by Marjorie Rice (here recording information on the 9 types of pentagons known to tile) which was the key to all her work.

In her own way, Marjorie had discovered a way to manage an enormous amount of information that would emerge in a thorough combinatorial search of possible combinations of angles (and sides) which yield tiling pentagons. Mathematicians use symbols for objectivity, conciseness and clarity—and good notation must be simultaneously suggestive and definitive as well. Marjorie's pictorial notation looks like hieroglyphics yet it records the possible combinations of angles with a simplicity not possible with more conventional mathematical notation. The notation eliminates completely the need to worry about repetition of cases caused by assigning different letters to the angles of a pentagon.

Beginning with two copies of a single pentagon stuck together, Marjorie considered how further copies of the pentagon could be added to these to create a tiling of the plane. The information on how corners of the pentagon met at a vertex of the tiling and how sides touched were all kept track of by using her symbolic notation. If it became clear that a certain combination was impossible (a tiling would not result), this case was eliminated; when a tiling seemed possible she sketched an actual example of such a pentagon and its tiling.

In order to make calculations quickly and test new pentagons to see whether or not their angles might add up so as to create vertices of a new pentagonal tiling, Marjorie arbitrarily divided 360° into units of 18°, marking the divisions on a small protractor. Then using these units, an angle of 36° was represented as 2, an angle of 108° as 6, and so on. "When constructing a pentagon for trial I usually began with the 2 angles that equal 180° numbered as 4 (72°) and 6 (108°) and adjusted them later on as

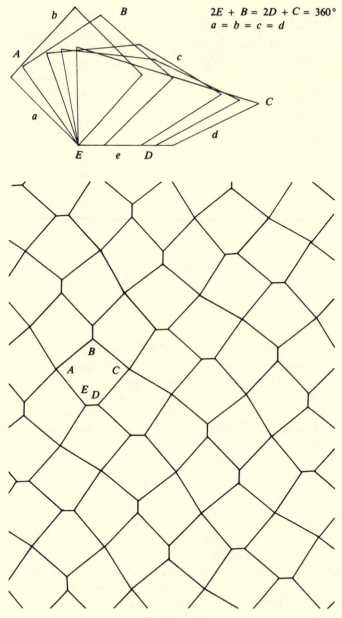

$$2E + B = 2D + C = 360°$$
$$a = b = c = d$$

FIGURE 6

The first discovery (February 1976) of a new type of tiling pentagon by Marjorie Rice. The range of shapes that the pentagon can assume is shown along with a tiling by one representative of this type. A tiling by a representative of this type whose sides are in golden ratio is the underlying grid for an Escher-like design of bees in clover, designed by Mrs. Rice. (See Figure 16A.)

required. This was the busy Christmas [1975] season which took much of my time but I got back to the problem whenever I could and began drawing little diagrams on my kitchen counter when no one was there, covering them up quickly if someone came by, for I didn't wish to have to explain what I was doing to anyone. Soon I realized that many interesting patterns were possible but did not pursue them further, for I was searching for a new type and a few weeks later, I found it."

In mid-February, 1976, Marjorie sent her discovery to Gardner with a sketch of the range of shapes which the pentagon could assume and tilings by two different representatives of this new type of pentagon tile (Figure 6). She wrote, "Here is a pentagonal tile that I believe really is different from any you had listed though similar to type 7 and 8. One of the enclosed examples in which the two sizes of line are in golden proportion makes a very pleasing arrangement, I think." Again, Gardner dispatched the discovery to several interested parties via MG^2 including Kershner and myself. It was verified to be indeed a new addition to the list of tiling pentagons; Kershner wrote to Marjorie to ask how she discovered it, and acknowledged that he had erroneously eliminated this as a possible type in his search. As happens with a great deal of reader correspondence, the discovery was not reported in Gardner's column but filed away for future reference. (Other items had also been received by Gardner in response to the James tessellation—at least one quilt and a beautifully woven rug were inspired by the design. See color plate V.)

When I examined the material that Marjorie had sent to Gardner and compared it with Kershner's types 7 and 8, it appeared that these three types of tiling pentagons (I named hers type 9) might all be examples in a still larger class of tiling pentagons. I made the following conjecture and sent it to Gardner: "Any pentagon having four equal sides and containing four different angles P, Q, R, S, such that $2P + Q = 360°$ and $2R + S = 360°$, tiles the plane." In less than two weeks I received a letter from Marjorie in which she showed that she had considered the conjecture and proved it false. "As the symbols below show, there are only 8 possibilities—[8 ways in which the corners of such a pentagon can come together so that only the equations $2P + Q = 360°$ and $2R + S = 360°$ are used]. Four of them *1, 5, 6*, and *8* can tile the plane, the other four seem to be impossible for the reasons illustrated. *6* seems only to work when two adjacent angles equal 180° thus making it a type 1. It does however give two interesting ways of assembling a special type kind of type 1 as shown on the enclosed sheet." (Figure 7).

This was my first correspondence from Marjorie and my first encounter with her notation and method of checking possibilities by construction. It was so far from the conventional ways that mathematicians use that I puzzled over the diagrams trying to figure out what she was saying and how these proved anything. Her "reasons" that some pentagons considered would not tile were little sketches, not algebraic or geometric

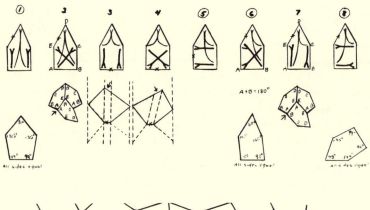

FIGURE 7

The pictorial "proof" that Schatlschneider's conjecture was false. Case *6*, having 4 equal sides and 2 adjacent angles whose sum is 180°, assembles into blocks in 2 distinct ways (note the change in the tiling across the dashed line).

arguments that mathematicians require for proof. Probably because it seemed so obvious to her, she had not enclosed any explanation of her pictorial notation. In her notation, the heavy chicken tracks $\succ$— represent the equations $2P + Q = 360°$, $2R + S = 360°$ which would bring together these corners at a vertex of the tiling, and the lighter curves connecting corners of a pentagon represent the remaining equation on how corners must come together at a vertex of the tiling. Recall that the sum of the interior angles of a pentagon is $540°$ so that this fact, together with the two equations involving P, Q, R, S imply that $Q + S + 2T = 360°$, where T is the 5th angle of the pentagon. She assumed no further equations were necessarily satisfied by angles of a pentagon and so a tiling by that pentagon could use only the angle relationships represented symbolically. Thus her cases 2 and 7 were eliminated because (as her arrow indicates) another equation on angles would have to be added if the pentagons were to tile in this way. Her cases 3 and 4 are impossible to construct—her sketches tell this pictorially. It was not hard for me to verify algebraically that no pentagon could satisfy the relationships on angles shown in these cases. Marjorie had constructed several examples of her case 6 and always 2 adjacent angles seemed to add up to $180°$. But this observation did not constitute a mathematical proof. In trying to prove her observation, I found that the assumed angle relationships did not force this fact. It was Kershner who later supplied an elegant proof, showing the usefulness of his generalized laws of sines and cosines. This provided a very striking illustration of an amateur's intuition and observation using elementary tools leading to a correct conclusion, but the necessity of more sophisticated, mathematical means and a trained mathematical mind to provide irrefutable proof.

The four cases which remained did indeed tile and were already known—her cases 1 and 8 were Kershner's types 8 and 7 respectively. Her case 5 was her new discovery (type 9) and case 6 was type 1. Thus the conjecture was completely disposed of.

Now that Marjorie had established direct contact with Kershner and with me, MG^2 was no longer an intermediary; however, Gardner was kept informed of the continuing work on the problem. No doubt Marjorie was encouraged by the praise for her work received in correspondence, but it was the problem itself which continued to entice her. Although she was busy with family events she kept returning to the puzzle, drawing tilings and considering and reconsidering possibilities in whatever snatches of time she could find. The problem was like a partially finished jigsaw puzzle laid out in a spare room—worked on intently for a while until a small satisfaction is achieved, then abandoned. It is not forgotten, however, and lures you back again and again to tempt you to add a few more pieces and see a little more of the pieced-together scene.

I asked Marjorie to write me all she had done on the problem and keep me informed of any new results since I had been asked to write an

P'
Λ

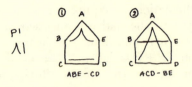

FIGURE 8

Symbolic representation of pentagons in which each angle is "used" once in a vertex of the tiling.

article on the pentagonal tiling problem for *Mathematics Magazine*. In March, 1976, I received her codified analysis of how she had considered ways in which a pentagon could tile. The diagram showed groups of pentagons considered—each group corresponded to a set of angle relationships satisfied by a pentagon. These angle relationships (and only these) were to be reflected in an associated tiling of the plane by the pentagon. How the angles came together then forced any conditions on the sides of a tiling pentagon. The first group, called "$p1$" (for pentagon-1), considered pentagons in which each angle was "used" once in a vertex of the tiling. Thus three different angles "came together" to sum to 360° and the two remaining angles "came together" to sum to 180°. Only 2 pentagons, types 1 and 2, were in this category (Figure 8). The next 12 groups (listed 1 through 12 under a heading "$p2$") considered pentagons in which each angle was used twice if different vertices of the tiling were listed. Group 12 in this listing is the collection of pentagons which Marjorie had written to me about earlier, in response to my conjecture. On this codified listing (Figure 9), Marjorie explained "There are 3 tests, the first is whether the group itself is possible (5 and 6 are not). Then sketches are made of combinations indicated by each member of a group to see if the proper angles will come together. If they combine successfully, it will be obvious which lines [sides of pentagons] must be equal lengths. The last test is whether it can be translated into specific angles—if it can, it will tile successfully." Marjorie had written "no" next to each symbolic pentagon in the list which would not tile in the way specified; next to those that would tile, she put the type number from Kershner's list. In addition, for each pentagon that tiled she produced an illustrative tiling showing that it could tile in the way specified (Figures 9a, 9b). In her listing, types 1, 2, 4, 6, 7, 8, 9 had occurred and she had 26 different tilings. Several of these tilings were new. It is not surprising that types 3, 5, and James's tile (which I call type 10) did not occur on this list because if all of the angle relationships which are satisfied at the vertices of tilings by these types are given, then each angle of a pentagon tile is used three times.

A few weeks after receiving this information, Marjorie sent an even more extensive list. "Regarding the 2-pentagon patterns I sent you earlier—I have found on rechecking that I had missed quite a few—have gone over them much more carefully and here is the revised list with examples." Her revised list showed the same 12 groups but many more cases considered. Now she had found 35 pentagons with angle relationships that led to plane tilings. Some angle combinations yielded two or more distinct tilings; thus there were a total of 45 sketched tilings accompanying this listing. Although no new types of tiles were discovered, the range of tilings was greatly expanded. Her letter indicated that she was already at work on the "$p3$" groups of pentagons—those whose angle relationships would use each angle of a pentagon three times. "The majority are quickly

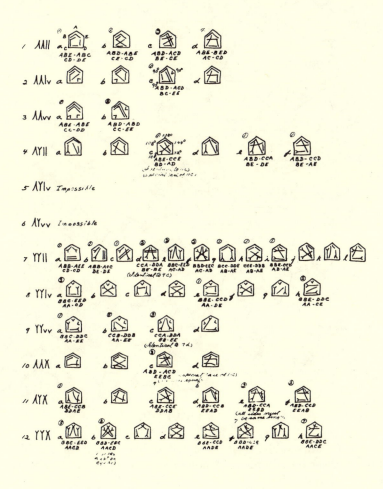

FIGURE 9

The 12 groups of pentagons considered which might form "2-pentagon" patterns. Each angle of such a pentagon is "used" twice in the list of angle relationships occuring at the vertices of the tiling. Sketches of tilings are shown in Figures 9A and 9B for each successful combination of angles. This listing and collection of tilings is the second one made by Marjorie Rice.

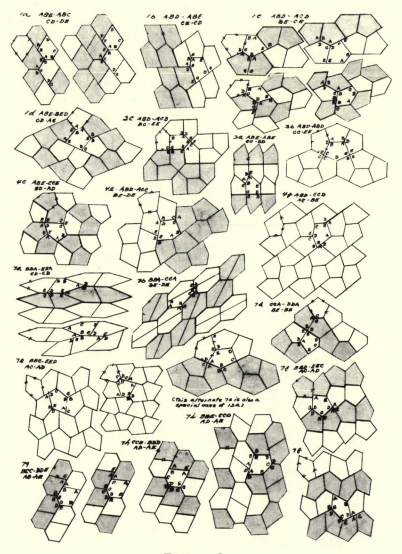

FIGURE 9A

seen to be impossible, so it doesn't look like such a formidable job to go through those remaining. ... Among them are types 3 and 5 and Mr. James's new one [type 10]."

In October, 1976, I received another bulging envelope from Marjorie. She had made a new listing of all of the pentagon tilings she had discovered thus far—58 in all. She had reorganized her listing; this time she had arranged the pentagons (and associated tilings) into 12 classes, each class corresponding to which sides of the pentagon must be equal. Six pages of illustrative tilings demonstrated the thoroughness of her work.

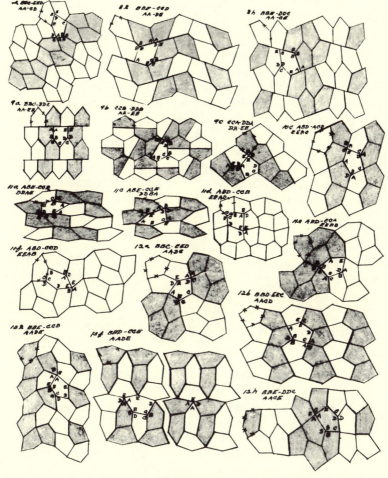

FIGURE 9B

Every one of the ten types of tiling pentagons occurred, and there were many new tilings. She closed her letter "this is as far as I can go with my limited knowledge, so I am through looking. Perhaps there are still others I have not come across."

In mid-November, I sent Marjorie a preprint of a paper by Branko Grünbaum and Geoffrey Shephard which showed the 24 tile-transitive tilings by pentagons of types 1 through 5 and also the first draft of my article on the pentagon problem for publication in *Mathematics Magazine*. This was an expanded version of a talk I had given at a Conference on Recreational Mathematics at Miami University in Oxford, Ohio. John H. Conway had been at that conference and showed great interest in the problem and the contributions made by James and Rice. He admitted that he had once set out to find all tiling pentagons but had abandoned the problem when it became too time-consuming. At the end of my article I raised some natural questions: "Is the list of pentagons which can tile in an edge-to-edge manner complete?" "Can we find the complete list of all equilateral pentagons which tile?"

Marjorie couldn't ignore the questions. I received another letter in December, 1976. "Had thought to spend no more time on pentagons but they weren't so easy to lay aside." This time she had tackled the questions raised in my article. In reply to the last question, she had sketched all of her tilings by equilateral pentagons and the angle relationships forced by the tilings. She had also tackled the first question. In the article, I had explained "block-transitive" tiling by pentagons, noting that all of the recent discoveries—Kershner's, James's, and hers were pentagons which could not tile isohedrally, but for which a block of two or three stuck-together pentagons was necessary to generate the tiling. This notion was new to Marjorie and she reexamined all of her tilings which began with two pentagons stuck together and noticed that "most of them consist of 4 tiles forming 2 hexagons which tile in one of 6 ways." Focusing in on one possible dissection of such a block into four congruent pentagons had led her to discover several new edge-to-edge tilings. Two weeks later (December 27, 1976—the Christmas season seemed to be her most creative time!), came the exciting news: "I have been working with this idea further and have some new patterns, and to my surprise and delight—2 new types, closely related." Indeed, she had discovered types 11 and 12, and the accompanying tilings were very striking (Figure 10).

The discovery of the new types had been an unexpected bonus in her methodical analysis of 2-block transitive designs. She had found that the double hexagons could be dissected into four congruent pentagons in nine distinct ways (Figure 11) and these dissections together with the variety of

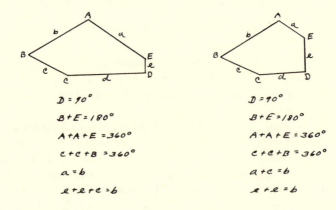

$D = 90°$

$B + E = 180°$

$A + A + E = 360°$

$c + c + B = 360°$

$a = b$

$e + e + e = b$

$D = 90°$

$B + E = 180°$

$A + A + E = 360°$

$c + c + B = 360°$

$a + c = b$

$e + e = b$

(angles of the 2 pentagons - 117°- 54°- 153°- 90°- 126°)

FIGURE 10

FIGURE 10 (cont'd)

Tiling pentagons of types 11 and 12 discovered by Marjorie Rice in December, 1976.

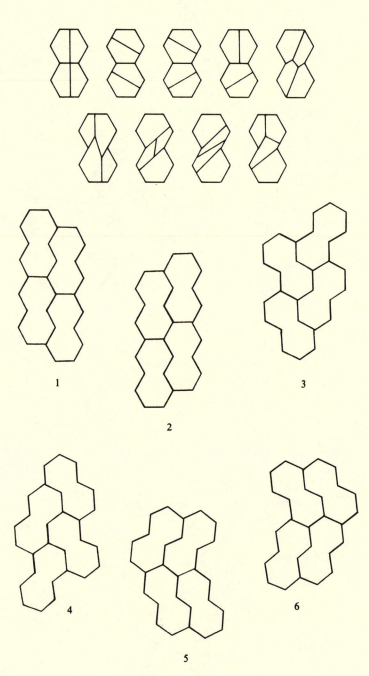

FIGURE 11

Consideration of how "double hexagon" blocks (represented here symbolically) could be dissected into four congruent pentagons and the ways in which such double hexagon blocks can tile, produced a wealth of new tilings by pentagons.

tilings by their blocks, led to over 50 different tilings by pentagons (the tilings were "2-block transitive"). Through the spring she continued to pursue the idea and found several dissections leading to 3-block pentagon tilings and even some 4-block pentagon tilings.

By now the original MG² circulation of information on the pentagon problem had grown and information was traveling to three continents. Another group of amateurs—11th year school children in New South Wales, Australia, had spent a week investigating the problem of discovering convex equilateral pentagons which tile and the ways in which they tile. Led by their teachers, George Szekeres and Michael Hirschhorn, they had made good progress on the problem. One particular equilateral pentagon was capable of a great variety of tilings (this was the special tile "case 6" of Marjorie Rice's first correspondence with me). Hirschhorn had discovered many unusual tessellations using this pentagon, including two beautiful central tessellations with just six-fold rotational symmetry (Figure 12).

In the summer of 1977 I provided Marjorie with the article "The 81 Isohedral Tilings of the Plane" by Branko Grünbaum and G. C. Shephard. The paper contained careful sketches illustrating the 81 distinct types of tilings and I felt that Marjorie might find among these some whose tiles could be dissected into congruent pentagons, thus creating new pentagonal tilings. Throughout the fall, she continued her previous work and also utilized the Grünbaum-Shephard article in a far more sophisticated manner than I had anticipated. "The isohedral tilings by Grünbaum and Shephard were of much interest. I have copied them into a four page version I could more easily use." She had, in fact, represented each of the

FIGURE 12

Michael Hirschhorn's central tessellation by an equilateral pentagon, fashioned into an engraved silver pendant.

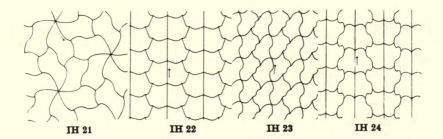

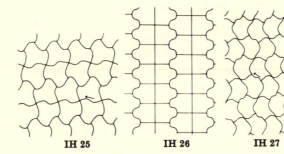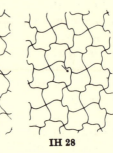

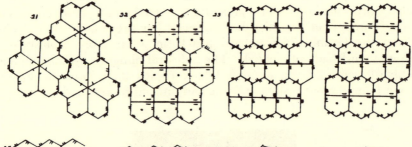

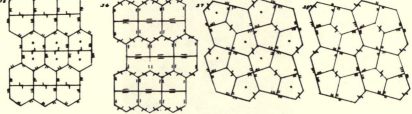

FIGURE 13

(Upper half) Several tilings by shaped tiles as they appear in "The 81 Types of Isohedral Tilings in the Plane", by Branko Grünbaum and G. C. Shephard (Mathematical Proceedings of the Cambridge Philosophical Society. September, 1977. pp. 190–191). (Lower half) The same tilings as redrawn in symbolic form by Marjorie Rice.

Grünbaum-Shephard tilings of shaped tiles by a symbolic marked tiling showing only the topological network of the tiling and the action of the symmetry group on the tiles (Figure 13). Using these, she had reanalyzed all of her previous tilings and discovered several new ones.

It was Christmas season, 1977, when she sent me a fat envelope containing her voluminous work. Again, there was a Christmas surprise. "Just a couple of weeks ago, this new type of pentagon turned up (and I thought there would be no more). This one (Figure 14) like type 4 has two opposite 90° angles but the requirements for the lengths of the sides are different." Her illustrative tiling by this new type 13 showed an interesting pattern of interlocked bow ties, made up of 4 of the pentagons. I had just received the galley proofs of my article on the pentagon problem and so was able to insert this latest discovery before publication. (The article, in the January 1978 issue of *Mathematics Magazine*, contains a fuller account of the mathematical details of pentagons which tile.) Still Marjorie's work on the problem did not end—she was determined to try to see if she could prove that all pentagons which could tile had been found. Although her attempt at proof was not complete, her thorough combinatorial check of all 2-block and 3-block patterns reduces the remaining task considerably. Perhaps by the time this story is published the question as to whether there are still other pentagons which tile will be answered. Hirschhorn is confident that all equilateral convex pentagons which tile have been found (these are described in the *Mathematics Magazine* article); his argument utilizes a computer in a proof by elimination of possible angle relationships.

What makes a person pursue a problem so steadfastly as Marjorie? She was not trained to do this, nor paid to do it, but obviously gained personal satisfaction in her patient and persistent search. No doubt her personal history is like that of many amateurs who are inspired by Gardner's writing.

She was born in 1923 in St. Petersburg, Florida, a first child. At age 5, she began school in Garden Valley School, a one-room country school with grades 1 through 8, having a total of about two dozen pupils. "My mother wished me to have a good start and had taught me well at home so I was placed in the second grade." She was a shy child, loved to read and "could easily become absorbed in a book or in my daydreams and forget the world around me." She did well in school; "arithmetic was easy and I liked to discover the reasons behind the methods we used." "I was interested in the colors, patterns and design of nature and dreamed of becoming an artist. ..." Her later years at the school "were enriched by two very fine teachers, Miss Keasey and Miss Timmons ... [who] helped make up for the deficiencies of a small country school."

"When I was in the 6th or 7th grade our teacher pointed out to us one day the Golden Section in the proportions of a picture frame. This immediately caught my imagination and though it was just a passing incident, I never forgot it. I've continued reading on a wide variety of

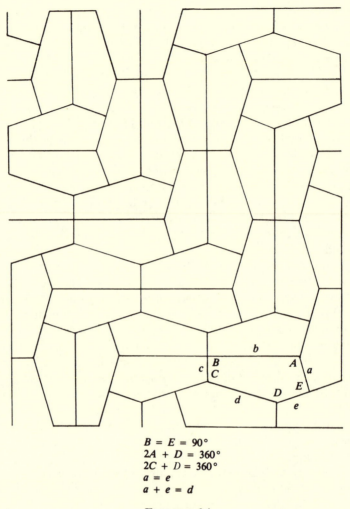

$$B = E = 90°$$
$$2A + D = 360°$$
$$2C + D = 360°$$
$$a = e$$
$$a + e = d$$

FIGURE 14

Tiling pentagon of type 13 discovered by Marjorie Rice in December, 1977.

subjects over the years and have been especially interested in architecture and the ideas of architects and planners such as Buckminster Fuller. I've come across the Golden Section again in my reading and considered its use in painting and design. A book that was especially helpful and inspiring to me in this regard was *The Geometry of Art and Life* by Matila Ghyka." Marjorie's interest in art continued—she became especially interested in textile design and the works of M. C. Escher. As she pursued the problem of pentagons and their tilings, she produced some beautiful geometric designs and imaginative Escher-like patterns (Figures 15, 16).

While in high school the family moved to Pine Castle, near Orlando,

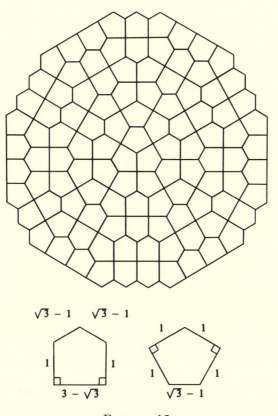

$\sqrt{3} - 1$ $\sqrt{3} - 1$

$3 - \sqrt{3}$ $\sqrt{3} - 1$

FIGURE 15

A beautifully symmetric tiling by Marjorie Rice which radiates outwards from the center and can be continued to fill the plane. Two pentagons, each having two 90° angles and three 120° angles, create the design.

Florida. At Orlando senior high Marjorie studied shorthand and typing to prepare for future employment and did poorly at both. She regretted that she could not take mathematics beyond the required general course. After graduating from high school at 16 she was employed first in the office of a laundry, then in a small printing firm until her marriage to Gilbert Rice in 1945. "During those years I frequented the public library and learned much about science, psychology and other subjects I had missed out on in school. I also started a correspondence course in commercial art. ..." After the Rices' first son was born they moved to San Diego, California. "These were very busy years for us both"—an understatement to be sure. The Rices raised five children and Gilbert started his own trade typesetting shop. Marjorie was drawn back into mathematics by her children. "When my oldest son, David, was in junior high, ... the 'new math' was just beginning to be used. ... I wanted to study his lessons and keep up with him

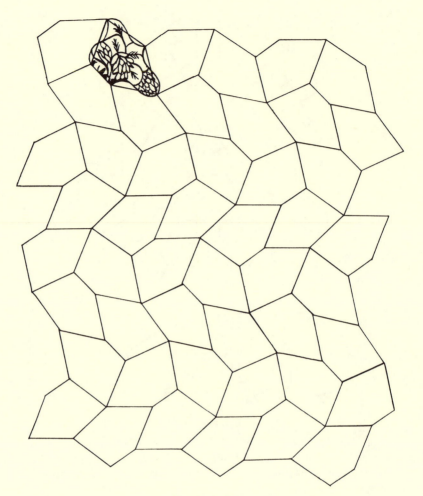

FIGURE 16A

Underlying grid for Bees in Clover (see Plate I.)

FIGURE 16

Three Escher-like plane-filling designs by Marjorie Rice are based on the geometric grids of some of her unusual tilings by pentagons. (M. C. Escher used a well-known tiling by pentagons as the underlying grid for some of his plane-filling designs.)

as he learned and he encouraged me to do this—but these were busy days. I soon fell behind and gave up the idea. My interest in his lessons continued however and I could often find solutions to his problems by unorthodox means, since I did not know the correct procedures. He shared with me the mathematical games he learned in class, such as Hex and three dimensional tic-tac-toe. ..."

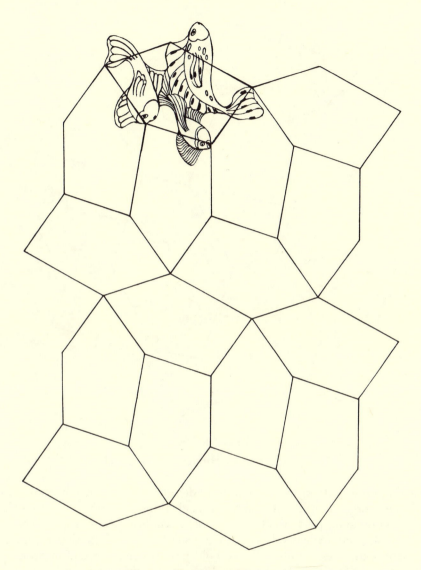

FIGURE 16B
Underlying grid for Fish (see Plate II.)

"I enjoy puzzles of all kinds, crosswords, jigsaw puzzles, mathematical puzzles and games, and have purchased books of mathematical puzzles over the years. Those of a geometric nature are a special delight. Thus when my son's *Scientific American* arrives I first turn to Martin Gardner's "Mathematical Games" section." Her absorbing fascination with such puzzles and keen perception of shapes, proportions and designs is tellingly revealed in her account of a recent trip. "In November, 1974, my husband and I started on a journey which would take us around the world. ... I had

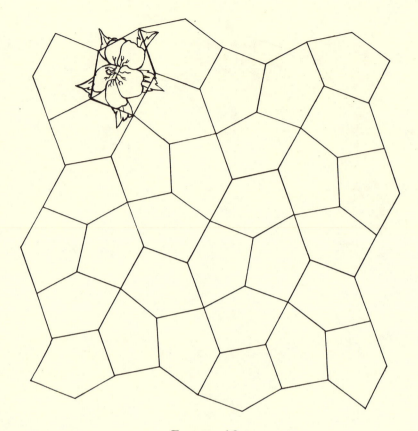

FIGURE 16C

Underlying grid for Hibiscus (see Plate III.)

much interest and curiosity concerning designs and proportions that were different and unfamiliar. I looked for such things wherever we traveled, taking notes of proportions of houses, doors, windows, division of space, the designs of grilles over windows. ... I especially enjoyed the delightful and colorful textile designs often in big bold patterns that were often worn in Ghana and Nigeria." "We kept our luggage to a minimum on this trip ... but I did slip in a small paperback book, *Work This One Out*, 105 puzzling brain-teasers by L. H. Longley-Cook. Thinking on these problems helped pass the time quickly when we had long periods of waiting. ..."

The mind and spirit are the forte of all such amateurs—the intense spirit of inquiry and the keen perception of all they encounter. No formal education provides these gifts. Mere lack of a mathematical degree separates these "amateurs" from the "professionals". Yet their dauntless curiosity and ingenious methods make them true mathematicians. Martin Gardner has awakened many such mathematicians.

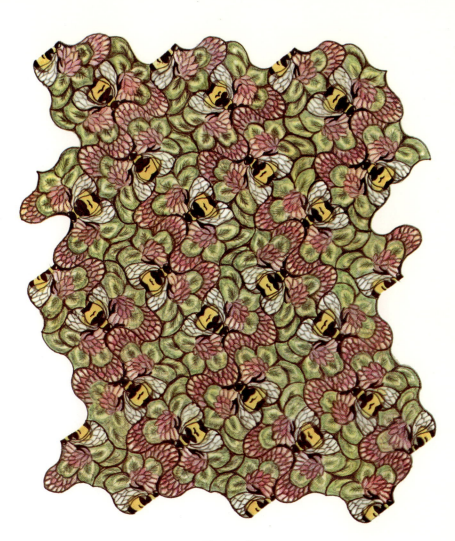

PLATE I

Bees in Clover by Marjorie Rice.

PLATE II

Fish by Marjorie Rice.

PLATE III

Hibiscus by Marjorie Rice.

PLATE IV

Bees and Wasps by M. C. Escher

PLATE V

A hand woven rug (created by Australian A. G. Bomford of Canberra, Australia and reproduced here with his permission) captures Richard James's design. The rug is 2.03 by 1.31 meters, 46,986 stitches.

Some Problems on Plane Tilings

Branko Grünbaum
UNIVERSITY OF WASHINGTON

G. C. Shephard
UNIVERSITY OF EAST ANGLIA

Although the art of tiling is as old as human history, the science of tiling seems to have been curiously neglected until recent times.

The advanced state of the art in the Middle Ages is demonstrated by the remarkable tilings on many mosques and other Saracenic buildings— the fourteenth-century Spanish palace known as the Alhambra is one of the prime examples (Figures 1 and 2). There is little doubt that it was a visit to the Alhambra that helped to inspire the Dutch artist M. C. Escher in some of his well-known creations (Figure 3).

The first treatment of tilings from a mathematical point of view seems to be that of the German astronomer Johannes Kepler (1571–1630) (Figure 4) in his book *Harmonice Mundi* published in 1619 (Figure 5). Kepler's geometrical discoveries were so overshadowed by his contributions to physical astronomy that, incredible as it may seem, they were mostly forgotten for nearly three hundred years. Apart from Kepler's book, very little work on tilings seems to have been done before the end of the nineteenth century; so the science of tilings, by which we mean the active investigation of their mathematical properties, is barely a hundred years

FIGURE 1

A view of part of the Alhambra palace at Granada in Spain. This dates from the fourteenth century and, like many other Saracenic buildings, is remarkable for the wealth of tilings used as decoration. A few of the tilings from Alhambra are shown in Figure 2.

FIGURE 2

Eight tilings from the Alhambra. These sketches were made by the Dutch artist M. C. Escher when he visited the Alhambra in 1936. The subject of tilings had a profound effect on the artist's later work.

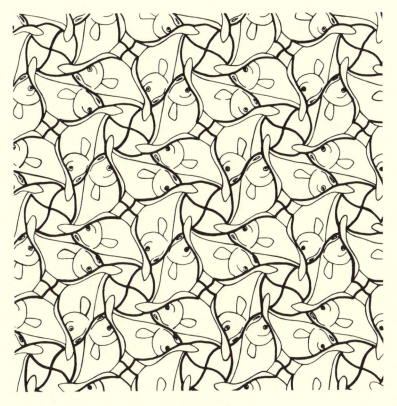

FIGURE 3

One of Escher's less well known tilings. In Escher's work there are many examples of tilings in which the tiles resemble animals or inanimate objects.

FIGURE 4

The famous astronomer Johannes Kepler, whose pioneering research of tilings was forgotten for almost three centuries.

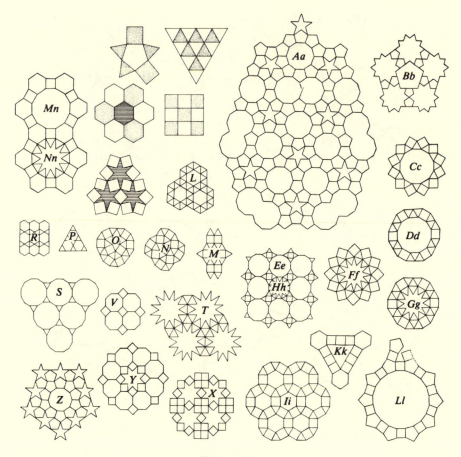

FIGURE 5

Kepler's book *Harmonice Mundi*, which appeared in 1619, contains the first mathematical treatment of tilings ever published. The plates from this book, reproduced above, show some of the tilings by regular and star polygons which Kepler investigated.

old. During this century a large amount of material has been published, much of it by crystallographers, engineers and other non-professional mathematicians. The subject still abounds with open problems; we shall draw attention to some of them in the following pages.

Recently, with an increased awareness of the educational merits of geometry, the investigation of tilings has come to be regarded as an appropriate mathematical activity in schools and colleges. It is interesting to note that periodicals intended for mathematics teachers seem to be devoting more and more space to this topic. This resurgence of interest can be partly attributed to Martin Gardner's beautiful articles in the *Scientific*

American. By reporting new discoveries, he has repeatedly stimulated the interest of his readers in the subject. In the following pages we shall have occasion to mention several of these articles and we hope that the gratitude which the mathematical community should feel towards Martin Gardner, the expositor *par excellence*, will be made apparent.

1   What is a *tiling*? From a mathematical point of view it is a family $\mathcal{T}$ $= \{T_1, T_2, \ldots,\}$ of closed sets T_i (the tiles) which cover the plane without gaps or overlaps of non-zero area. As it stands, this definition is much too general, and throughout the article we shall restrict attention to tilings for which the tiles T_i, $i = 1, 2, \ldots$, are topological disks; that is, each is obtained from a circular disk by a continuous deformation. In particular, each tile is connected and simply-connected. Thus tiles which are fragmented into two or more disjoint parts, or have holes in them, are not permitted. Some of the commonest tilings are those in which the tiles are of a limited number of different shapes. This situation is conveniently described in the following way. Let $\mathcal{S} = \{P_1, P_2, \ldots, P_k\}$ be a (finite) family of closed sets such that every tile T_i in $\mathcal{T}$ is congruent to one of the P_i $(i = 1, \ldots, k)$. Then $\mathcal{S}$ is called a set of *prototiles* for $\mathcal{T}$ and we say that $\mathcal{S}$ *admits* the tiling $\mathcal{T}$. If $\mathcal{S}$ contains exactly k distinct sets (by which we mean that no two are congruent) and all of these are used in $\mathcal{T}$, then $\mathcal{T}$ will be called *k-hedral*.

When $k = 1$ we call the tiling *monohedral*. Such tilings are very familiar. Examples are provided by the regular tilings (Figure 6) which have been known from time immemorial. The uniform tilings (Figure 7), known to Kepler (see Figure 5) and probably discovered by him, 2-hedral (*dihedral*) or 3-hedral (*trihedral*). Many other *k*-hedral tilings with small

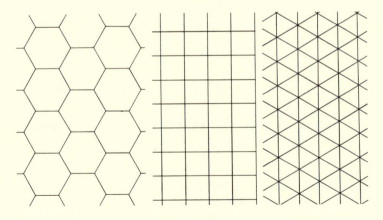

FIGURE 6

The three regular tilings. These familiar tilings have been known since time immemorial.

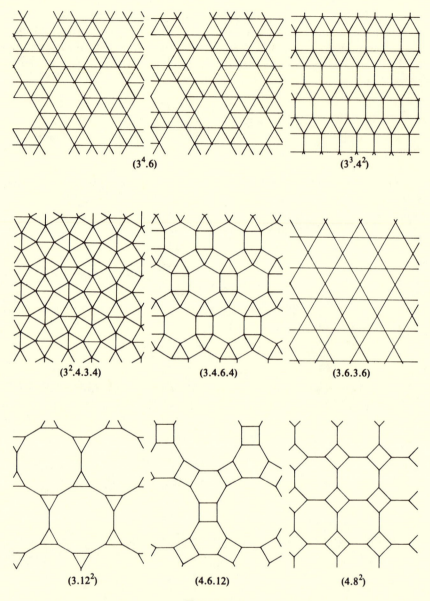

$(3^4.6)$ $(3^3.4^2)$

$(3^2.4.3.4)$ $(3.4.6.4)$ $(3.6.3.6)$

(3.12^2) $(4.6.12)$ (4.8^2)

FIGURE 7

The eight uniform tilings that are not regular. In each, the tiles are regular polygons and there exist symmetries of the tiling which map any chosen vertex onto any other vertex. One of the tilings (denoted by $(3^4.6)$) occurs in two mirror-image forms as shown. It is believed that as a group, these tilings were discovered by Kepler in the early seventeenth century.

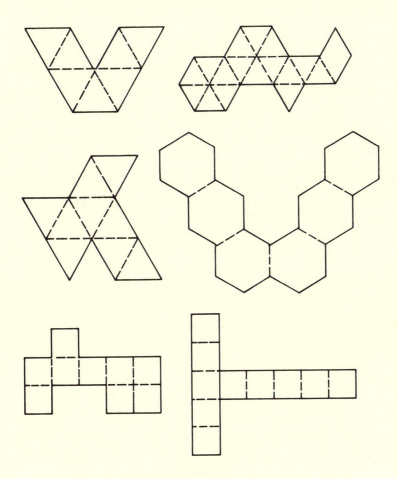

FIGURE 8

Are these prototiles of monohedral tilings? It is by no means easy to answer this question.

values of k are known, such as those shown in Kepler's drawings K, T, X and Aa of Figure 5.

It might be thought that monohedral tilings, in which all the tiles are the same shape, are mathematically trivial and therefore, hardly worth investigating. But this is far from true. A few of the great variety of possibilities are shown in Figures 21, 23, 28 and 29. To convince the reader of the difficulties, we show a few prototiles in Figure 8. For some of these it is not known with certainty whether or not they admit monohedral tilings, and in every case it is by no means easy to settle the question. Problems of this type were discussed by Martin Gardner in his article in *Scientific American* (August 1975). The basic question is the following:

PROBLEM 1 Is there any well-defined procedure (or algorithm), not employing trial-and-error, for testing whether or not a given prototile P admits a monohedral tiling of the plane?

This is a very difficult problem, and to discuss all its ramifications would take us deep into the subject of mathematical logic. Although this is not appropriate here, in Section 5 we will briefly note the connection between this problem and that of finding aperiodic tiles. It is worth mentioning that one of the few conditions known for testing whether a given prototile admits a tiling or not is the so-called *Conway criterion* (see Figure 9). This is of surprisingly wide application, but it is only a *sufficient* condition for a tiling to exist and is not *necessary*. Therefore, it does not provide a solution to Problem 1.

The depth of the problem as well as the extent of our ignorance is shown by the fact that we even do not know all the shapes of polygons that

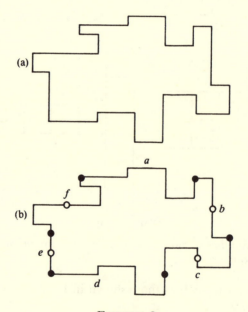

FIGURE 9

Is it possible to tile the plane with the tile shown in (a)? This question is answered in the affirmative by the *Conway criterion* which says that a prototile admits a monohedral tiling if it is possible to divide its boundary into six parts (as indicated by the black circles in (b)) in such a way that parts a and d are translates of each other, while the other four parts b, c, e and f, are curves each of which has a center of symmetry (indicated by an open circle. In the tiling four of the six adjacents of the given tile are obtained by a 180° rotation of the given tile about the open circles, and the other two are translates of the given tile. The criterion applies to any tile—not just to polygons like the tile shown here.

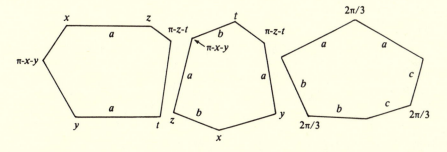

FIGURE 10

The three types of hexagon that admit monohedral tilings of the plane; in fact each will also tile isohedrally. These types were discovered by K. Reinhardt in 1918. Relations between angles (marked x, y, z, t) and between sides (marked a, b, c) are indicated.

admit monohedral tilings. Every triangle and every quadrangle admit such tilings, as do three families of hexagons (see Figure 10). (A *family* is defined as a set of polygons satisfying certain stated relationships between its sides and angles.) But the number of families of pentagons is still uncertain. At the present time, thirteen families of pentagons that admit tilings are known (Figure 11), but whether or not this list is complete is still not settled. Here again Martin Gardner made a contribution to the subject by publishing, in *Scientific American* (July 1975), what was thought at that time to be a complete list of pentagons—namely, those discovered by K. Reinhardt in 1918 and by R. B. Kershner in 1968. Several readers pointed out that the list was incomplete, and the new tiles that turned up are included in Figure 11. This leads us to our second question.

PROBLEM 2   Do there exist any families of convex pentagons, other than those shown in Figure 11, that admit monohedral tilings of the plane?

Recently D. C. Hunt and M. D. Hirschhorn claim to have proved that a list of *equilateral* pentagons that tile the plane, compiled by Doris Schattschneider in 1978, is complete. However, their proof has not yet been published.

2   Even if a prototile admits a monohedral tiling of the plane, there is no *a priori* method of deciding in how many different ways it will tile. Let us say that a tile is *monomorphic* if it admits a unique tiling. The commonest example of a monomorphic tile is the regular hexagon—the only tiling possible is the regular tiling shown in Figure 6. On the other hand, a square is not monomorphic. In fact, by "sliding" rows of squares relative to one another, it is evident that a square admits an uncountable infinity of different tilings. In Figure 12 we show some less familiar examples of monomorphic tiles, and in considering these an important question of definition

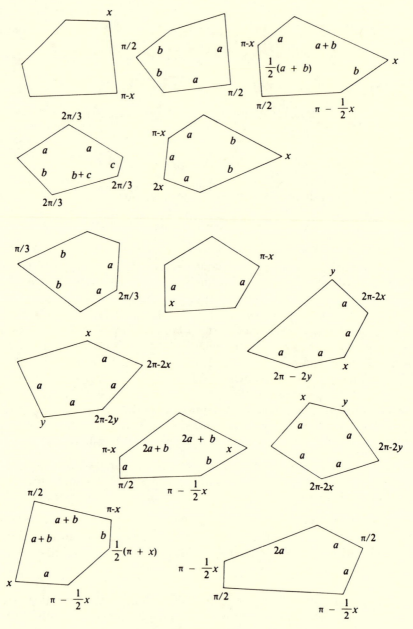

FIGURE 11

The thirteen types of pentagon that are known at present to admit monohedral tilings of the plane; the first five types also tile isohedrally. The list is taken from Schattschneider's paper (*Mathematics Magazine*, 51 (1978), pp. 29–44) in which further details and examples of the tilings can be found.

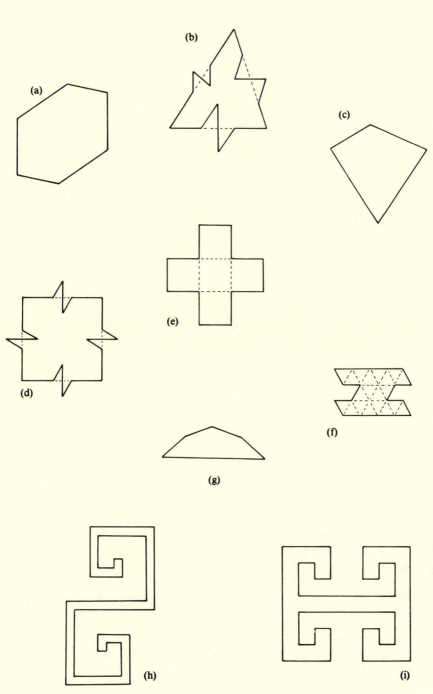

FIGURE 12

Nine examples of monomorphic prototiles.

FIGURE 13

The prototiles (e) and (f) of Figure 12 admit the tilings shown above. Since the two possible tilings with each are mirror images of one another, the tiles are regarded as monomorphic.

arises. The tiles (e) and (f) of Figure 12 each appear to admit two tilings (Figure 13), but in each case these are mirror-images of each other. Should we count these as different? The most appropriate answer is a matter of opinion, but for various reasons we prefer not to distinguish between mirror images and so we assert that the tiles (e) and (f) are monomorphic.

Do there exist tiles which are *dimorphic*, that is, tiles which admit precisely two monohedral tilings? The answer is that such tiles exist and examples are shown in Figure 14. A *trimorphic* prototile, and the three corresponding tilings are shown in Figure 15; this example is unique in the sense that all other known trimorphic prototiles differ trivially from this one. The discovery of dimorphic and trimorphic prototiles is not easy, so we propose the following.

PROBLEM 3   Find additional, essentially different, examples of dimorphic and trimorphic prototiles.

PROBLEM 4   Do *r*-morphic prototiles exist for any finite value of $r \geqslant 4$?

The following is a more technical question of considerable theoretical interest.

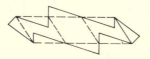

FIGURE 14

Two dimorphic prototiles and the corresponding tilings.

PROBLEM 5   Does there exist a tile which only admits a countable infinity of monohedral tilings?

Thus Problems 4 and 5 are asking whether the large gap in values between $r = 3$ and an uncountable infinity, can be bridged.

If we try to extend these ideas to k-hedral tilings (those in which k shapes of tiles are used) then new considerations enter. Let us consider k-hedral r-morphic prototiles ($k \geqslant 2$, $r \geqslant 1$). Their existence was established by H. Harborth in 1977. He pointed out that a suitable rhomb and a tile which is a union of rhombs (see Figure 16) solved the problem for $k = 2$, and hence for $k \geqslant 2$. (This was done by cutting either of the tiles into an appropriate number of pieces in such a way that these pieces can only be fitted together again by reconstituting the original tiles.) However, the construction of Figure 16 does not solve the problem completely; it does so only if we insist that in all the tilings, copies of every prototile must be used. Without this condition the example fails because a rhomb on its own admits an uncountable infinity of tilings. We therefore pose the following question.

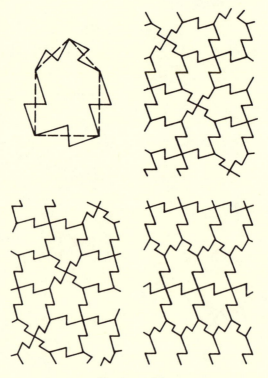

FIGURE 15

A trimorphic prototile and the corresponding tilings.

(a) (b)

(c) (d)

FIGURE 16

Harborth's construction for a pair of prototiles which admit exactly r tilings (assuming that both kinds of tile occur in each tiling). In this diagram, $r = 4$. One tile is a rhomb with angles $2\pi/p$ and $(p - 2)\pi/p$ where $p = 6r - 7$, and the other is obtained by fusing together $2r - 2$ rhombs as shown.

PROBLEM 6   For every value of $k \geqslant 2$ and $r \geqslant 1$ does there exist a set $\mathscr{S}$ of k prototiles such that exactly r distinct tilings are admitted by $\mathscr{S}$ even if copies of all the prototiles in $\mathscr{S}$ need not be used?

An example that solves Problem 6 in the case $k = r = 2$ is shown in Figure 17.

3   It is well known that in 1900, D. Hilbert posed a series of problems that have had a major effect on the development of mathematics. One of these—the eighteenth problem—was concerned with tilings. The following is a brief account of that problem.

To state the problem we need to use the word *isohedral*. This is defined as follows. Every tiling $\mathscr{T}$ has a group $S(\mathscr{T})$ of *symmetries*, that is, rigid motions of the plane that map $\mathscr{T}$ onto itself. (See, for example, Figure 18, where several different kinds of symmetry are indicated.) An informal way of thinking of a symmetry (not without reservations as to its mathematical validity) was described by Fourrey in 1907. He suggested tracing the tiling

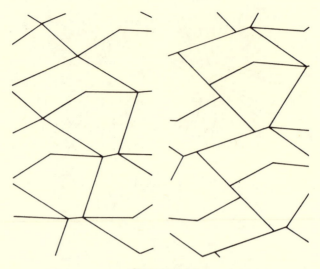

FIGURE 17

Two dihedral tilings using the same two pentagonal prototiles. These are the only tilings possible, so the pair of pentagons forms a dimorphic set. Notice that here *neither* prototile on its own admits a tiling.

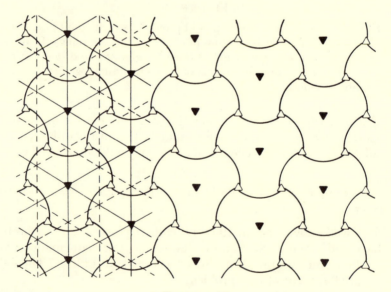

FIGURE 18

A tiling with, in the left part of the figure, its symmetries (except translations) marked. Triangles represent centers of 3-fold rotational symmetry, solid lines represent lines of reflection, and dashed lines represent glide-reflections. Any vector connecting two of the solid triangles is a translational symmetry.

FIGURE 19

Two monohedral tilings using the same prototile. The right tiling is isohedral because, given any two tiles, there exists a symmetry of the tiling which maps one onto the other. The left tiling is not isohedral since, for example there is no symmetry which maps the tile A onto the tile B.

onto a transparent sheet and then every symmetry of the tiling corresponds to a motion of the sheet (including the possibility of turning it over) such that the tracing again fits exactly over the tiling. A tiling $\mathscr{T}$ is *isohedral* if, given any two tiles T_i, T_j of $\mathscr{T}$, there exists a symmetry of $\mathscr{T}$ that maps T_i onto T_j. Every isohedral tiling must be monohedral—the distinction between the two concepts is illustrated in Figure 19.

Hilbert's eighteenth problem can be stated in the following way. Does there exist a prototile which admits a monohedral tiling but no isohedral tiling? In fact the problem was originally posed in a three-dimensional context, and there is reason to suppose that in the plane case Hilbert believed the answer to be trivially in the negative. But he was wrong! In 1935 H. Heesch found a counterexample (see Figure 20), that is, a tile which admits an infinity of tilings, but none of these tilings is isohedral. Since then, other examples have been found; some of the convex pentagons of Figure 11 have this property.

At first it may appear that the difference between monohedral and isohedral tilings is slight—but this is far from the truth. For example, if we restrict attention to isohedral tilings, then Problems 1, 2, 4 and 6 can be solved. In particular, in connection with Problem 2, H. Heesch and

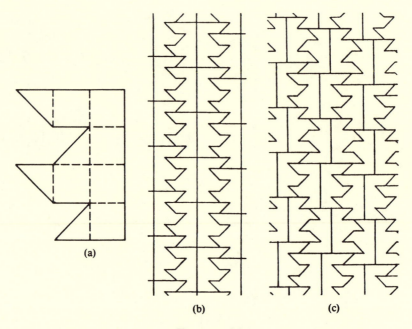

FIGURE 20

Heesch's tile, shown here, was the first to be discovered that admits a monohedral tiling but no isohedral tiling. Many other examples are now known—for example several of the pentagons in Figure 11 have this property. In (a) we show how the tile is built up from squares and half-squares, and in (b) and (c) we show examples of the (infinitely many) possible tilings using this prototile.

O. Kienzle gave in 1963 a complete list of polygons that admit isohedral tilings. This is possible because *all* isohedral tilings can be described. It can be shown that there are 81 types of which 47 can be realized using polygons. Some examples are shown in Figure 21. The actual method of classification into types is too technical to describe here; it has only recently been satisfactorily resolved in a general context. (This fact probably explains why many of the older papers dealing with classification problems contain many obscurities and errors.)

4   One of the most remarkable theorems concerning tilings is the *Extension Theorem*. A special case of this is known as Wang's Theorem (named for its discoverer), but the general case has only recently been proved and no proof has yet been published. Suppose that $\mathscr{S}$ is a given (finite) set of prototiles, each one a topological disk, and suppose that, however large R is, we are able to tile over a circular disk D_R of radius R using the tiles of $\mathscr{S}$. Then the Extension Theorem asserts that $\mathscr{S}$ admits a tiling of the plane. The phrase *tile over* needs clarification. It means that it is possible to construct a set (or *patch*) of tiles in such a way that their overlaps have zero area, yet they completely cover a region which includes D_R. For this theorem to be true, it is essential that the prototiles be bounded and finite in number. For example, in Figure 22 we show a single prototile copies of which will tile over arbitrarily large disks, but which does not admit a tiling of the plane. In this case, of course, the theorem does not apply since the prototile is unbounded.

Some of the consequences of this theorem are surprising. For example, it implies that if we can tile over a quarter-plane using the tiles of $\mathscr{S}$, then we can tile the whole plane. At first sight it may appear obvious that if one can construct larger and larger patches of tiles, without limit, then eventually one will arrive at a tiling of the plane. But this becomes much less clear when one realizes that as R increases, it may be necessary to

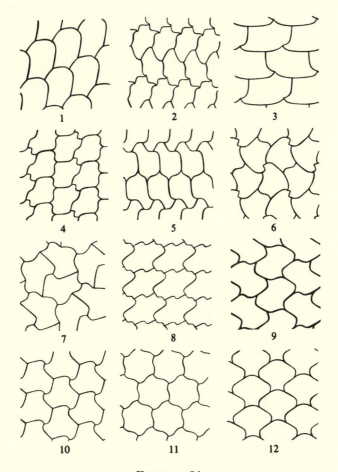

FIGURE 21

Twelve examples of isohedral tilings.

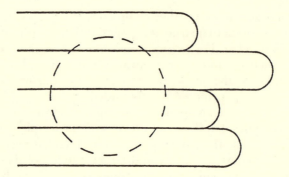

FIGURE 22

Here each tile is a semi-infinite strip ending in a semicircle. Such tiles will clearly tile over arbitrarily large circular disks, but will not admit a tiling of the plane. This example shows that for the truth of the Extension Theorem it is necessary to assume that the tiles are bounded.

continually rearrange the tiles and it may happen that *none* of the patches used to cover the disks can appear as part of the final tiling.

A related problem is known as Heesch's problem. It is known that there exist prototiles P which do not admit a tiling of the plane, yet P can be completely surrounded by copies of itself (see Figure 23). By *completely surround P* we mean that we can construct a ring R of tiles around P (leaving no gaps) in such a way that each point of P is at a distance greater than some fixed positive number from every point of the plane that is outside R. By the phrase *P is completely surrounded twice* we mean that *P is surrounded by a ring R*, and then R is surrounded by a second ring R'.

PROBLEM 7   Does there exist a prototile P which does *not* admit a tiling of the plane yet can be completely surrounded twice by copies of P?

More generally it is clear what we mean when we say that a tile P can be surrounded r times, and if r is the largest such integer it is called the *Heesch number* for the tile P.

PROBLEM 8   Do any tiles exist with Heesch number $r = 3, 4, 5, \ldots$?

5   A tiling $\mathcal{T}$ is called *periodic* if there exist two non-parallel translations which are symmetries of $\mathcal{T}$. Examples of periodic tilings appear in Figures 13, 15 and 18; in each case one can think of the tiling as being made up of a small patch of tiles repeated in a lattice arrangement. A set of prototiles $\mathcal{S}$ is called *aperiodic* if it admits a tiling of the plane, but no such tiling is periodic. Until recently no aperiodic sets were known; the first example was discovered by R. Berger in 1966, and other examples have been found

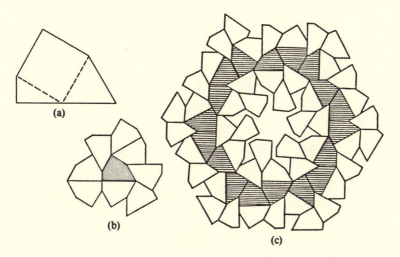

FIGURE 23

Heesch's tile shown in (a) has the remarkable property that it can be completely surrounded by copies of itself, yet admits no tiling of the plane. Even so, extensive patches can be built up, such as that in (c). The shape of the tile is indicated in (a)—it consists of the union of a square, an equilateral triangle, and a half equilateral triangle as shown by the dashed lines.

by Robinson, Penrose, Ammann and others. Figure 24 shows a tiling using the first set of aperiodic tiles discovered by R. Penrose.

For a delightful account of Penrose's aperiodic tiles known as kites and darts, the reader should consult Martin Gardner's article in the *Scientific American* of January 1977. These tiles have many remarkable and unexpected properties, not all of which have been adequately explained. Since 1977, several new sets of aperiodic tiles have been discovered by Robert Ammann, to whom we are grateful for permission to reproduce the tiles shown in Figure 25.

The subject of aperiodic tilings is much too big to deal with here. The history of the subject and its connection with mathematical logic is a fascinating chapter in the history of tilings, a chapter which is still not fully written. Here we can do no more than mention these few facts. However, no article attempting to survey the properties of tilings would be complete without mentioning the main outstanding problem in this area.

PROBLEM 9   Does there exist a single prototile P which forms an aperiodic set?

In other words, P may admit many monohedral tilings of the plane, but no such tiling is periodic. It can be shown that this problem is, surprisingly, related to Problem 1, or at least H. Wang has shown a connection between aperiodicity and the *tiling problem*, that is, the determination of an

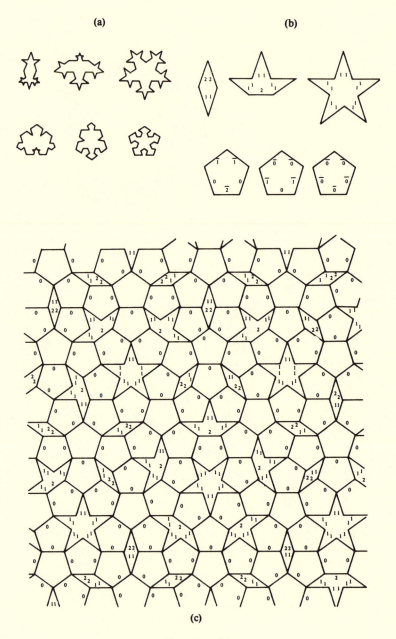

(a)

(b)

(c)

FIGURE 24

The first set of aperiodic tiles to be discovered by Roger Penrose. The prototiles are shown in (a), and in (b) we indicate how the projections and indentations can be replaced by numbers which give an equivalent "matching condition": 0, 1, 2 must fit against $\overline{0}$, $\overline{1}$, $\overline{2}$ respectively. In (c) we show an example of the tiling.

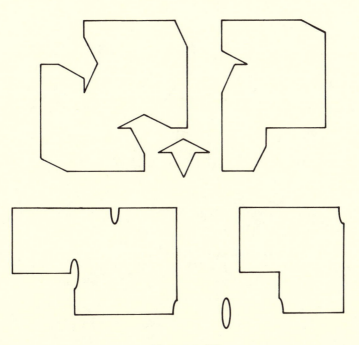

FIGURE 25

Two sets of aperiodic tiles recently discovered by R. Ammann. Each set of three tiles contains a small "key tile" which fits into indentations in the larger tiles, and so restricts the ways in which the latter can abut against each other.

algorithm for deciding whether a given set $\mathscr{S}$ of prototiles admits a tiling. Wang's results only apply to a very special kind of "square tile with colored edges", and we do not know to what extent similar considerations apply to tiles of general shapes. However, it seems likely that an affirmative solution to problem 9 would imply a negative answer to Problem 1.

6   We conclude with some curiosities. The first of these is the *enclosure problem*. In 1934 K. Reinhardt asked whether it is possible for two tiles— each congruent to a prototile P—to enclose another copy of P. In 1936 this problem was solved by H. Voderberg, who produced the example shown in Figure 26, which admits a periodic tiling. This prototile has the strange property that not only do some tiles enclose a third, but other pairs of tiles enclose two copies of P. We can say that Voderberg's tile has the 2-enclosure property—more generally, a tile P is said to have the r-enclosure property if two copies of P enclose a region equal to the union of r non-overlapping copies of P. It is remarkable that however large r is chosen, tiles with the r-enclosure property exist. A construction for such tiles is shown in Figure 27. However, these tiles do not solve the following (probably not very difficult) question.

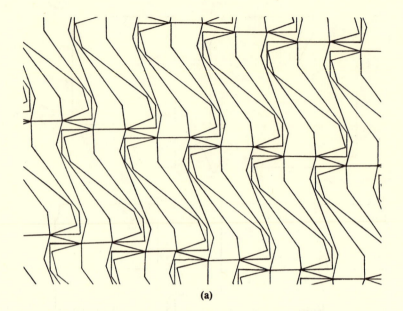

(a)

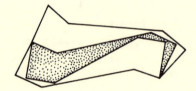

(b)

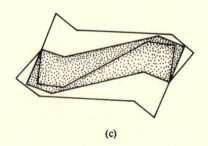

(c)

FIGURE 26

Voderberg's tile. This has the remarkable property that two copies of tile can enclose a third (b) or even two other copies (c). In (a) a periodic tiling is shown using the Voderberg tile as prototile.

PROBLEM 10   Does there exist a tile P with the r-enclosure property $(r \geqslant 3)$ and which admits a tiling of the plane?

In his paper Voderberg remarks, almost casually, that his tile admits a tiling with "spiral form" (see Figure 28). The appearance of this is sufficiently striking for it to have attracted considerable attention. In January 1977, Martin Gardner published a diagram of Voderberg's spiral, and also Michael Goldberg's explanation of how it is constructed. Once the method is explained, it is obvious that the spiral appearance has nothing to do with the enclosure property; moreover, it is easy to find many different tiles that admit spiral tilings.

However, all spirals that can be constructed by Goldberg's method necessarily have an *even* number of arms, that is, rows of tiles spiralling

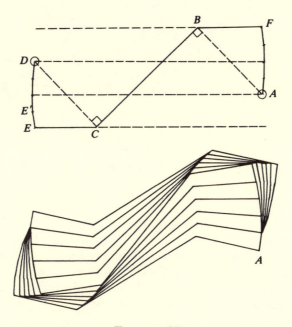

FIGURE 27

The construction of a tile with the r-enclosure property (in the figure, $r = 8$). First construct a zigzag line ABCD with right angles at B and C and the points ABCD on four equidistant parallel lines, shown dashed. DE is an arc centered at A and FA is an arc centered at D. EE′ is one fourth of the arc ED. Denote AFBCE by S and let S′ be the result of rotating S about A until E coincides with E′. Then the tile bounded by S, S′ and EE′ has the 8-enclosure property as shown in the right-hand diagram. For values of r other than 8 the same construction is carried out with E′ chosen so that the arc length DE is $[\frac{1}{2}(r + 1)]$ times that of EE′.

FIGURE 28

Voderberg's spiral tiling. This uses the same prototile as the tiling in Figure 26.

outwards from the center. Recently spirals with an odd number of arms have been found and examples appear in Figure 29. The tile used here has been called *versatile* since it admits many other unusual tilings (see Figure 30). However, there are still several open problems. For example, in Figure 29 we notice that about half the copies of the prototile are reflections of the others. Is this necessarily so?

> PROBLEM 11   Does there exist a monohedral spiral tiling with an odd number of arms, which uses only direct copies (no reflections) of the prototile?

Perhaps we should mention that the subject of spiral tilings, while aesthetically attractive, has a big mathematical disadvantage—so far we have not been able to say exactly what a spiral tiling is! Is it a genuine mathematical concept, or is it merely psychological? We conclude with the following problem, which is probably more suitable for general discussion than for a mathematical investigation:

> PROBLEM 12   Give a precise definition of a spiral tiling.

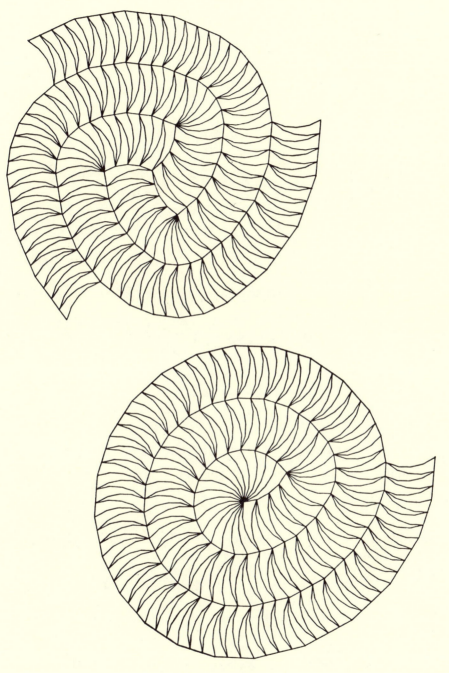

FIGURE 29

Monohedral spiral tilings with three and with one arms. The prototile
is known as a "versatile".

FIGURE 30

An ornamental spiral tiling using the same prototile as the spirals of Figure 29.

References and Further Reading

In addition to the articles by Martin Gardner quoted in the text, the following books and papers will be of interest. They are listed by the sections of the paper to which they are relevant.

Introduction.   Several collections of Escher's works have been published; the most extensive is *The World of M. C. Escher* (Abrams, New York, 1971). A very interesting account of Escher and his tilings is given in B. Ernst's book *The Magic Mirror of M. C. Escher* (Random House, New York, 1976). A discussion of the tilings to be found in the Alhambra appears in E. Müller, "Gruppentheoretische Ornamente aus der Alhambra in Grenada" (ETH dissertation, Zürich, 1944). Kepler's book *Harmonice Mundi*, originally published in Linz in 1619, has been reprinted in *Kepler's Complete Works*, edited by M. Caspar (*Gesammelte Werke*, Band VI, Beck, München, 1940 and by Culture et Civilisation, Bruxelles, 1968). These texts are in Latin; a German translation, *Weltharmonik*, by M. Caspar has also been published (Oldenbourg, München, 1967).

1   The three types of hexagons that admit tilings of the plane were determined by K. Reinhardt in his thesis "Über die Zerlegung der Ebene in Polygone", Frankfurt University, 1918 (Noske, Leipzig, 1918). Kershner's paper is "On paving the plane", *American Mathematical Monthly* 75 (1968), 839–844; our list of pentagons is taken from D. Schattschneider, "Tiling the plane with congruent pentagons", *Mathematics Magazine* 51 (1978), 29–44. The announcement by M. D. Hirschhorn and D. C. Hunt that the list of equilateral pentagons is complete also appears in the *Mathematics Magazine* 51 (1978), p. 312.

2   The subject of k-morphic tilings is considered in the authors' "Patch-determined tilings", *Mathematical Gazette* 61 (1977), 31–38, and Harborth's example in "Prescribed numbers of tiles and tilings", *Mathematical Gazette* 61 (1977), 296–299.

3   Hilbert's famous problems were printed (English translation) in "Mathematical Problems", *Bulletin of the American Mathematical Society* 8 (1902), 437–479 and reprinted in "Mathematical Developments Arising from Hilbert Problems", *Proc. Sympos. Pure Math.*, Vol. 28, (American Math. Soc., Providence, R.I., 1976). The book by Heesch and Kienzle is *Flächenschluss* (Springer-Verlag, Berlin-Göttingen-Heidelberg, 1963). A recent treatment of isohedral tilings is the authors' "The eighty-one types of isohedral tilings in the plane", *Math. Proc. Cambridge Philos. Soc.* 82 (1977), 177–196.

4   A proof of the Extension Theorem will appear in the authors' book mentioned below. Heesch's problem appears in his book *Reguläres Parkettierungsproblem* (Westdeutscher Verlag, Köln-Opladen, 1968).

5   Accounts of aperiodic tiles appear in R. M. Robinson's article "Undecidability and non-periodicity of tilings of the plane", *Inventiones Math.* 12 (1971), 177–209, and R. Penrose's papers "The role of aesthetics in pure and applied mathematical research", *Bull. Inst. Math. Appl.* 10 (1974), 266–271, and "Pentaplexity", *Eureka* 39 (1978), 16–22. The most up-to-date account is Martin Gardner's article cited in the text, and a more detailed exposition will appear in the author's forthcoming book. The connection between aperiodicity and the tiling problem is discussed in Robinson's paper quoted above, and also in H. Wang, "Proving theorems by pattern recognition II", *Bell System Techn. Journal* 40 (1961), 1–42.

6   Voderberg's papers are "Zur Zerlegung der Umgebung eines ebenen Bereiches in kongruente", *J.-Ber. Deutsch. Math.-Verein.* 46(1936), 229–231, and "Zur Zerlegung der Ebene in kongruente Bereiche in Form einer Spirale", *ibid.* 47(1937), 159–160. Goldberg's explanation of the structure of spiral tilings appears in "Central tessellations", *Scripta Math.* 21(1955), 253–260. A short article "Spiral tilings and versatiles" by the authors appeared in *Mathematics Teaching* 88(1979), 50–51.

The above are only a few references to the wealth of published material on tilings in the plane. Further information and problems will appear in the authors' book *Tilings and Patterns*, to be published by W. H. Freeman and Company, San Francisco.

Angels and Devils

H. S. M. Coxeter
UNIVERSITY OF TORONTO

About forty years ago, Abraham Sinkov and I wrote twin papers on the subject of groups determined by the periods of two generators S, T and of their commutator $S^{-1}T^{-1}ST$ [Coxeter 1936; Sinkov 1936], never dreaming that twenty years later M. C. Escher would be using such groups (unconsciously) as symmetry groups for a carved ball and four other works of art [Escher 1971, Figures 112, 115, 226, 235, 244, 247; MacGillavry 1976, p. 18]. Those works have been reproduced here by kind permission of the Escher Foundation, Haags Gemeentemuseum, The Hague.

Patterns in the Euclidean Plane

On ordinary wallpaper, a motif is repeated by *translations* in two directions in accordance with the symmetry group called **p1**. Theoretically, a more interesting pattern could be obtained by other symmetry operations such as a *half-turn* which turns the motif upside down (so that b would yield q), or a *reflection* which changes a left hand to a right hand (or b to d or p). A half-turn is a *rotation* of period 2, but we might also use a rotation of period 3 or 4 or 6. By combining these "isometries" in every possible way, E. S. Fedorov proved in 1891 that there are just seventeen planar symmetry groups involving translations in two directions. Eleven of the seventeen were unconsciously discovered long ago by the Moors in their decoration of the Alhambra. Some of these eleven, and an additional five, were used by the Bakuba and Benin tribes in Africa (south of the Sahara) in their pottery, weaving, and basket-making [Crowe 1971; 1975]. The remaining one, called **p31m**, occurs in a Chinese pattern [Fejes Tóth 1964, p. 40, Plate II.1].

Escher's patterns are more interesting, because their motifs are mainly animals, ingeniously fitted together so as to cover the plane without leaving any gaps. For instance, in Figure 1, every point of the plane belongs either to an angel or to a devil or to the curve that separates one from the other. The pattern, regarded as continuing so as to fill and cover the whole plane, is symmetrical by quarter-turns (rotations through $\pi/2$) about each point where wing-tips of four angels meet wing-tips of four devils. It is symmetrical also by reflections in certain vertical and horizontal lines, and by any product of such rotations and reflections. More economically, the whole symmetry group is generated by one rotation of period 4 (say S) and one reflection T. For instance, if T is the reflection in

FIGURE 1

Escher's *Sketch for "Angels and Devils"*.

one of the vertical mirrors, and S is a quarter-turn whose center is as near as possible to that mirror, then S will transform T into the reflection $T_1 = S^{-1} TS$ in one of the horizontal mirrors. The powers S^k of the rotation S $(k = 0, 1, 2, 3)$ will transform T into reflections $S^{-k} TS^k$ in all the sides of a square.

The rotations S^k form the cyclic group $\mathfrak{C}_4$ of order 4, generated by S. The two reflections T and T_1 generate the dihedral group $\mathfrak{D}_2$ of order 4, in which the product $T_1 T$ generates a subgroup $\mathfrak{C}_2$ (because the half-turn $T_1 T$ has period 2).

The relations

$$S^4 = 1 \quad \text{and} \quad T_1^2 = T^2 = (T_1 T)^2 = 1$$

are abstract definitions or *presentations* for the groups $\mathfrak{C}_4$ and $\mathfrak{D}_2$, respectively, in the sense that every relation satisfied by the generator S, or by the two generators T_1 and T, is an algebraic consequence of these simple relations. Writing $S^{-1} TS$ for T_1, we deduce the presentation

$$S^4 = T^2 = (S^{-1} TST)^2 = 1$$

for the infinite group generated by the rotation S and the reflection T.

This infinite group **p4g** [Coxeter and Moser 1972, p. 47] is the special case $[4^+, 4]$ of the group

$$[l^+, 2p],$$

which has the presentation

$$S^l = T^2 = (S^{-1} TST)^p = 1 \qquad (l \geq 2, p \geq 1).$$

Here S is a rotation of period l (that is, through an angle $2\pi/l$) and T is a reflection whose mirror is in such a position that the product $T_1 T$, where $T_1 = S^{-1} TS$, has period p. In other words, the powers of S transform the mirror for T into the sides of a regular l-gon, and two adjacent sides (such as the mirrors for T_1 and T) form an angle π/p. It follows that such a group $[l^+, 2p]$ occurs whenever the plane can be tessellated with regular l-gons, $2p$ round each vertex. Such a tessellation is denoted by $\{l, 2p\}$ [Coxeter and Moser 1972, p. 102]. For instance, $\{4, 4\}$ is the ordinary tiling of squares (the "squared paper" pattern). The only other possibility in the Euclidean plane is $\{3, 6\}$, indicated by the heavy lines in Figure 2.

The crystallographer Caroline MacGillavry [1976, Plate 8], impressed by Escher's rediscovery of most of Fedorov's symmetry groups, pointed out that Escher, like the Africans, had missed the group

$$\textbf{p31m} \cong [3^+, 6]$$

(misnamed **p3m1** in [Coxeter 1969, p. 413] and [Coxeter and Moser 1972, pp. 49, 136]). At her request, Escher filled this gap by creating a new picture of red bees and yellow-green wasps (Plate IV). Here the edges of

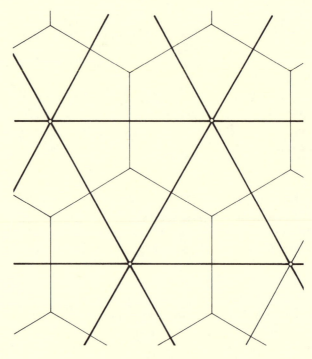

FIGURE 2

A fragment of the dual tesselations $\{3, 6\}$ (dark) and $\{6, 3\}$ (light).

the underlying tesselation $\{3, 6\}$ are the obvious lines of symmetry of the insects. These edges are perpendicularly bisected by the edges (appearing as light lines in Figure 2) of the dual tessellation $\{6, 3\}$, whose vertices are the points where the "elbows" of three bees meet those of three wasps. The generator S of $[3^+, 6]$ is the rotation through angle $2\pi/3$ about such a point, and we easily verify the relations

$$S^3 = T^2 = (T_1 T)^3 = 1,$$

where $T = S^{-1} TS$.

We might stretch the meaning of the symbol $[l^+, 2p]$ to include

$$[2^+, \infty] \quad \text{and} \quad [\infty^+, 2]$$

as the fifth and sixth of the seven frieze groups [Coxeter 1969, p. 48]. The former, generated by a half-turn and a reflection, is the symmetry group of the sine curve and of the frieze

$$\ldots \vee \wedge \vee \wedge \vee \ldots;$$

the latter, generated by a translation and a reflection (the translation being along the direction of the mirror), is the symmetry group of the frieze

$$\ldots D D D D D \ldots.$$

FIGURE 3

Escher's *Sphere with Fish.*

Carved Balls

A more interesting extension of the Euclidean tessellations arises when we replace the ordinary plane by a non-Euclidean plane—either the sphere or the hyperbolic plane.

The surface of a sphere may be regarded as a plane whose lines are the great circles. This idea occurred to Abû'l Wafâ (940–998) [see Woepke 1855, pp. 352–357]. Rotation about a diameter of the sphere is regarded as rotation about either of the points where this diameter penetrates the surface. A regular tetrahedron $\{3, 3\}$, inscribed in the sphere, is symmetrical by rotations of period 3 about its vertices. These rotations generate the tetrahedral group of order 12, which is denoted by $\mathfrak{A}_4$ because it is the alternating group of degree 4: the group of even permutations of the 4 vertices. Four more tetrahedra may be inscribed in the same sphere so that the whole set of 20 vertices belong to a regular dodecahedron $\{5, 3\}$ [Coxeter and Moser 1972, p. 35]. Adjoining to $\mathfrak{A}_4$ a rotation of period 5, we obtain the *icosahedral* group of order 60, which is denoted by $\mathfrak{A}_5$ because it is the alternating group of degree 5: the group of even permutations of the 5 tetrahedra.

FIGURE 4

Polyhedron with Flowers.

Escher used $\mathfrak{A}_4$ as the symmetry group for one of his carved globes (Figure 3), and $\mathfrak{A}_5$ for another (Figure 4). In the latter, the twelve flowers are chiral, like periwinkles, and the tips of the petals are at the vertices of the dodecahedron. In fact, this globe is evidently based on the above-mentioned compound polyhedron (Figure 4a), which is denoted by

$$\{5, 3\} \, [5\{3, 3\}] \, \{3, 5\}$$

because its twenty vertices belong to a dodecahedron $\{5, 3\}$ while its twenty faces lie in the same planes as the faces of the dual icosahedron $\{3, 5\}$.

FIGURE 4A

J. F. Petrie's drawing of the five tetrahedra.

More directly relevant to our discussion of groups $[l^+, 2p]$ is his other carved globe (Figure 5), because it is symmetrical by a rotation S, of period 3, about a point where wing-tips of three angels meet wing-tips of three devils. It is also symmetrical by a reflection T whose mirror is one of three mutually perpendicular planes. Thus S and T generate the group $[3^+, 4]$, and the underlying spherical tesselation $\{3, 4\}$ is the octahedron whose faces are the spherical triangles cut out on the sphere by these three planes of symmetry. $[3^+, 4]$ is sometimes called the *pyritohedral* group because it is also the symmetry group of a crystal of iron pyrites—a not-quite-regular dodecahedron. Abstractly, this group of order 24 is the direct product

$$\mathfrak{A}_4 \times \mathfrak{C}_2$$

[Coxeter and Moser 1972, pp. 3, 39]. In fact, the defining relations

$$S^3 = T^2 = (S^{-1} T S T)^2 = 1$$

are satisfied by the permutations

$$S = (a\,b\,c), \ T = (a\,b)\,(c\,d)\,(e\,f),$$

FIGURE 5
Sphere with Angels and Devils.

which generate the direct product of the $\mathfrak{A}_4$ generated by

$$(a\,b\,c) = S, \ (a\,b)\,(c\,d) = STSTS$$

and the group of order 2 generated by

$$(e\,f) = (ST)^3.$$

For the sake of completeness, we may mention also the trivial groups

$$[2^+, 2p] \cong \mathfrak{D}_{2p} \quad \text{and} \quad [l^+, 2] \cong \mathfrak{C}_l \times \mathfrak{C}_2$$

(of orders $4p$ and $2l$), which resemble $[2^+, \infty]$ and $[\infty^+, 2]$, only now the friezes are wrapped round a cylinder. In other words, $[2^+, 2p]$ is the symmetry group of a p-gonal *antiprism* [Coxeter 1969, p. 149].

Circle Limits

In 1958 I sent Escher a reprint which contained a drawing like Figure 6. In reply, he thanked me and said that "[the illustration gave him] quite a shock. Since a long time I am interested in patterns with 'motives' getting smaller and smaller till they reach the limit of infinite smallness. The question is relatively simple if the limit is a point in the center of a pattern. Also a line-limit is not new to me, but I was never able to make a pattern in which each 'blot' is getting smaller gradually from a center towards the outside circle-limit, as shows your figure. I tried to find out how this figure was geometrically constructed, but I succeeded only in finding the centers and radii of the largest inner-circles. If you could give me a simple explanation how to construct the following circles, whose centers approach gradually from the outside till they reach the limit, I should be immensely pleased and very thankful to you! Are there other systems besides this one to reach a circle-limit?"

In response to Escher's letter, I told him that $\{4, 6\}$ and $\{6, 4\}$ are two of infinitely many regular tessellations $\{p, q\}$ consisting of congruent regular p-gons fitting together, q at each vertex. If p and q are too big for the tessellation to exist on a sphere or in the Euclidean plane, it requires a *hyperbolic* plane, in which the regular p-gon has a smaller vertex angle. In

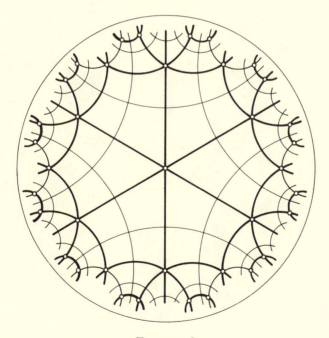

FIGURE 6

A fragment of the dual tessellations $\{4, 6\}$ (dark) and $\{6, 4\}$ (light).

one of Poincaré's models, the "straight lines" of the hyperbolic plane appear as arcs of circles orthogonal to a boundary circle Ω drawn in the Euclidean plane. Angles are represented faithfully but distances are distorted, with the points of Ω itself being infinitely far away. Although the regular quadrangles formed by the dark lines in Figure 6 get "smaller gradually from the center towards the outside circle-limit", we enter the spirit of hyperbolic geometry when we stretch our imagination in order to pretend that these quadrangles are all regular and congruent.

Escher's sketch-books show that he diligently pursued these ideas before completing his *Circle Limit IV* (Figure 7), whose symmetry group (in the sense of hyperbolic geometry) is $[4^+, 6]$:

$$S^4 = T^2 = (S^{-1} \, TST)^3 = 1.$$

The generator S is (as in Figure 1) a quarter-turn about one of the points where wing-tips of four angels meet wing-tips of four devils—that is, about

FIGURE 7

Escher's *Circle Limit IV*.

a vertex of $\{6, 4\}$ (drawn with light lines in Figure 6). The other generator T is the reflection in one of the dark lines in Figure 6 (each containing infinitely many edges of $\{4, 6\}$). If the mirror for T is one of the straight lines through the center, S transforms it into one of the dark arcs, and the hyperbolic reflection $T_1 = S^{-1} T S$ appears in the Poincaré model as the *inversion* in the circle that carries that arc. Figure 7 differs from Figure 1 in that the rotation $T_1 T$ (about the tips of the angels' feet) is of period 3 instead of 2.

Escher used the analogous group $[3^+, 8]$ for his *Circle Limit II* (Figure 8), which is just as interesting mathematically although Bruno Ernst [1976, p. 109] dismissed it with a little joke. The underlying tessellation $\{3, 8\}$ (Figure 9) has for vertices the centers of all the crosses. The points where the three colors come together, being the centers of the triangular faces of $\{3, 8\}$, are the vertices of the dual tessellation $\{8, 3\}$ [Coxeter 1979, p. 23, Figure 5]. On the other hand, the centers of the

FIGURE 8

Escher's *Circle Limit II*.

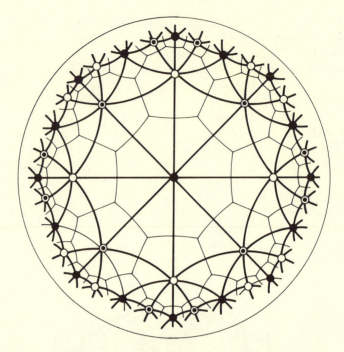

FIGURE 9

A fragment of the dual tessellations {3, 8} (dark) and {8, 3} (light).

crosses of one color are the vertices of a tessellation of quadrangles, {4, 8}. (In Figure 9 the vertices of {3, 8} are black, white or dotted accordingly as the corresponding crosses are black, white or grey.) In this sense, Escher may be said to have anticipated my discovery of the regular compound tessellation

$$\{3, 8\}\ [3\{4, 8\}]2\{8, 3\}$$

[Coxeter 1964, pp. 156–157] which consists of 3 superposed tessellations {4, 8} whose vertices belong to a single {3, 8} while their faces have the same centers as the faces of the dual {8, 3}, each used twice.

Conclusion

For any two integers l and p ($l > 2$, $p > 1$) there is a group $[l^+, 2p]$ generated by a rotation S of period l and a reflection T of period 2, whose commutator $S^{-1}TST$ is a rotation of period p. The kind of plane involved is spherical, Euclidean or hyperbolic depending on whether the number $(l - 2)(p - 1)$ is less than 2, equal to 2, or greater than 2. The finite group $[3^+, 4]$ (of order 24) is the symmetry group of Escher's *Sphere with Angels and Devils*. He used also the two Euclidean groups $[4^+, 4]$ and $[3^+, 6]$, and it is interesting that, among the infinitely many hyperbolic groups $[l^+, 2p]$ with

$$(l - 2)(p - 1) > 2,$$

he instinctively chose the two simplest: $[4^+, 6]$ and $[3^+, 8]$.

References

1 Coxeter, H. S. M. 1936. The groups determined by the relations $S^l = T^m = (S^{-1}T^{-1}ST)^p = 1$ *Duke Math. Journal.* 2: 61–73.

2 ———. 1964. Regular compound tessellations of the hyperbolic plane. *Proc. Royal Soc. London* A 278: 147–167.

3 ———. 1969. *Introduction to Geometry.* 2nd ed. New York: Wiley.

4 ———. 1979. *The non-Euclidean symmetry of Escher's picture 'Circle Limit III'.* Leonardo 12: 19–25, 32.

5 ——— and Moser, W. O. J. 1972. *Generators and Relations for Discrete Groups* 3rd ed. Berlin: Springer.

6 Crowe, D. W. 1971. The geometry of African art I. *J. Geom.* 1: 169–182.

7 ———. 1975. The geometry of African art II. *Historia Math.* 2: 253–271.

8 Ernst, Bruno. 1976. *The Magic Mirror of M. C. Escher.* New York: Random House.

9 Escher, M. C. 1971. *The World of M. C. Escher.* New York: Abrams.

10 Fejes Tóth, L. 1964. *Regular Figures.* New York: Pergamon.

11 MacGillavry, Caroline. 1976. *Fantasy and Symmetry—The Periodic Drawings of M. C. Escher.* New York: Abrams.

12 Sinkov, Abraham. 1936. The Groups Determined by the Relations $S^l = T^m = (S^{-1}T^{-1}ST)^p = 1$. *Duke Math. Journal* 2: 74–83.

13 Woepke. F. 1855. Recherches sur l'histoire des sciences mathématiques chez les orientaux, d'après des traités inédits arabes et persans. *J. Asiatique* 5: 309–359.

Three-Dimensional Tiling

Packing Problems and Inequalities

D. G. Hoffman

AUBURN UNIVERSITY

Four rectangular farms, each 7 miles by 8 miles, are to be carved out of a square county, 15 miles on a side. How is this to be done? (Try it yourself before peeking at the answer in Figure 1a on page 213.

Notice that the county has area $15^2 = 225$ square miles, and each farm has an area of 56 square miles. Hence there will be $225 - 4(56) = 1$ square mile left over.

Here is a more general puzzle. Suppose that x and y are positive numbers. Will four rectangle farms, each x miles by y miles, fit into a square county, $x + y$ miles on a side? Now the area of this county is $(x + y)^2$ square miles, and each farm has area xy square miles. So unless

1 $$4xy \leq (x + y)^2,$$

we don't have a chance of solving this puzzle. In other words, if the puzzle has a solution, then the combined area of the four farms cannot be greater than the area of the county, so that inequality **1** must hold.

This more general puzzle does have a solution, and I'm sure you've found it if you solved the first puzzle, where $x = 7$ and $y = 8$. Figure 1 below gives the answer to the first puzzle in (a), and the answers to the general puzzle in (b), (c), and (d), depending on the relative sizes of x and y. The arithmetic mean, or average of two numbers x and y is half their sum:

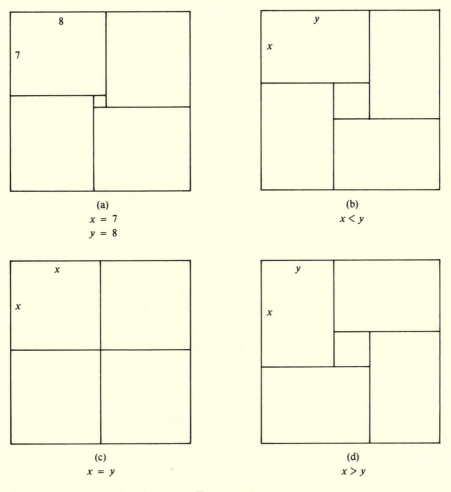

FIGURE 1

2 $$\text{A.M.} = \frac{1}{2}(x+y)$$

There is another "average" of x and y that is sometimes useful when x and y are positive. It is called the geometric mean of x and y, and it is the square root of their product:

3 $$\text{G.M.} = \sqrt{xy}$$

For example, if $x = y$, then A.M. $= \frac{1}{2}(x+x) = x$, and G.M. $= \sqrt{xx} = x$, so these two means coincide if $x = y$. However, with $x = 4, y = 16$, we have A.M. $= 10$, while G.M. $= 8$. It turns out that the geometric mean is never bigger than the arithmetic mean; that is,

4
$$\sqrt{xy} \le \frac{1}{2}(x+y).$$

Here is an easy proof, using our four farms puzzle. Figure 1 proves that the puzzle does have a solution, therefore inequality **1** must be true. If we take the square root of both sides of **1**, we get $2\sqrt{xy} \le x+y$, and dividing both sides by 2 gives precisely inequality **4**, concluding the proof.

Puzzles in which a number of objects are to be fit into a container are called packing problems. Such puzzles are well known to Gardner readers, recreational mathematicians, and smugglers.

The moral of the above example is that any solution to a packing problem yields a proof of an inequality. If all the objects fit into the container, then their combined area (or volume) must be no greater than the area (or volume) of the container. Perhaps the reverse is true! That is, can we start with an inequality known to be true, and make an amusing packing problem out of it? Here is an example.

The arithmetic mean of three numbers, x, y and z, is their average, or one third their sum:

5
$$\text{A.M.} = \frac{1}{3}(x+y+z).$$

The geometric mean of three positive numbers x, y and z is the cube root of their product:

6
$$\text{G.M.} = \sqrt[3]{xyz}.$$

The inequality **4**, for two numbers, is also true for three numbers:

7
$$\sqrt[3]{xyz} \le \frac{1}{3}(x+y+z).$$

To make this into a packing problem, multiply both sides by 3, and then cube both sides. The result is

8
$$27xyz \le (x+y+z)^3.$$

We notice that xyz is the volume of a brick with dimensions x by y by z, while $(x+y+z)^3$ is the volume of a perfect cube, $x+y+z$ on a side. So here is our packing problem:

Will 27 identical bricks, each with dimensions x by y by z, fit into a cubical box, $x+y+z$ on a side?

Of course, if this packing problem has a solution, that solution would prove that **8**, (and therefore **7**) is a true inequality.

We mentioned above that **7** is in fact true for any positive numbers x, y and z. This fact in itself does *not* guarantee that the bricks will fit in, but only that their combined volume is no bigger than the volume of the box.

(The volume of my fishing pole is less than the volume of my briefcase, but the fishing pole won't fit in my briefcase!)

At this point you might want to obtain a set of 27 bricks and try this puzzle. Let me advise you on what dimensions you should choose for x, y and z. There is one basic problem in choosing x, y and z. Many values of x, y and z will yield puzzles that are far too easy! For example, $x = 1, y = 1$, and $z = 100$ is ridiculous; 27 sticks, each $1''$ by $1''$ by $100''$ will fit into a $102''$ by $102''$ by $102''$ box with enough room left over for a tuba. Another extreme is $x = y = z$; here the 27 bricks are cubes that fit snugly into the box. To be certain of avoiding all such trivialities, make sure that the three dimensions are different, and that the sum of the two larger ones is less than three times the smallest.

In other words;

9 $$0 < x < y < z$$

10 $$y + z < 3x.$$

If you make a set of bricks violating either **9** or **10**, there might (and probably will) be solutions that are easy to find.

The only other considerations in choosing the dimensions x, y and z, besides **9** and **10**, depend only on what materials you have at hand, and the convenience of construction. No values of x, y and z satisfying **9** and **10** produce a puzzle easier or harder than any other such values, with one possible exception: a puzzle with $y = \frac{1}{2}(x + z)$ might be just a bit harder! Here is a good example: $x = 4, y = 5, z = 6$. So, will 27 4 by 5 by 6 bricks fit into a 15 by 15 by 15 box?

Let me assure you that this puzzle does have a solution. In fact, I learned from both J. H. Conway and William Cutler, that there are exactly 21 solutions, not counting rotations and reflections of the box, provided of course that **9** and **10** hold.

A solution seems hard to find. I tried for hours to solve this puzzle merely using pencil and paper. Finally, I gave up and called my friend David Klarner, who has a power saw. He says that as he sanded the blocks one by one, he stacked them into the solution shown below without changing one block. Thus, he earned the distinction of being the first person to solve my puzzle. (See the photos on the following page.) I've seen solution times ranging from about 20 minutes to a few days. If you have built a set, and you want to put your blocks back into the box (as you should, if you are done playing with them), I enclose a solution in Figure 2 below, with $x = 4$, $y = 5$, $z = 6$. The actual dimensions don't matter of course, you can use Figure 2 for any x, y and z satisfying **9** and **10**, just as Figure 1b works for any $x < y$.

The 27 bricks are arranged in three layers of nine bricks each, and Figure 2 is a cross section through each of the three layers. Let us examine this puzzle more closely. We are assuming that x, y and z are numbers satisfying **9** and **10**.

1

2

The first to build a set of blocks, and to find a packing of them was
D. A. Klarner. Shown above is the first set built, each block having
dimensions 7 by 8 by 10, being packed by a chip off the old block, Carl
Klarner. The photos were taken by Kara Lynn Klarner.

3

4

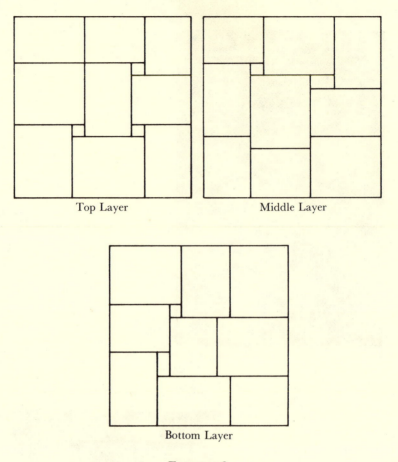

Top Layer   Middle Layer

Bottom Layer

FIGURE 2

A perfect cube, $x + y + z$ on a side, can be cut into 64 perfect cubes, each $\frac{1}{4}(x + y + z)$ on a side by *nine cuts* (see Figure 3a). These nine cuts determine nine planes, which we shall call the special planes (see Figure 3b). These nine planes fall into three sets of three mutually parallel planes. Within each set, the middle plane is $\frac{1}{2}(x + y + z)$ from the other two, and these other two are each $\frac{1}{4}(x + y + z)$ from the outside of the box. Two special planes that are not parallel meet in a line. Altogether, there are 27 such lines, and we call them the special lines (see Figure 3c). Three special planes, of which no two are parallel, meet at a point. Altogether, there are 27 such points, we call them the special points (see Figure 3d). Suppose now that this cube has been packed with 27 x by y by z bricks.

Consider now one of the three sets of three mutually parallel planes. They divide the cube into four slabs, each with thickness $\frac{1}{4}(x + y + z)$. Can any one of the 27 bricks in our packing lie entirely within one of these four slabs, miraculously untouched by any of the three knife cuts?

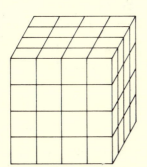

FIGURE 3A

The cube cut.

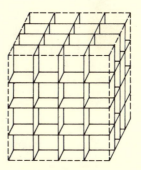

FIGURE 3B

The nine special planes.

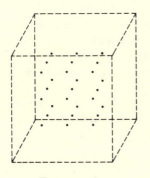

FIGURE 3C

The 27 special lines.

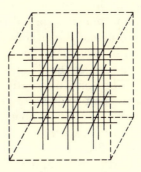

FIGURE 3D

The 27 special points.

Since $y + z < 3x$, we have $x + y + z < 4x$, or $\frac{1}{4}(x + y + z) < x$. Thus the smallest dimension of our brick, x, is bigger than the width $\frac{1}{4}(x + y + z)$ of each slab. Unlike the magician's lovely assistant, no brick has escaped all three of the parallel knife cuts.

To summarize what we have just proved: each brick in the packing is cut by at least one of the three special planes in any of the three parallel classes of special planes. As a consequence, the special point in which these three special planes meet must be inside the brick. We have shown that each of the 27 bricks in the packing has at least one special point inside it. But there are only 27 special points altogether. Hence each brick has exactly one special point inside it, and each special point is on the inside of exactly one brick. (We have also proved, by the way, that there is no hope of getting 28 or more bricks in the box, as there are not enough special points to go around.)

Each special line has three special points on it, so much of it lies on the inside of the three bricks containing these three special points. Can there be gaps, as in Figure 4?

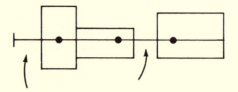

FIGURE 4

Gaps on a special line.

We will show that there can be no gaps. For each special line, let us measure how much of its total length of $x + y + z$ is on the inside of bricks in the packing. Let us denote the sum of these 27 numbers, one for each of the 27 special lines, by the symbol ℓ. Each of the 27 bricks has a special point inside it, and this special point has three special lines through it, in three perpendicular dimensions. Hence each brick must gobble up at least $x + y + z$ units worth of special lines. Hence $27(x + y + z) \leq \ell$. On the other hand, each of the 27 special lines is of length $x + y + z$, so $\ell \leq 27 (x + y + z)$. The only alternative is that $\ell = 27(x + y + z)$, and that each special line is completely within the three bricks on its three special points. In particular, there can be no gaps, as in Figure 4. Hence each special line looks like Figure 5, where $a + b + c = x + y + z$, and each of a, b and c is one of x, y or z. Since $x < y < z$, there are only two ways this could happen. Either a, b and c are x, y and z in some order, or $a = b = c = y$. Further, this second case cannot happen unless $y + y + y = x + y + z$, or $y = \frac{1}{2}(x + z)$.

We can prove that the second case, $a = b = c = y$, cannot happen at all. Each of the 27 special lines is divided into three segments by the three bricks containing this line. Altogether, there are $27 \cdot 3 = 81$ such segments. Moreover, these 81 segments must consist of exactly 27 x's, 27 y's, and 27 z's, since each of the 27 bricks cuts off three segments—one of each length. The paragraph above points out that each special line has either one or three segments of length y. But there are only 27 segments of length y to go around, so each special line has exactly one segment of length y. Hence $a = b = c = y$ cannot happen.

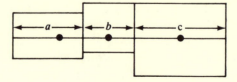

FIGURE 5

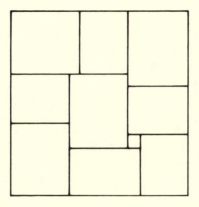

FIGURE 6

We have shown that each of the 27 special lines is entirely inside three bricks of the packing, which divide it into three segments, of three different lengths, x, y and z. This is quite a useful fact to know. For example, if you attempt to solve the puzzle by putting the first nine bricks in a layer at the bottom of the box as in Figure 6, you will be unable to complete the puzzle. Do you see why?

Another useful fact to know is that no special plane can have a corner like Figure 7 below, since whichever brick goes in the center of this plane must abut both A and B, and yet C is in the way. Knowing these facts certainly helps to solve the puzzle, but there are many more difficulties and false starts to be encountered.

Notice that each of the nine special planes contains nine special points, and hence cuts across nine bricks. Thus, this special plane has been packed with nine rectangles, each either x-by-y, x-by-z or y-by-z, It is easy to prove then, the following fact: these nine rectangles comprise exactly three x-by-y's, exactly three x-by-z's and exactly three y-by-z's. For three examples, see Figure 2 above. We have proved, by the way, that

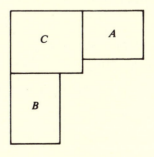

FIGURE 7

11 $$3(xy + xz + yz) \le (x + y + z)^2,$$

since the left hand side is the total area of the nine rectangles, and the right hand side is the area of the square they pack.

If conditions **9** and **10** hold, there are exactly 78 ways, not counting rotations and reflections, in which a square, $x + y + z$ on a side, can be packed with three x by y rectangles, three x by z rectangles, and three y by z rectangles.

The inequalities **4** and **7** are special cases of what is known as the *arithmetic mean, geometric mean inequality*. This states that if $x_1, x_2, \ldots, x_n$ are positive numbers, then

12 $$\sqrt[n]{x_1 \cdot x_2 \cdot \ldots \cdot x_n} \le \frac{1}{n}(x_1 + x_2 + \cdots + x_n).$$

The left hand side is called the geometric mean of the numbers, while the right hand side is the familiar arithmetic mean.

To make **12** into a puzzle, multiply both sides by n, then raise both sides to the nth power. The result is

13 $$n^n(x_1 \cdot x_2 \cdot \cdots \cdot x_n) \le (x_1 + x_2 + \ldots + x_3)^n.$$

Now $x_1 \cdot x_2 \cdot \cdots \cdot x_n$ is the n-dimensional *volume* of an n-dimensional brick with dimensions $x_1, x_2, \cdots, x_n$; and $(x_1 + x_2 + \ldots + x_n)^n$ is the n-dimensional *volume* of a perfect n-dimensional cube; each side of the cube has length $x_1 + x_2 + \cdots + x_n$. So will n^n such bricks fit into the n-dimensional cube?

Let me assure those readers unaccustomed to thinking in four or more dimensions, that such thought, contrary to rumor, does not cause, significant brain damage if practiced moderately. Here is some explanation, if the above paragraph made you shudder.

Just as a rectangle has two dimensions (a length and a width) and a shoe box has three, an n-dimensional brick or box has n dimensions, $x_1, x_2, \cdots, x_n$ and each dimension x_i is a positive number. Let us denote such a brick, or box by the vector $(x_1, x_2, \cdots, x_n)$. Thus $(3,5)$ denotes a 3-by-5 rectangle, $(4,6,2,6)$ denotes a 4-dimensional 4-by-6-by-2-by-6 "box", and both $(2,3,9)$ and $(2,9,3)$ denote 2-by-3-by-9 ordinary 3-dimensional boxes. (In one dimension, $n = 1$, (x_1) denotes a line segment x_1 units long).

A measure of how much "n-dimensional stuff" there is in the box $(x_1, x_2, \cdots, x_n)$ is the number $x_1 \cdot x_2 \cdot \cdots \cdot x_n$, the product of its dimensions. It is called the hypervolume of the box. Thus for $n = 1$, hyper-volume is length, for $n = 2$, hyper-volume is area, and for $n = 3$, hyper-volume is usual volume.

Here is a question which, as far as I know, has not been resolved:

14   For which positive integers n is it true that for any positive numbers $x_1, x_2, \cdots, x_n$, n^n bricks, each with dimensions $(x_1, x_2, \cdots, x_n)$, will fit into a perfect n-dimensional cube, $x_1 + x_2 + \ldots + x_n$ on a side?

The problem is trivial for $n = 1$, and Figures 1 and 2 show how to do $n = 2$ and 3 for *any* (x_1, x_2) or (x_1, x_2, x_3). How about $n = 4$ or more?

Let me remind the reader that the inequality **13** does not assure us that the bricks will fit, it only fails to discourage us from trying. (If **13** did not hold, we would be certain that the bricks would not fit into the box).

Let us call those dimensions n, for which a packing exists, the *good* dimensions. Thus we know that 1, 2 and 3 are good. J. Selfridge has told me that R. Robinson has proved the following theorem:

15 If m and n are good dimensions, so is mn.

In particular, since 2 and 3 are good, so are 4, 6, 8, 12 and 181398528 $= 2^{10} \cdot 3^{11}$. Thus $4^4 = 64$ four-dimensional bricks, each with dimensions (x_1, x_2, x_3, x_4), will fit into a perfect 4-dimensional cube, $x_1 + x_2 + x_3 + x_4$ on a side.

One might suspect that all positive dimensions n are good. If this is so, it need only be proved for all prime numbers n, since any positive integer is a product of prime numbers, and solutions can be "multiplied together" by **15**. The smallest dimension in doubt is thus 5.

I hope some ambitious readers will attempt to resolve this question: Will $5^5 = 3125$ 5-dimensional bricks, each x_1-by-x_2-by-x_3-by-x_4-by-x_5, fit into a perfect 5-dimensional cube, $x_1 + x_2 + x_3 + x_4 + x_5$ on a side?

There is only one little stumbling block in understanding Robinson's proof of **15**, and that is the fact **16** below. Once **16** is understood you will see that it is easy to prove **15**.

Before we state this fact, let us give an example. We showed in Figure 1 that four $(7,8)$ rectangles pack a $(15,15)$ square. Hence four $(7,8,100)$ bricks will pack a box $(15,15,100)$. Similarly, four 5-dimensional bricks, each $(7,8,100,14,47)$ will pack a 5-dimensional $(15,15,100,14,47)$ box.

16 Suppose that k n-dimensional bricks, with dimensions
$$(x_{1,1}, x_{1,2}, \cdots, x_{1,n}), (x_{2,1}, x_{2,2}, \cdots, x_{2,n}), \cdots, (x_{k,1}, x_{k,2}, \cdots, x_{k,n})$$
fit into a box of dimensions $(y_1, y_2, \cdots, y_n)$. Then the k $(n+t)$-dimensional bricks with dimensions
$$(x_{1,1}, x_{1,2}, \cdots, x_{1,n}, z_1, z_2, \cdots, z_t),$$
$$(x_{2,1}, x_{2,2}, \cdots, x_{2,n}, z_1, z_2, \cdots, z_t), \cdots,$$
$$(x_{k,1}, x_{k,2}, \cdots, x_{k,n}, z_1, z_2, \cdots, z_t),$$
will pack a box of dimensions
$$(y_1, y_2, \cdots, y_n, z_1, z_2, \cdots, z_t).$$

We are now ready to prove Robinson's theorem (15). We know then that m and n are good dimensions, and we have to show that mn is also a good dimension. In other words, we must find a way to pack $(mn)^{mn}$ (mn)-dimensional bricks, each with dimensions
$$(x_{1,1}, x_{1,2}, \cdots, x_{1,n}, x_{2,1}, x_{2,2}, \cdots, x_{2,n}, x_{3,1}, \cdots, x_{m-1,n}, x_{m,1}, x_{m,2}, \cdots, x_{m,n}),$$
into a perfect (mn)-dimensional box, with each side having length t, where t is the sum of the mn-dimensions $x_{1,1}, \cdots, x_{m,n}$. The proof will be easier to

follow if we don't string out the mn dimensions of a brick in one line, but write them in matrix form instead:

$$\begin{pmatrix} x_{1,1}, & x_{1,2}, & \dots, & x_{1,n} \\ x_{2,1}, & x_{2,2}, & \dots, & x_{2,n} \\ \vdots & \vdots & & \vdots \\ x_{m,1}, & x_{m,2}, & \dots, & x_{m,n} \end{pmatrix}$$

To begin, divide the $(mn)^{mn}$ bricks into groups of m^m bricks each. Thus there will be $(mn)^{mn}/m^m = m^{m(n-1)}n^{mn}$ such groups. Let $s_1 = x_{1,1} + x_{2,1} + \cdots + x_{m,1}$. Since m is a good dimension, m^m bricks, each with dimensions $(x_{1,1}, x_{2,1}, \dots, x_{m,1})$, will pack a box of dimensions $(s_1, s_1, \dots, s_1)$. Using **16**, we see that the m^m bricks in each of the $m^{m(n-1)}n^{mn}$ groups can pack a box of dimensions

$$\begin{pmatrix} s_1, & x_{1,2}, & \dots, & x_{1,n} \\ s_1, & x_{2,2}, & \dots, & x_{2,n} \\ \vdots & \vdots & & \vdots \\ s_1, & x_{m,2}, & \dots, & x_{m,n} \end{pmatrix},$$

making $m^{m(n-1)}n^{mn}$ such boxes. Divide these new boxes into groups of m^m boxes each. There will be $m^{m(n-1)}n^{mn}/m^m = m^{m(n-2)}n^{mn}$ groups. The m^m boxes in each group can be packed into a box of dimensions

$$\begin{pmatrix} s_1, & s_2, & x_{1,3}, & \dots, & x_{1,n} \\ s_1, & s_2, & x_{2,3}, & \dots, & x_{2,n} \\ \vdots & \vdots & \vdots & & \vdots \\ s_1, & s_2, & x_{m,3}, & \dots, & x_{m,n} \end{pmatrix}$$

where $s_2 = x_{1,2} + x_{2,2} + \dots + x_{m,2}$. (Here we have used the fact that m^m bricks, each of dimensions $(x_{1,2}, x_{2,2}, \dots, x_{m,2})$, will fit into a box of dimensions $(s_2, s_2, \dots, s_2)$; and then we used fact **16** again.)

Continue following this process n times, once per column. At each stage, the boxes produced at the previous stage are divided into groups of m^m boxes each, and the boxes in each group are assembled into a new box. The dimensions of these new boxes, (one per group), are obtained by replacing every entry of the column in question by the sum of the column entries. The result is this: the $(mn)^{mn}$ original bricks have all been packed into $m^{m(n-n)}n^{mn} = n^{mn}$ boxes, each box with dimensions

$$\begin{pmatrix} s_1, & s_2, & \dots, & s_n \\ s_1, & s_2, & \dots, & s_n \\ \vdots & \vdots & & \vdots \\ s_1, & s_2, & \dots, & s_n \end{pmatrix},$$

where $s_i = x_{1,i} + x_{2,i} + \cdots + x_{m,i}$ for $i = 1, 2, \dots, n$.

Now we do the same thing, this time working with the rows instead of the columns. Divide the n^{mn} boxes into groups of n^n boxes each. There will be $n^{mn}/n^n = n^{(m-1)n}$ such groups. Since n is a good dimension, n^n bricks, each with dimensions $(s_1, s_2, ..., s_n)$, will fit into a box with dimensions $(t, t, ..., t)$, where $t = s_1 + s_2 + \cdots + s_n$. (Note that t is the sum of all the mn original brick dimensions $x_{1,1}, x_{1,2}, ..., x_{m,n}$, as defined above, just after **16**. So t is the length of each side of the mn-dimensional cube we are trying to pack.) Thus, using **16**, we see that the n^n boxes in each group will fit into a new box with dimensions

$$\begin{pmatrix} t, & t, & ..., t \\ s_1, & s_2, & ..., s_n \\ \vdots & \vdots & \vdots \\ s_1, & s_2, & ..., s_n \end{pmatrix},$$

and there will be $n^{(m-1)n}$ such boxes. Repeat this for each of the m rows, with the result that everything can be packed into $n^{(m-m)n} = 1$ box of dimensions

$$\begin{pmatrix} t, t, ..., t \\ t, t, ..., t \\ \vdots \vdots \quad \vdots \\ t, t, ..., t \end{pmatrix}$$

which is precisely our goal. Thus all $(mn)^{mn}$ original bricks fit in, and mn is a good dimension. This completes the proof.

We have seen several examples in this chapter of a nice connection between inequalities and packing problems: sometimes known inequalities yield interesting packing problems, and conversely, sometimes an inequality can be proved by solving a packing problem.

The proof above illustrates a further connection. In some of the standard proofs of the arithmetic mean-geometric mean inequality, one step is to prove that if it is true for m numbers, and if it is true for n numbers, then it is true for mn numbers. This proof can be stated so that each step of the proof corresponds to a step in Robinson's proof above! So the moral is this: sometimes a known proof of an inequality can be used to find a solution to a packing problem.

Mathematics is rich with inequalities and surely many of these can yield interesting packing problems. May I recommend to the reader the excellent book *Geometric Inequalities* by N. D. Kazarinoff. (New Mathematical Library, Vol. 4, Random House). This (and in fact all the books in New Mathematical Library series) can be readily understood by anyone with a reasonable background in high school mathematics, and an open mind.

When next accosted by an inequality, use whatever force is necessary to change it to a packing problem—you might be pleased by the result!

Can Cubes Avoid Meeting Face to Face?

———◁✑▷———

Raphael M. Robinson

University of California

In this paper, we shall study the tiling of space by congruent cubes. By saying that the cubes form a tiling of space, we mean that they cover space without overlapping. We will assume that the edges of the cubes are parallel to the coordinate axes. There is no loss of generality in using unit cubes, and we shall usually do so.

We will start by looking at the (easier) two-dimensional case. Here we are concerned with tilings of the plane by congruent squares with sides parallel to the coordinate axes. The simplest tiling is the standard tiling shown in Figure 1. In this tiling, each square meets edge to edge with four

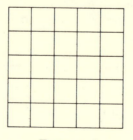

FIGURE 1
Standard tiling of the plane.

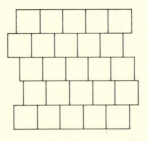

FIGURE 2

Tiling of the plane with displaced rows.

other squares. We say that two squares meet edge to edge only if they share a complete edge with each other. We can also tile a plane by sliding different rows of the standard tiling by different amounts, as in Figure 2. In this way, we can form tilings in which each square meets edge to edge with only two other squares. Instead of displacing rows, we could displace columns, as in Figure 3. It is easy to see that in any tiling of the plane by congruent squares, the squares must lie either in rows or in columns, or perhaps in both. Thus, in every such tiling, each square must meet at least two other squares edge to edge. (A lattice point in the plane or in space is a point (x,y) or (x,y,z) with integer coordinates.) If we place a unit square with its center at each lattice point in the plane, then we will obtain a standard tiling of the plane.

We now turn to the three-dimensional case. If we place a unit cube with its center at each lattice point in space, then we will obtain a standard tiling of space. In this tiling, each cube meets face to face with six other cubes. (We say that two cubes meet face to face only if they share a complete face with each other.) This tiling may be decomposed into layers parallel to the xy-plane. Modified tilings can be formed by sliding different layers by arbitrary amounts in the x- and y-directions. Within each layer, we may either slide the rows in the x-direction or the columns in the y-direction by

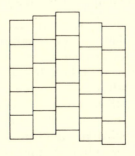

FIGURE 3

Tiling of the plane with displaced columns.

various amounts, just as in the plane case. In some layers we may slide rows, and in others we may slide columns. In this way, we can obtain tilings in which each cube meets only two other cubes face to face.

A similar construction can be carried out, starting with layers parallel to any coordinate plane. But the cubes do not necessarily fall into layers; in fact, we shall give several examples of tilings which are not layered. In describing these, it is convenient to use the word *column* to denote a stack of cubes lined up in the direction of any one of the coordinate axes.

The simplest example of an unlayered tiling can be obtained by starting with the standard tiling and choosing one column in each coordinate direction in such a way that the three columns have no cubes in common, and then displace each column along itself.

An extension of this idea will produce a tiling in which a certain cube avoids meeting any other cube face to face. To do this, start with the standard tiling, and select a cube which is to avoid meeting other cubes face to face. The six adjacent cubes may be incorporated into six columns—two in each coordinate direction—which may be slid independently without interfering with each other or the central cube. For example, the cubes adjacent to the central cube in the x-direction may be slid in the y-direction; the cubes adjacent to the central cube in the y-direction may be slid in the z-direction; and the cubes adjacent to the central cube in the z-direction may be slid in the x-direction. In each case, an entire column is moved in the indicated direction. The central cube then meets no other cube face to face.

In a similar fashion, we can construct a tiling in which infinitely many cubes avoid meeting other cubes face to face. The tiling described below was discovered by Hans Jansen in 1909. Start with the standard tiling obtained by placing a unit cube with its center at each lattice point (x,y,z). Then shift the cubes with centers (x,y,z) by $\frac{1}{2}$ unit

in the x-direction if y is even and z is odd,

in the y-direction if z is even and x is odd,

in the z-direction if x is even and y is odd.

The forms of the cross sections $z = c$, where c is an even or odd integer, are shown in Figure 4. These planes pass through the centers of most cubes. However, in the case of cubes which were displaced in the z-direction, the cross section passes through faces of the cubes. This is indicated in Figure 4 by cross-hatching. Arrows indicate the direction of the motion of the other cubes. The cubes with x, y, z all even or all odd remain fixed, and therefore, are unmarked. They do not meet any other cubes face to face. Obviously, the cubes which have avoided meeting any other cubes face to face constitute one-fourth of all the cubes. On the other hand, given any tiling of space by congruent cubes, it is easy to see that if a certain face of a given cube is not shared by another cube, then there must be two cubes touching

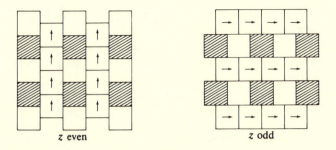

FIGURE 4

Cross sections of the Jansen tiling of space.

this face which meet each other face to face. Thus, while some cubes can avoid meeting other cubes face to face, not all cubes can do so.

We now consider multiple tilings of the plane by congruent squares or multiple tilings of space by congruent cubes. We assume that each point of the plane or space is covered a finite number of times, and that each point not on the boundary of any square or cube is covered the same number of times. If this number is k, then we speak of a k-fold tiling. The tilings considered previously are 1-fold tilings, or simple tilings. Multiple tilings of the plane are not very interesting. It is not hard to see that any k-fold tiling of the plane can be formed by superposing k simple tilings. Hence every square will meet at least two other squares edge to edge.

In space, something new happens. Not all multiple tilings can be formed by superposing simple tilings. Furthermore, it is possible for all cubes to avoid meeting other cubes face to face. This was proved in 1974, when I discovered a 25-fold tiling of space by congruent cubes of which no two meet face to face. In describing this tiling, it will be convenient to use cubes whose edge is 5 instead of using unit cubes. This change of scale obviously does not affect the problem. The cubes used will all have centers at lattice points. We will use exactly those centers (x, y, z) satisfying one of the following four conditions:

$$x \equiv y \equiv z \pmod 2, \qquad x + y + z \equiv 0 \pmod 5;$$

$$x + 1 \equiv y \equiv z \pmod 2, \qquad x + y + z \equiv 1 \pmod 5;$$

$$x \equiv y + 1 \equiv z \pmod 2, \qquad x + y + z \equiv 2 \pmod 5;$$

$$x \equiv y \equiv z + 1 \pmod 2, \qquad x + y + z \equiv 3 \pmod 5.$$

Here we have used the congruence notation $a \equiv b \pmod m$ to mean that a and b differ by a multiple of m. Thus the first line above states that the coordinates x, y, z are either all even or all odd, and that their sum is a multiple of 5. In general, the first condition of each line depends only on whether the three coordinates are even or odd, while the second condition gives the remainder when $x + y + z$ is divided by 5. The first condition will be called the parity condition. All of the points satisfying one of the parity conditions will be said to form a parity class. We shall show that all of the above cubes together form a 25-fold tiling of space, and that no two of them meet face to face.

To show that we have a 25-fold tiling, it is necessary to check that every lattice point is covered 25 times. Now the centers of cubes with edge 5 containing a lattice point (x',y',z') are exactly the lattice points (x,y,z) in a cube with edge 5 whose center is (x',y',z'). We start by considering the case $(x',y',z') = (0,0,0)$. Thus, we are given the cube with edge 5 whose center is at the origin. The lattice points in this cube are the 125 points (x,y,z) with each coordinate assuming the values $-2, -1, 0, 1, 2$. These form a $5 \times 5 \times 5$ array. We look first at the parity class defined by $x \equiv y \equiv z$ (mod 2). This consists of two parts: there are 27 points in a $3 \times 3 \times 3$ array with x,y,z all even, and 8 points in a $2 \times 2 \times 2$ array with x,y,z all odd.

We now form the sections $x + y + z = c$, which are perpendicular to a diagonal of the cube. The sections of the $3 \times 3 \times 3$ array will contain $1,3,6,7,6,3,1$ points, as shown in Figure 5. The sections of the $2 \times 2 \times 2$ array will contain $1,3,3,1$ points. Considering all 35 points together, we see that they will lie in equally spaced planes containing

$$1,0,3,1,6,3,7,3,6,1,3,0,1$$

points. Combining those in which $x + y + z$ (mod 5) have the same value, we obtain the totals shown below.

	1	0	3	1
6	3	7	3	6
1	3	0	1	
7	7	7	7	7

Thus, there are exactly 7 lattice points in the given cube with $x \equiv y \equiv z$ (mod 2) that have any prescribed value for $x + y + z$ (mod 5).

All of the lattice points in the given cube form the $5 \times 5 \times 5$ array mentioned above. Again we consider the sections $x + y + z = c$. Three of the sections are shown in Figure 6. The numbers of points in all of the sections are listed below; they are combined according to the value of $x + y + z$ (mod 5).

FIGURE 5

Oblique sections of a $3 \times 3 \times 3$ array.

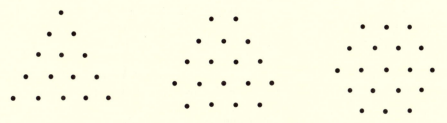

FIGURE 6

Oblique sections of a 5 × 5 × 5 array.

	1	3	6	10
15	18	19	18	15
10	6	3	1	
25	25	25	25	25

Thus each value occurs 25 times. This could also be seen from the fact that x and y may be assigned arbitrary values from the list $-2, -1, 0, 1, 2$, and then z is a uniquely determined value, when the value of $x + y + z$ (mod 5) is given. Each value of $x + y + z$ (mod 5) occurs 7 times for points with $x \equiv y \equiv z$ (mod 2), and therefore 18 times for other points. Now the other three parity classes are permuted by permuting x, y, z cyclically; in other words, the lattice points of the other three parity classes are interchanged by a rotation of 120° about the line $x = y = z$. Thus, in each of these parity classes, any prescribed value of $x + y + z$ (mod 5) must occur exactly 6 times in the given cube.

Now consider the cube of edge 5 with center at any lattice point (x', y', z'). If we move this point to the origin by a translation, then a lattice point (x, y, z) lying in the given cube will go into a lattice point $(x-x', y-y', z-z')$ lying in the cube of edge 5 centered at the origin. If the point (x', y', z') is given, then the parity class of this point is determined by the parity class of (x, y, z), and the reverse is also true. In particular, we will have

$$x - x' \equiv y - y' \equiv z - z' \text{ (mod 2)}$$

just when (x, y, z) lies in the same parity class as (x', y', z'). From the results proven above, we see that there are 7 such points (x, y, z) in the given cube with a prescribed value for $x + y + z$ (mod 5). But in any one of the other three parity classes, there will be only 6 points in the given cube with a prescribed value for $x + y + z$ (mod 5).

We thus see that the cubes whose centers satisfy any one of the four conditions in the definition of the tiling will cover all lattice points which satisfy the same parity condition 7 times, and other lattice points 6 times. Thus the four sets of cubes together will cover all lattice points 25 times, and therefore, furnish a 25-fold tiling. All of this is independent of the values assigned to $x + y + z$ (mod 5) in the four cases.

We now check that no two cubes meet face to face. If two cubes did so meet, then the center of the second cube would be obtained from the center of the first cube by changing one coordinate by 5 units. This would change the parity class, but not the value of $x + y + z \pmod 5$—which would be impossible since different values of $x + y + z \pmod 5$ were used for the four parity classes. Thus we have a 25-fold tiling in which no two cubes meet face to face.

Notice that in this tiling, formed with cubes of edge 5, the entire tiling is unchanged by a translation of 10 units along any coordinate axis, or by the translation through a vector with components $(5,5,5)$. It can be shown that these properties are consequences of the fact that no two cubes meet face to face.

It is difficult to make any drawing which gives an idea of this tiling. The best we can do is to take cross sections, and indicate the centers of the cubes used. In Figure 7, we show two of the cross sections, those in the planes $z = 0$ and $z = 5$. The portion drawn indicates the sections of the unit cubes with centers at lattice points (x, y, z), where x and y vary from 0 to 9. The dots indicate the centers of cubes of edge 5 used in the 25-fold tiling. Each of the two diagrams is repeated both horizontally and vertically. Other cross sections have a similar appearance.

In order to have a multiple tiling of space, it is necessary that 5×5 squares drawn in the same position in both sections contain the same number of dots. In order that no two cubes meet face to face, it is necessary that the dots always lie in different positions in the two sections, and that no two dots are 5 units apart either horizontally or vertically in either section.

The same 25-fold tiling described above was published recently in [1, section 12] with a somewhat different proof. (The paper [1] was devoted mainly to the corresponding problem in higher dimensions, but with the additional condition that the centers of the cubes form a lattice.) For some other related results and historical remarks, see Stein [2].

It is an open question whether there is a k-fold tiling of space by congruent cubes with $k < 25$ for which no two cubes meet face to face.

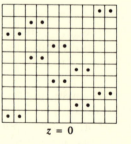

$z = 0$

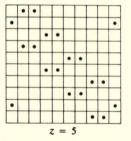

$z = 5$

FIGURE 7

Cross sections of the 25-fold tiling of space.

References

1   Robinson, R. M. 1979. Multiple tilings of n-dimensional space by unit cubes. *Mathematische Zeitschrift* 166: 225–264.

2   Stein, S. K. 1974. Algebraic tiling. *American Mathematical Monthly* 81: 445–462.

POSTSCRIPT   Since this article was written, Basil Gordon has shown that it is possible to take $k = 2$. Indeed, there is a 2-fold tiling of space by cubes of edge 2, each composed of eight unit cubes of the standard tiling, in which no two cubes meet face to face.

Packing Handed Pentacubes

C. J. Bouwkamp

TECHNISCHE HOGESCHOOL EINDHOVEN

To refresh the reader's memory, there are 29 distinct three-dimensional *pentacubes*; a pentacube being a solid built from five *unit cubes* such that neighbor cubes have a full face in common and any component cube has at least one neighbor. Of them, the 12 *planar* pentacubes, also called solid pentominoes, are well known in the field of recreational mathematics. Together they have a volume count of 60 units, and boxes of various sizes such as $(2 \times 3 \times 10)$, $(2 \times 5 \times 6)$, $(3 \times 4 \times 5)$ can be packed with them [1, 2, 3]. Among the non-planar pentacubes there are 5 that have at least one plane of symmetry; each of them is its own mirror image. The remaining 12 pentacubes consist of 6 pairs, each pair containing a pentacube and its mirror image. The pentacubes of a pair are in the same relation to each other as one's left and right hands; I shall, therefore, call them a pair of *handed* pentacubes. The complete set of handed (or *chiral*) pentacubes is depicted in Figure 1, together with their identifying labels 1 through 12. Again, their total volume count is 60 units and, as with the planar pentacubes, it is natural to ask whether they can pack a $3 \times 4 \times 5$ box, for example.

I was confronted with this question in July 1973 through a letter from Dr. G. V. Baddeley of the School of Chemistry, University of New South Wales, Australia. His principal interest is in stereochemistry and he wrote "to ask if [I] have a solution to the 3-4-5 block formed from the chiral pentacubes and if so, whether ... the solution is unique".

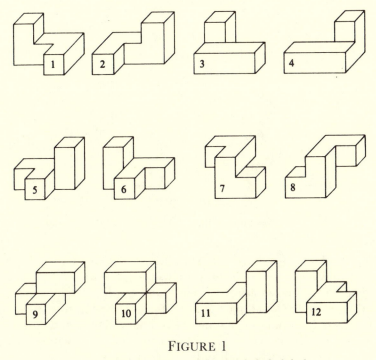

FIGURE 1

The handed pentacubes and their labels.

Parenthetically, if the reader is lucky enough to have access to a set of handed pentacubes, I challenge him to find a solution on his own before reading further. Although the handed pentacubes are much more difficult to handle than the planar ones, he should not give up too soon, in view of the overwhelming number of distinct solutions that are now known to exist.

A very clever, yet simple way to get many solutions at the same time was dicovered by J. M. M. Verbakel of Philips Research Laboratories at Eindhoven. He partitioned the set of handed pentacubes into four sets of three each, so as to be able to form four simultaneous blocks, here called *pianos*. One of Verbakel's constructions is shown in Figure 2, where the two pianos at the left are the mirror images (or *duals*) of the two at the right. It is obvious that the four pianos can be assembled into two simultaneous 2 × 3 × 5 blocks (and in three different combinations) and hence into a 3 × 4 × 5 block, which settles the existence question raised by Baddeley. Moreover, the solution is not unique at all. As a by-product of Verbakel's construction it is clear that 2 × 3 × 10 and 2 × 5 × 6 blocks can be formed from the handed pentacubes as well, again in more than one way (as in the case of the planar pentacubes).

By computer I found all possible partitions of the set of handed pentacubes in four simultaneous pianos. There are 8 different partitions modulo rotation and translation (or 6 modulo reflection, rotation, and

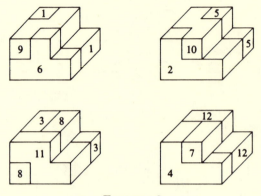

FIGURE 2

Four simultaneous pianos built from the handed pentacubes.

translation). They are listed in Table 1. The first four of them are attributed to Verbakel and these are all *self-dual*. Partitions 5 and 6 are each other's dual, and so are partitions 7 and 8. (The reader is invited to realize the remaining seven partitions of Table 1 in solid blocks (indeed a neat exercise in manipulating handed pentacubes). He should observe that there are two different pianos $(5, 6, 9)$ and two different pianos $(5, 6, 10)$.)

For the time being I shall only consider *compound* blocks $3 \times 4 \times 5$; that is, blocks that are built from *rectangular* blocks of smaller size. (By the way, it is easy to see that a compound $3 \times 4 \times 5$ block must necessarily be built from two $2 \times 3 \times 5$ sub-blocks.) By computer, I determined the components of all compound $3 \times 4 \times 5$ blocks. The results are listed in

1	(1, 6, 9) (2,5,10) (3,8,11) (4,7,12)
2	(1, 6, 9) (2,5,10) (3,8,12) (4,7,11)
3	(1, 7, 9) (2,8,10) (3,5,11) (4,6,12)
4	(1, 7, 9) (2,8,10) (3,6,12) (4,5,11)
5	(1, 6, 9) (2,8,10) (3,5,11) (4,7,12)
6	(1, 7, 9) (2,5,10) (3,8,11) (4,6,12)
7	(1, 8,11) (2,9,12) (3,4, 7) (5,6,10)
8	(1,10,11) (2,7,12) (3,4, 8) (5,6, 9)

TABLE 1

The eight partitions of the set of handed pentacubes each generating
four simultaneous pianos.

Type	Pentacubes in first block	Number of Solutions for		Total ($S =$ self-dual)
		first block	second block	
1	(1,2,5, 6, 9,10)	1	3	3 S
2	(1,2,5, 7, 9,10)	1	1	1
3	(1,2,7, 8, 9,10)	3	4	12 S
4	(1,2,7, 8,11,12)	1	1	1 S
5	(1,2,7,10,11,12)	1	2	2
6	(1,3,4, 5, 9,11)	3	1	3
7	(1,3,4, 6, 7,10)	1	1	1
8	(1,3,4, 6, 8, 9)	1	2	2
9	(1,3,4, 7, 8,11)	1	2	2
10	(1,3,5, 6, 8,10)	1	1	1
11	(1,3,5, 6, 9,11)	4	1	4
12	(1,3,5, 7, 9,11)	2	2	1 + 2 S
13	(1,3,5, 8,10,12)	1	1	1 S
14	(1,3,6, 7, 9,12)	1	1	1 S
15	(1,3,6, 7,10,12)	2	2	1 + 2 S
16	(1,3,6, 8, 9,11)	1	1	1 S
17	(1,3,6, 8, 9,12)	2	2	1 + 2 S
18	(1,3,6, 8,11,12)	2	1	2
19	(1,4,5, 7, 9,11)	1	1	1 S
20	(1,4,6, 7, 9,10)	1	1	1
21	(1,4,6, 7, 9,11)	1	1	1 S
22	(1,4,6, 7, 9,12)	3	3	3 + 3 S
23	(1,5,6, 8, 9,10)	1	1	1
24	(1,5,6, 8,10,11)	2	1	2
				Total 28 + 30 S

TABLE 2

The 24 types of compound $3 \times 4 \times 5$ blocks

Table 2. I distinguish 24 types according to the six different pentacubes that pack the first block (which contains pentacube 1). Also indicated is the number of different $2 \times 3 \times 5$ blocks that can be formed from these pentacubes, as well as the number of solutions for the second block with the complementary pentacubes not indicated. The total number of distinct (modulo reflection and rotation—and, of course, modulo translation— implicit in the sequel) combinations of each type is indicated in the last column, where the letter S stands for self-dual. These numbers add up to $28 + 30S$, which means that there are $2 \times 28 + 30 = 86$ different (modulo rotation) combinations of two simultaneous $2 \times 3 \times 5$ blocks. Some, but not all blocks involved are built from pianos. By way of an example, Figure 3 shows type-3 blocks with three solutions for the first block and four solutions for the second block. As usual, each block is represented by two rectangular cross-sections, one for the bottom (left) and one for the top

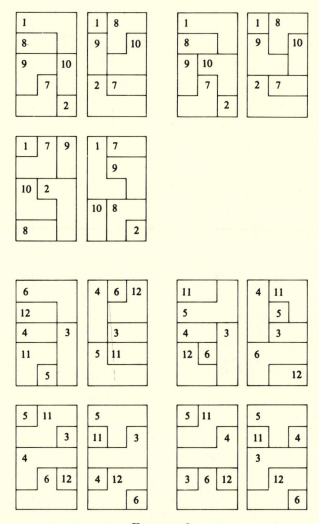

FIGURE 3

The basic 2 × 3 × 5 blocks of type 3 for the construction of compound 3 × 4 × 5 blocks.

(right) layer. It is clear that only three of the seven blocks are built from pianos. It is remarkable that all seven blocks are symmetric; that is, each of them has a *center* of symmetry.

The total number of different (modulo reflection and rotation) compound 3 × 4 × 5 blocks turns out to be 372. Of them, 28 have a *plane* of symmetry and as many have a *center* of symmetry. The set of 28 with a plane of symmetry is *complete*, because a 3 × 4 × 5 block with a plane of symmetry is necessarily compound. For the actual construction of compound blocks with symmetry, I refer the reader to Figure 4 showing a 2

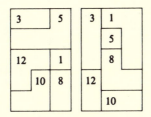

FIGURE 4

This block and its dual can be assembled into a 3 × 4 × 5 block with either center or plane of symmetry each in two different ways.

× 3 × 5 block built from the pentacubes (1,3,5,8,10,12). The corresponding dual block can be built from the complementary pentacubes (2,4,6,7,9,11). The two blocks can be joined (in two ways) to form a 3 × 4 × 5 block with a plane of symmetry. They can also be positioned so that the resulting block has a center of symmetry—again—in two ways. Many other examples of 3 × 4 × 5 blocks with either type of symmetry can be found with piano partitioning as in Figure 2.

I now turn to 3 × 4 × 5 blocks that are *simple*; that is, not compound. An easy way to prove their existence is to refer to Figure 5 showing the construction of simple blocks with the help of four simultaneous pianos. Since the pianos in Figure 5 can be permuted in several ways and since there are eight different partitions into pianos (see Table 1), it is evident that simple blocks are numerous as well.

Although I often tried to find a "random" solution by hand I was never successful—it was kind of unfair to challenge the energetic reader above to get a solution of his own. Yet, I felt sure that the few hundred solutions known so far represented only a small fraction of the total set of solutions. I was right!

To program the problem for the computer to obtain one or a few solutions is one thing, to get all of the solutions by computer is quite another. To eliminate rotations and reflections is not easy in this case, and solutions showing symmetry should be so indicated in the output. Both by

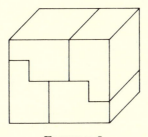

FIGURE 5

Four simultaneous pianos assembled into a 3 × 4 × 5 block that is not compound.

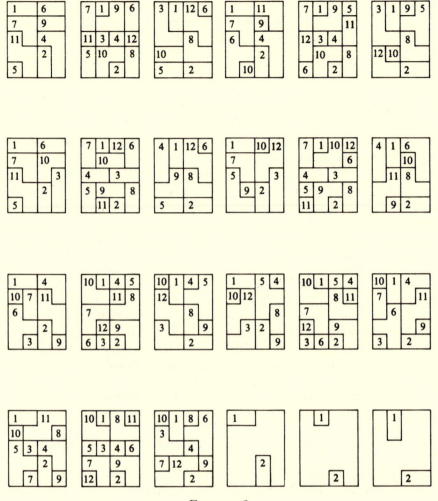

FIGURE 6

Five simple and two compound 3 × 4 × 5 blocks each with a center of symmetry. With pentacubes 1 and 2 fixed as indicated the set is complete.

the general program as well as by a special program I found that the total number of different 3 × 4 × 5 blocks with central symmetry is 742. This number includes the 28 compound blocks mentioned before. Note that their contribution to the total is less than four per cent.

Figure 6 shows all possible solutions with pentacubes 1 and 2 fixed as indicated in the lower right corner. A solution is represented by the three cross-sections of (from left to right) bottom, middle and top layers. Five of the blocks are simple and two are compound (type 15).

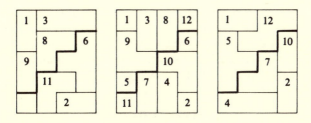

FIGURE 7

Block with center of symmetry that can be decomposed into two "steps". The two steps can be reassembled into a $3 \times 4 \times 5$ block in various ways.

A beautiful block among those with center of symmetry is shown in Figure 7. It is composed of two "steps" which can be assembled into different blocks in various ways. There are many more blocks of this structure.

The last example of a block with center of symmetry is given in Figure 8. It is peculiar in more than one way. As the reader will observe, pentacubes $(7,8,9,10)$ are not indicated in the 3-layer diagram of the block. As a matter of fact, there are two possibilities to fill in the missing labels in connection with the congruence of blocks built from pentacubes $(7,8)$ and $(9,10)$ in the lower part of Figure 8. It so happens that the two resulting blocks are each other's mirror image (modulo rotation). Blocks of this structure are very scarce; only ten of them exist.

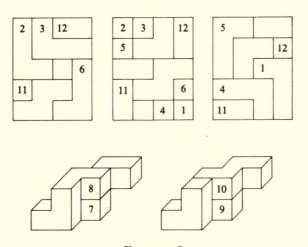

FIGURE 8

Example of a block that is turned into its mirror image by translation of four of the twelve pentacubes. For details, see text.

It is about time to close this paper and to disclose the total number of solutions of the $3 \times 4 \times 5$ block formed from the twelve handed pentacubes. Believe it or not, the total number is 29162 modulo reflection and rotation, and 57554 modulo rotation. They were obtained in computer slack time during a period of about two years.

References

1 Bouwkamp, C. J. *Catalogue of solutions of the rectangular 3 × 4 × 5 solid pentomino problem.* 1967. The Netherlands: Technische Hogeschool Eindhoven, Department of Mathematics, Eindhoven.

2 ――――. Packing a rectangular box with the twelve solid pentominoes. 1969. *J. Combinatorial Theory* 7: 278–280.

3 ――――. Catalogue of solutions of the rectangular 2 × 5 × 6 solid pentomino problem. 1978. *Kon. Ned. Akad. Wetensch., Proc., ser A* 81: 177–186.

My Life Among The Polyominoes

———— ✤ ————

David A. Klarner

STATE UNIVERSITY OF NEW YORK

$\mathbf{M}$y life as a mathematician was probably launched by my father. He had a scientific mind, and he was a very constructive man with the attitude that he could do anything. I also had an excellent high school teacher, Nemo Debely, who nurtured my interest in science and mathematics; in particular, he encouraged me to read Martin Gardner's columns in The Scientific American about the time they began to appear in the late fifties. Inspired by my father's example, I had the attitude toward mathematics that I could do it. When Deb gave me a classical problem in geometry, number theory, combinatorics, or whatever, I set to work thinking I would solve it. Of course, I never managed to solve Fermat's Last Theorem, the four color conjecture (now a theorem), or to prove that there is an infinite set of prime pairs, but I learned a lot by trying. Mixed into this naive period was my first contact with Martin Gardner.

Imagine my thrill as a high school student when Martin Gardner replied to my letters and put me in contact with other people who had interests similar to mine. (For example, it was through Martin that I first made contact with Solomon Golomb who later incorporated many of my results and problems in his book *Polyominoes*.)

During my induction into the mathematical world, I had a series of wonderful teachers: Nemo Debely in high school, James E. Householder in college, Leo Moser in graduate school, and Dick de Bruijn in my post doctoral period. Martin Gardner belongs in this list too, but his role differs from the others. For twenty years Martin has been an abiding inspiration, and he helped me share my mathematical hobbies with many people.

In this article I have attempted to present some of my mathematical findings dealing with polyominoes—a subject that seems too technical for a popular account. I intend, nevertheless, to present things so that non-specialists will understand them. The first few sections deal with enumeration problems. The technical papers covering this subject comprise scores of printed pages intersperced with ugly formulas; I have left out the proofs and oversimplified things throughout. The last few sections give a very brief account of some exact packing (tiling) results. While this subject is easier to understand, and probably more attractive to the average reader, it has been covered pretty well in Martin's columns and books already. Thus, I will not try to outdo the master Gardner.

Growing Polyominoes

Polyominoes dwell amid the integer points of the Cartesian plane. They are composed of *cells*, that is, unit squares that have their vertices at integer points. An *n*-omino is a union of *n* cells whose interior is connected. Solomon Golomb describes an *n*-omino as a set of *n* cells of a large chessboard where a rook path through the cells of the *n*-omino connect any two cells of it. Figure 1 indicates a unit cell, three pentominoes (5-

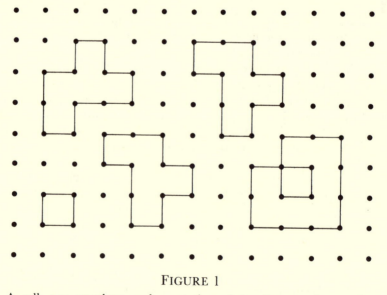

FIGURE 1

A cell, two translates and a rotation of the *F* pentomino, and a heptomino with a hole in it.

FIGURE 2

The 19 translation types of tetrominoes.

ominoes), and a heptomino (7-omino). The heptomino shown in Figure 1 underscores the fact that *n*-ominoes may not be simply connected; that is, they may have holes in them.

Another point underscored by Figure 1 is that for each *n* there are an infinite number of *n*-ominoes even though there are only a finite number of shapes involved. Various types of *n*-ominoes are defined on this infinite set by notions of equivalence given in terms of certain groups of isometries of the plane. First of all, *translation types* are defined by saying two *n*-ominoes are equivalent if some translation of the plane maps one onto the other. For example, translation types of the tetrominoes (4-ominoes) are shown in Figure 2. *Rotation types* arise when two *n*-ominoes are equivalent if there is a sequence of rotations of the plane which maps one onto the other. Since a translation can be accomplished by composing two appropriate rotations, translational equivalence is a refinement of rotational equivalence. In terms of Figure 3, this means that the representatives of rotational types may be selected from the set of representatives of translation types. An intuitive notion of a rotation type of *n*-omino is that it is a one-sided puzzle piece; that is, it can't be turned over but it can be moved around freely otherwise. Finally, two *n*-ominoes may be considered equivalent if they are

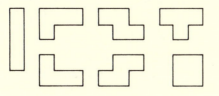

FIGURE 3

The 7 rotation types of tetrominoes.

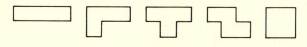

FIGURE 4

The 5 reflection types of tetrominoes.

congruent. Since every isometry is a composition of reflections, two *n*-ominoes are congruent only when there is a sequence of reflections which maps one onto the other. Congruent *n*-ominoes form reflection types. The 5 reflection types of tetrominoes are shown in Figure 4.

Computers have been used to generate complete sets of *n*-ominoes for small values of *n* (say, $n \leq 24$). But the number of *n*-ominoes increases so rapidly with *n* that there would not be enough storage space in our solar system to list all the *n*-ominoes, even for fairly modest values of *n* (say $n \geq 30$). Ronald Rivest has suggested an ingenious algorithm to generate the rooted translation type of *n*-ominoes. To root a translation type of *n*-omino, one of the *n* cells is distinguished from the others by endowing it with an umbilicus. There are *n* times as many rooted translation types of *n*-ominoes as there are translation types. The basic idea of Rivest's algorithm is to define a growth process which creates each rooted translation type of

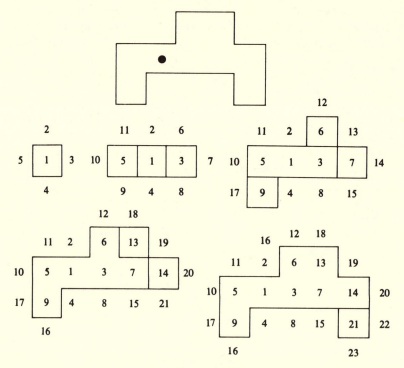

FIGURE 5

Stages of Rivest's growth process for a 9-omino.

n-omino just once. To grow a rooted translation type n-omino, begin with the umbilicus which is labelled 1. Then, start with the cell above the umbilicus and label the cells surrounding it $2,3,4,5$ in a clockwise fashion. Cells adjacent to cells of a polyomino will be called *border cells*. The growth process is a sequence of stages. Each stage consists of a polyomino which has its border cells labelled with the smallest positive integers. A larger polyomino is grown from this one by adding labelled border cells which belong to the n-omino being grown. New unlabelled border cells usually appear and these are labelled with the smallest positive integers not already used as labels. These label assignments are made in a clockwise fashion beginning above and nearest to the umbilicus. It is conceivable that one must wind around the umbilicus more than once to assign all the border cells their labels. Figure 5 shows the stages of growth involved in creating a particular rooted 9-omino.

Rivest's growth scheme defines a family tree for rooted translation-type n-ominoes. The set of numbers $\rho(A)$ assigned to the cells of a rooted n-omino A is its *name*. For example, the rooted 9-omino shown in Figure 3 has $\{1,3,5,6,7,9,13,14,21\}$ as its name. A rooted polyomino B is said to be *descended* from a rooted polyomino A if A's name is part of B's name; that is, $\rho(A) \subseteq \rho(B)$. Part of the rooted translation type polyomino family tree is shown in Figure 6.

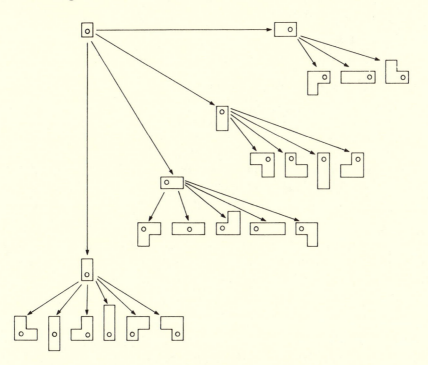

FIGURE 6

The family tree for rooted translation type polyominoes.

If n is greater than 1, then $3n - 1$ is the largest integer used in the name of any rooted n-omino. This is obvious when you reason as follows: the largest integer used to label a border cell of a rooted monomino (1-omino) is 5. Addition of a cell to a rooted polyomino increases its border cells by 3 (at the most). Hence, the largest integer used to label the border cells of a rooted n-omino is, at most, $3n + 2$. When $n > 1$, a rooted n-omino is grown by adding a border cell to some rooted $(n - 1)$-omino. The integer used to label this border cell is, at most, $3(n - 1) + 2 = 3n - 1$. Thus, since a rooted n-omino is grown by adding border cells with larger and larger labels, the largest integer used to label a cell is $3n - 1$. This shows that every rooted translation type n-omino has a name which is an n-element subset of the set $\{1, 2, \ldots, 3n - 1\}$. Of course, 1 is in every name, so rooted translation type n-ominoes correspond one-to-one with some of the $(n - 1)$-element subsets of the $(3n - 2)$-element set $\{2, 3, \ldots, 3n - 1\}$. The number of these subsets is $\binom{3n - 2}{n - 1} = (3n - 2)!/(2n - 1)!(n - 1)!$. If $t(n)$ denotes the number of translation type n-ominoes, then $nt(n)$ is the number of rooted translation type n-ominoes. The foregoing argument shows that

1
$$t(n) \leq \frac{1}{n}\binom{3n - 2}{n - 1}.$$

But it is easy to verify that

2
$$\frac{1}{(n + 1)}\binom{3n + 1}{n} \bigg/ \frac{1}{n}\binom{3n - 2}{n - 1} = \frac{3(9n^2 - 1)}{2(2n^2 + 3n + 1)} < \frac{27}{4},$$

For $n = 1, 2, \ldots$, so a simple induction implies

3
$$t(n) < \left(\frac{27}{4}\right)^{n - 1} \quad \text{or} \quad [t(n)]^{1/n} < \frac{27}{4}.$$

Methods for obtaining exponential lower bounds for $t(n)$ will be described in the next section. The best published lower bound for $t(n)$ first appeared in my Ph.D. thesis. It was shown there that

4
$$(3.72)^n < t(n) \text{ or } 3.72 < [t(n)]^{1/n}$$

holds for all sufficiently large n.

These bounds show that the number of translation type n-ominoes is growing exponentially. Note that if $r(n)$ denotes the number of rotation type n-ominoes, and $c(n)$ denotes the number of congruence type n-ominoes, then

5
$$t(n)/8 \leq c(n) \leq r(n) \leq t(n).$$

Thus, the number of n-ominoes of each type grows at the same exponential rate. This means that even the most efficient algorithm designed to generate all translation type n-ominoes could not do its job for very large n

(say, $n = 30$) because there wouldn't be space enough in all the world to store the output.

Counting Polyominoes

The set of translation type n-ominoes is too large to list, even for moderate values of n. This does not rule out the possibility that one might be able to compute the number of elements in this set. Recall that $t(n)$, $r(n)$, and $c(n)$ denote the number of translation, rotation, and congruence type n-ominoes respectively. Since $t(n)$, $r(n)$, and $c(n)$ are bounded above exponentially, by 8^n say, it follows that each of these numbers consists of not more than n decimal digits. Sometimes there are good algorithms to compute terms of sequences which grow exponentially. For example, the decimal representation of 2^n can be computed using about $2 \log_2 n$ multiplications. Also, there is an algorithm which only requires on the order of $\log_2 n$ basic operations to compute the nth term of the Fibonacci sequence $1, 1, 2, 3, 5, 8, 13, 21, \ldots$ (The nth term of the Fibonacci sequence is formed by summing the two terms immediately preceding it.) Again, the Fibonacci sequence is growing exponentially, so its nth term has about n decimal digits. Is there a good algorithm to compute $t(n)$? To put this even more vaguely, is there a *formula* for $t(n)$?

In a sense, there *is* a formula for $t(n)$. For example, Rivest's growth process could be used to form the set of all rooted translation type n-ominoes, the elements of this set could be counted, the total divided by n, and the result would be $t(n)$. Unfortunately, this algorithm requires about $nt(n)$ steps, and $nt(n)$ is exponentially large. As far as I know, all known algorithms for computing $r(n)$, $t(n)$, or $c(n)$ have this problem; that is, the computation times for the algorithms are bounded below by elementary functions of $r(n)$, $t(n)$, or $c(n)$ instead of being bounded above by some polynomial in n. It is conceivable that there is no good algorithm to compute these numbers, but the recently-born theory of computational complexity seems far from the level of development required to settle this problem.

Surprisingly, even the very elegant algorithm discovered by R. C. Read runs in exponential time. Read's algorithm computes $t_k(n)$, the number of translation type n-ominoes located in a strip k cells wide. This algorithm can be adapted to compute $t(n)$, $r(n)$, and $c(n)$, and Read was able to find these numbers by hand computation for about a dozen values of n. Figure 7 shows translation type n-ominoes located in a strip 2 cells wide for $n = 1, 2, 3, 4$. Here is a table of values of $t_2(n)$:

n	1	2	3	4	5	6	7	8	9	10	11	12
$t_2(n)$	2	3	6	11	20	37	68	125	230	423	778	1431

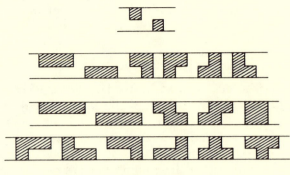

FIGURE 7

Translation type n-ominoes located in a strip 2 cells wide.

Perhaps after studying the table of values of $t_2(n)$, the reader will notice that from the fourth term onward, $t_2(n)$ is equal to the sum of the three terms immediately preceding it; that is,

1′ $$t_n(n + 3) = t_2(n + 2) + t_2(n + 1) + t_2(n)$$

for $n = 1, 2, 3, \ldots$. This formula indicates that the sequence $[t_2(1), t_2(2), t_2(3), \ldots]$ satisfies a third order difference equation. Read proved the more general result that for each k the sequence $[t_k(1), t_k(2), \ldots]$ satisfies a difference equation whose order depends on k. Implicit in Read's proof is an algorithm to compute this difference equation. He also formulated his computations in terms of powers of a square matrix M_k. To compute $t_k(n)$, we must compute $M_k, M_k^2, \ldots, M_k^{n-k+1}$. Unfortunately, the size of M_k is of order 3^k, and it is necessary to compute M_k^j with k and j both nearly half of n. This means Read's algorithm runs in exponential time. I suspect a computer implementation of Read's algorithm would make it possible to compute $t(n)$, $r(n)$, and $c(n)$ for many new values of n.

Other subsets of translation type n-ominoes have been considered. For example, Murray Eden, one of the pioneers in this subject, considered $b(n)$, the number of translation type n-ominoes which have each of its rows a connected bar of cells. Such n-ominoes look like board piles (Figure 8). Here is a table of values of $b(n)$:

n	1	2	3	4	5	6	7	8	9	10
$b(n)$	1	2	6	19	61	196	529	1517	4666	14827

Eden was able to show that $(3.14)^n < b(n)$ for all sufficiently large n. Leo Moser suggested improving Eden's bound to me when I came to the University of Alberta to start my graduate work. I was able to show that

2′ $$b(n + 4) = 5b(n + 3) - 7b(n + 2) - 4b(n + 1)$$

for $n = 1, 2, \ldots$. From this it follows that $(3.20)^n < b(n)$ for all sufficiently large n. About ten years later, Donald Knuth suggested that board piles

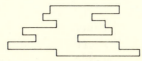

FIGURE 8

A typical board pile.

were row-convex n-ominoes, and the *convex n-ominoes* should be defined as board piles that are both row and column convex. Ronald Rivest and I found a way to compute $d(n)$ (the number of convex translation type n-ominoes) and we proved that $(d(n))^{1/n}$ tends to 2.309138 ... . Ed Bender derived an asymptotic formula for $d(n)$ based on our formulas.

Estimating the Number of n-ominoes

In previous sections, it was indicated that $t(n)$ (as well as $r(n)$ and $c(n)$) — the number of translation type n-ominoes, grows exponentially with n. Furthermore, it was shown all known algorithms for computing $t(n)$ run in exponential time. It may be a consequence of old age and growing pessimism, but I think it is possible that there is no good formula for $t(n)$. In fact, the number $t(n)$, for n (say, about 30) may be "unknowable". What does this mean? The claim that $t(30)$ is equal to a certain decimal number (with less than 30 digits) must be accompanied with a proof. Suppose there is no short proof? That is, it may be that all proofs of this fact are so long that they cannot be generated even by the largest and fastest computer in less than a thousand years of continuous work — a bleak prospect.

When faced with a difficult problem, it is often best to try a simpler (yet still related) problem. In my Ph.D. thesis, I showed that $(t(n))^{1/n}$ tends to a limit θ as n tends to infinity. Furthermore, the sequence $(t(1), [t(2)]^{1/2}, [t(3)]^{1/3}, ...)$ increases to θ, so that θ can be approximated from below by computing $[t(n)]^{1/n}$ for growing values of n. Unfortunately, this involves computing $t(n)$ and there is no known good algorithm to do this computation. Also, this sequence of lower bounds seems to be increasing very slowly to θ, so that a very large value of n would be required to improve the lower bound $3.72 < \theta$ which I established by a completely different method.

Rivest and I devised a method for computing a decreasing sequence of upper bounds for θ. The computer cost for this decreasing sequence of upper bounds grows exponentially, and we were not able to show that the sequence actually converges to θ. Our best upper bound is $\theta < 4.65$. Recently, Pat Woodworth and I refined this scheme, and it seems likely that the new method gives rise to a decreasing sequence of upper bounds which does conveige to θ. So far we have not implemented our method on a computer.

Thus, the number θ is also elusive; not one digit of the decimal expansion of θ is known! It seems possible that there is no good computation for θ, and that apart from the first two or so, the decimal digits of θ are unknowable. (Sometimes in talks on this subject, I have facetiously discussed θ as "Klarner's constant". None of the digits are known and they may be unknowable! If $\theta = 4$ could be proved, will school children be taught to count, "one, two, three, Klarner's constant, five, ..."?)

A table of the known values of $c(n)$ (the number of congruence type n-ominoes) concludes this section. These numbers (and a wealth of other information concerning polyominoes) were computed by D. H. Redelmeier. There is an outstanding conjecture that the sequence of ratios $c(2)/c(1)$, $c(3)/c(2)$, ... increases; that is,

$$\frac{c(n)}{c(n-1)} < \frac{c(n+1)}{c(n)} \qquad \text{for } n = 2, 3, \ldots.$$

If this conjecture is true, then this sequence increases to its limit which is equal to the Klarner constant θ. Furthermore, each of these ratios would be a lower bound for θ; in particular, $c(24)/c(23) < \theta$, that is, $3.8977 \ldots < \theta$ which beats the current lower bound $3.72 \ldots < \theta$. Of course, this improvement depends on the truth of the conjecture.

n	$c(n)$	n	$c(n)$	n	$c(n)$	n	$c(n)$
1	1	7	108	13	238591	19	742624232
2	1	8	369	14	901971	20	2870671950
3	2	9	1285	15	3426576	21	11123060678
4	5	10	4655	16	13079255	22	43191857688
5	12	11	17073	17	50107909	23	168047007728
6	35	12	63600	18	192622052	24	654999700403

TABLE 1

The number of different n-ominoes.

Polyomino Jigsaw Puzzles

Martin Gardner's 1958 article on polyominoes introduced me to the subject when I was a student in high school. Featured in the article were puzzles and proofs invented by Solomon Golomb. In particular, it was pointed out that the twelve reflection type pentominoes (simply called pentominoes from now on) can be used to form various rectangles with area 60. The shapes are (3×20), (4×15), (5×12), and (6×10). Years later, C. J. Bouwkamp used a computer to make complete lists of solutions of all box-filling problems using the twelve solid pentominoes. Besides the 2-dimensional boxes already listed, Bouwkamp also filled boxes with shapes

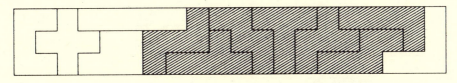

FIGURE 9

Two ways to fill a 3 × 20 rectangle with pentominoes. The shaded area
can be rotated 180° about its center to obtain the second solution.

(2 × 3 × 10), (2 × 5 × 6), and (3 × 4 × 5). Back in 1958, Martin's readers
did not have the benefit of Bouwkamp's lists or his ingenious computer
programs. In fact, I sought long and hard to find the two ways of filling a 3
× 20 rectangle with pentominoes, and was one of many enthusiasts who
sent his solutions to Martin. The two solutions (which are unique) are
indicated in Figure 9.

In preparing this article, it occurred to me that no one had considered
problems dealing with the rotation type pentominoes, that is, the eighteen
"one-sided" pentominoes shown in Figure 10. But it turns out that Golomb
mentions this set in his book *Polyominoes* by asking that these pieces be
used to tile a 9 × 10 rectangle. Probably the other problems discussed in
this section occurred to Golomb also. If the reader plans to make a set of
eighteen one-side pentominoes, use cardboard—preferably, black on
one side and white on the other. Then when playing with the pieces, they
must be kept one side up, say, black.

Since there are eighteen rotation type pentominoes, they might be
used to fill rectangles with area 90. It turns out that four shapes are
possible, namely, (3 × 30), (5 × 18), (6 × 15), and (9 × 10). The problems of

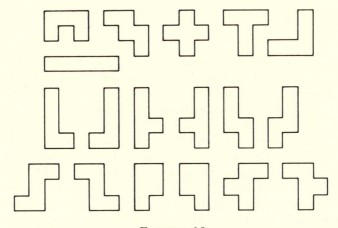

FIGURE 10

The eighteen rotation type pentominoes.

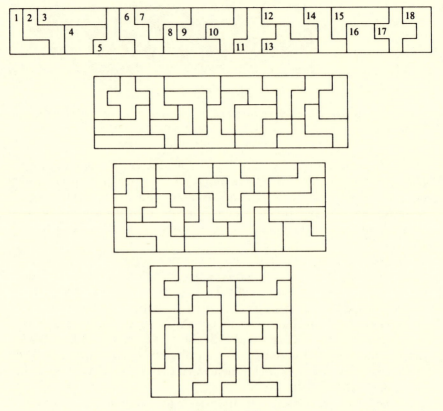

FIGURE 11

Rectangles tiled with one-sided pentominoes.

filling these rectangles become easier as one progresses from the slender to the fat rectangles. For example, the 3 × 30 rectangle took me about two hours to solve, while the 9 × 10 only required about two minutes. This varying order of difficulty probably reflects the fact that there are few solutions to the 3 × 30 problem, but there are thousands of solutions to the 9 × 10 problem. No doubt, Chris Bouwkamp will be the first person to quantify this statement! Solutions to these problems are shown in Figure 11.

It may seem difficult to tile a 3 × 30 rectangle with one-sided pentominoes. However, once one tiling has been found, it is often possible to transform this solution into new ones via a *rotation* or an *interchange*. To illustrate a rotation, consider the block of pieces numbered 6 through 13 in the tiling shown in Figure 11. This block can be rotated 180° about its center to obtain the first solution shown in Figure 12. Again, by using the tiling shown in Figure 11, two blocks consisting of pieces 6 through 11 and 15 through 17 can be interchanged to reach the second solution shown in Figure 12. There are five different tilings obtainable by transforming the

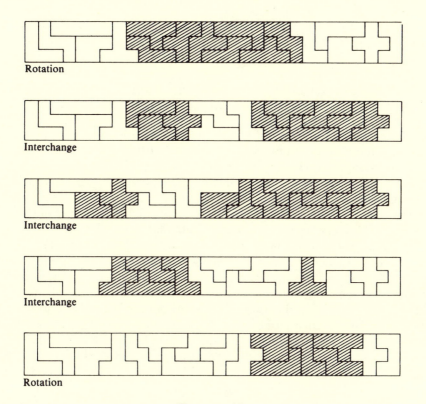

Rotation

Interchange

Interchange

Interchange

Rotation

FIGURE 12

Tilings of the 3 × 30 rectangle with one-sided pentominoes derived from the tiling shown in Figure 9.

tiling shown in Figure 11; two of these involve rotations, and the other three involve interchanges. Of course, these transformations can be applied to the tilings shown in Figure 12 also, and nine new tilings result. This process can be continued to find many new tilings. Perhaps the reader will enjoy seeing how many tilings can be derived from these by applying a sequence of rotations and interchanges. Of course, any tiling can be reflected to reach a new solution, but such reflections really shouldn't be considered distinct even though in a technical sense they are.

Early in my life among the polyominoes I tried to find elegantly defined sets of jigsaw pieces, and then turned my attention to seeing what could be done with them. This sort of enterprise has just been illustrated with the one-sided pentominoes. The elegance of this set of pieces is, in part, the simplicity of its definition. Over the years I have asked if even simpler sets of puzzle pieces might give rise to complex problems. My own inventions involved using several copies of one particular polyomino or polycube. (A polycube is the three-dimensional version of a polyomino.)

The problem has always been to characterize the rectangular boxes which can be exactly packed with copies of one piece. An example of this is given in the next section.

The N-pentacube

An N-pentacube is shown in Figure 13—the way it is positioned indicates where its name comes from. Also, it is clear that the N-pentacube is really just the N-pentomino with its five unit squares replaced with unit cubes. It is fairly easy to prove that the N-pentomino cannot tile (exactly pack) a rectangle. In fact, it is impossible to tile one end of a rectangle with N-pentominoes. Hence, the N-pentacube cannot tile a box having one of its sides equal to one unit. Can any rectangular boxes be tiled with N-pentacubes? It turns out the answer is yes, and I will begin by characterizing the boxes which can be tiled having one side equal to two units.

If an $a \times b \times c$ box has been exactly packed with pentacubes, then the volume of each pentacube (five) must divide the volume abc of the box. This means that one of the integers a, b or c is a multiple of 5. Consider the special case $a = 2$, then either b or c is a multiple of 5 (say c). It is easy to show that the N-pentacube cannot tile the end of a $2 \times b \times c$ box for $b = 1, 2, 3$ even when the third side c is a multiple of 5. However, it is possible to tile boxes with dimensions $(2 \times 4 \times 5)$, $(2 \times 5 \times 5)$, $(2 \times 6 \times 5)$, and $2 \times 7 \times 5$ as Figure 14 shows. These boxes can be stacked to form larger boxes which can also be tiled. It is fairly easy to show that every integer b greater than 3 can be expressed in the form $b = 4w + 5x + 6y + 7z$ with w, x, y, z nonnegative integers. Thus, the 2×5 ends of w $(2 \times 4 \times 5)$ boxes, x $(2 \times 5 \times 5)$ boxes, y $(2 \times 6 \times 5)$ boxes, and z $(2 \times 7 \times 5)$ boxes, can be matched together to form a $2 \times b \times 5$ box tiled with N-pentacubes. Then k of these boxes can be matched together along their $2 \times b$ sides to form a $2 \times b \times 5k$ box. The conclusion is that the set S_2 of all $2 \times b \times c$ boxes which can be tiled with N-pentacubes consists of all those boxes having b greater than 3 and c a multiple of 5. Furthermore, the four bricks shown in Figure 14 serve as a basis for S_2; that is, every $2 \times b \times c$ box which can be tiled with the N-pentacube can be tiled with the four bricks shown in Figure 14.

The set S_2, consisting of all 2-layer boxes which can be tiled with N-pentacubes, contains boxes which can be stacked to form boxes with an

FIGURE 13
The N-pentacube.

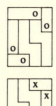

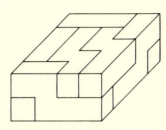

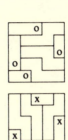

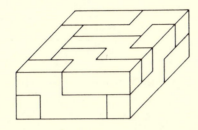

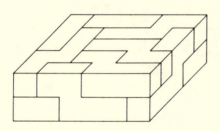

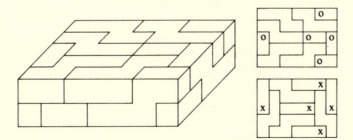

FIGURE 14

Four prime bricks for the N-pentacube. Cells marked with a small circle in the top layer are attached to cells marked with a small cross in the bottom layer.

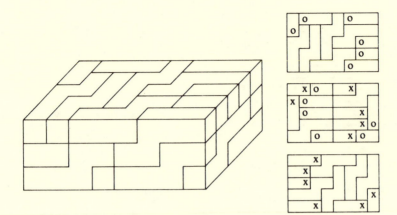

FIGURE 15

A 3 × 5 × 8 box tiled with $\mathcal{N}$-pentacubes.

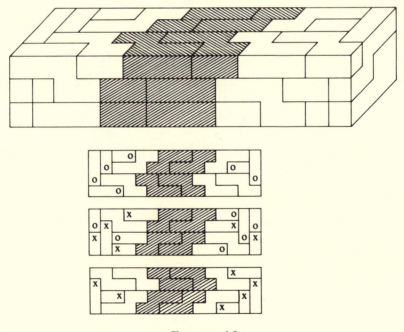

FIGURE 16

A 3 × 4 × 15 box tiled with $\mathcal{N}$-pentacubes. If the shaded pieces are removed and the remaining ends pushed together, a tiling of a 3 × 4 × 10 box results.

even number of layers. This means the four boxes shown in Figure 14 are a basis for boxes having dimensions $2a \times b \times 5c$ with $a,c = 1,2,3,\ldots$ and $b = 4,5,6,\ldots$. None of these boxes have just 3 layers. Is it possible to tile 3-layer boxes with N-pentacubes? So far, I have discovered tilings for a $3 \times 5 \times 8$ box (shown in Figure 15), and tilings for $3 \times 4 \times 5k$ boxes for $k = 2,3,\ldots$ (shown in Figure 16). All $3 \times 4 \times 5k$ boxes with k greater than 1 can be tiled with $3 \times 4 \times 10$ and $3 \times 4 \times 15$ boxes. It has been shown that the $3 \times 4 \times 5$ box cannot be tiled with N-pentacubes. It is not known which boxes with dimensions $3 \times 5 \times n$ can be tiled with N-pentacubes, but there is an algorithm to settle this question.

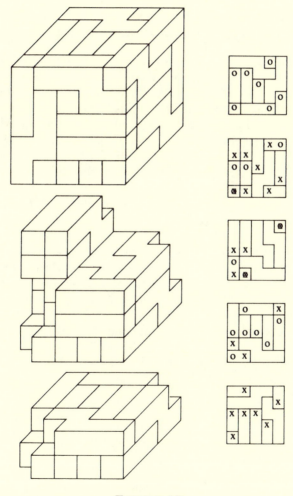

FIGURE 17

A cube of side 5 tiled with N-pentacubes.

Every 4-layer box which can be tiled with $\mathcal{N}$-pentacubes can be tiled with 2-layer or 3-layer boxes already tiled themselves. What about 5-layer boxes? The key problem here is to tile a cube with side 5. I discovered the tiling shown in Figure 17 after considerable effort; in fact, I nearly fainted when the last piece fell into place!

There is enough of a variety of boxes tiled with $\mathcal{N}$-pentacubes that it can be shown that every $a \times b \times c$ box can be tiled provided abc is a multiple of 5 and that a, b, c are all sufficiently large. As the discussion in the next section indicates, we should expect something like this. The boxes that can be tiled but cannot be cut into smaller boxes (that can be tiled) are called primes. The set of primes is finite, but unknown in the case of the $\mathcal{N}$-pentacube.

Tiling Boxes with Bricks

One of the many benefits of my two-year stay in the Netherlands working with de Bruijn, was meeting Frits Göbel, another dweller among the polyominoes. At the time, Frits had a notebook which stored information, problems, and conjectures concerning this subject. He recalled a problem set by Golomb which I had solved in the American Mathematical Monthly. In a nutshell the result is this: Every rectangle that can be tiled with L-tetrominoes can be tiled with 2×4 and 3×8 rectangles. Furthermore, the 2×4 and 3×8 rectangles can themselves be tiled with L-tetrominoes. Another result of this sort was established by David Walkup; namely, every rectangle that can be tiled with T-tetrominoes can be tiled with a square of side 4. Also, the square of side 4 can be tiled with T-tetrominoes. (Tilings of 2×4 and 3×8 rectangles with L-tetrominoes together with a tiling of the 4×4 square with the T-tetromino are shown in Figure 18.) Based on these and similar findings, Frits conjectured that the set of rectangles R which can be tiled with a given polyomino has a finite basis B; that is, there is a finite subset B of R, having the property that every rectangle in R can be tiled with the rectangles in B. Frits also conjectured that if the two rectangles A and B have areas with no common prime divisor, then all rectangles with both sides large enough can be tiled with A and B. He wanted to use this result to prove the first conjecture, but it turns out that the second conjecture is false.

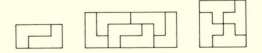

FIGURE 18

Prime boxes for the L-tetromino and the T-tetromino.

Stimulated by Göbel's conjectures, I found a necessary and sufficient condition for an $a \times b$ rectangle to be tiled with $p \times p$ and $q \times q$ squares. Either the rectangle can be tiled with just one size of square, or it can be cut into two smaller rectangles each of which can be tiled with just one size of square. This means one of the following holds true: 1) Both a and b are divisible by p. 2) Both a and b are divisible by q. 3) Either a or b (say a) is divisible by both p and q while $b = px + qy$ with x,y, non-negative integers. This means that if $p = 2$, $q = 3$, none of the squares 5×5, 25×25, 125×125, ..., $5^k \times 5^k$, ... can be tiled with 2×2 and 3×3 squares. Of course, this destroys Göbels' second conjecture, and one approach to proving the first conjecture. The first conjecture is true however; in fact, I was able to prove a much stronger result.

Suppose R is an infinite set of rectangles with integer sides. The rectangles are specialized in that they are located in the plane and oriented. That is, an $a \times b$ rectangle has its side of length a parallel to the y-axis and the other side parallel to the x-axis. Thus, if $a \neq b$, then $a \times b$ and $b \times a$ rectangles are different. Rectangles relocated by translations are considered equivalent. Next, the tiling is also specialized by being restricted to a divide and conquer approach; that is, one is only allowed to tile by cutting a rectangle into two smaller rectangles and then to tile these rectangles by the same method. For example, my theorem about tiling rectangles with two sizes of squares asserts that the rectangle can be tiled if and only if it can be tiled by the divide and conquer method. Also, divide and conquer tiling does not permit tilings of the sort shown in Figure 19.

My theorem deals with tiling oriented rectangles using the divide and conquer method. It asserts that every infinite set R of oriented rectangles contains a finite subset B such that every element of R can be tiled with elements of B using the divide and conquer method. Furthermore, an analogous result holds in k-dimensional space.

There is a very nice theorem about tiling boxes with bricks discovered by Dick de Bruijn. He defines a harmonic brick to be one with dimensions $a \times ab \times abc$ with a,b,c integers. (A similar definition is made for higher dimensional bricks.) The result is that a box can be tiled with these bricks if

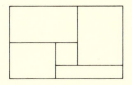

FIGURE 19

A tiling which cannot be accomplished by the divide and conquer method.

and only if the box can be tiled with all the bricks parallel; that is, the box has dimensions $ap \times abq \times abcr$. This means a harmonic brick tiles a box if and only if the box can be tiled by the divide and conquer method. Another theorem which I proved asserts that copies of a rectangle tile a larger rectangle if and only if the rectangle can be tiled by divide and conquer. This made me suspect that, in general, a brick tiles a box in k-dimensions if and only if the tiling can be done by divide and conquer. David Singmaster came up with a counter-example. A $1 \times 3 \times 4$ brick tiles a $5 \times 5 \times 12$ box, but there is no way to tile the box by divide and conquer! It follows from my solution to Göbels' first conjecture that all but a finite set of boxes which can be tiled with a given brick can be tiled by divide and conquer.

Happy Birthday Martin!

Fun and Problems

Disappearances

I wonder how magicians make their rabbits disappear;
Enchanted words like "hocus pocus" can not interfere
with laws of science and facts of mathematics that are clear.
The prestidigitators, making use of devious schemes,
(although they never tell you how) transport things as in dreams:
At times suspended, banished, null and void — or so it seems.
There must be something secret, yes, a trick that will involve
— when done with sleight of hand — a force that's able to *dissolve*.

N.B.: When the right-hand portions of this eight-line poem are interchanged,
a seven-line poem results. Which line disappears?

—D. E. Knuth

I wonder how magicians make their rabbits disappear;
Enchanted words like "hocus pocus" can transport things as in dreams:
with laws of science suspended, banished, null and void — or so it seems.
The prestidigitators, making use of devious schemes, involve
(although they never tell you how) a force that's able to *dissolve*.
At times I wonder how magicians make their rabbits disappear;
There must be something secret, yes, a trick that will not interfere
— when done with sleight of hand — and facts of mathematics that are clear.

Noneuclidean Harmony

---✦---

Scott Kim
STANFORD UNIVERSITY

Euclid

One of the outstanding achievements of the ancient Greeks was the formation of a deductive system of tonal harmony constructed on the principal sounds of everyday experience. This system was set forth in the 13 books of Euclid's *Elements of Harmony*, otherwise known as the "Harmonic Series". Starting from just 5 axioms:

1 Given two notes, there is an interval that joins them. (Axiom of Arpeggiation)

2 An interval can be prolonged indefinitely. (Axiom of the Fermata)

3 A progression can be constructed when its key center and one of its chords are given. (Axiom of the Sequence)

4 All tritones are equal. (The Devil's Axiom)

5 If a melodic line accompanying two melodic lines a fifth apart makes the interior intervals on the same side less than two tritones, the two melodic lines, if extended indefinitely, resolve on that side where the intervals are less than the two tritones. (The Fifth postulate)

Euclid was able to derive such theorems as:

The sum of the intervals of a triad is equal to two tritones.

He also investigated Apollonian progressions, and obtained the result that every point inside a circle of fifths has an invertible counterpoint.

Euclid, however, was clearly not satisfied with the statement of his fifth postulate. It lacked the concise elegance of the previous four. Indeed, Euclid delayed the entry of the fifth postulate in his "Elements" as long as possible, while employing the first four postulates to their full potential. At one point he considered appending yet another rule, to be called the "augmented fifth postulate", but rejected it as merely a "minor sixth."

During the following centuries, many people tried their hands at vindicating Euclid of every flaw. Perhaps the fifth postulate was redundant. Perhaps all pieces composed with the aid of the fifth postulate could be composed with only the first four. Some attempted to show this by deceptive cadence—the method of "resolution theorem-proving."

The 17th century Italian music theoretician, Gerolamo Saccheri, composed whole suites of pieces contradicting the fifth postulate. He expected to find a cross-relation; none appeared. Saccheri's work constitutes one of the earliest examples of Noneuclidean composition. His training, however, prevented him from realizing the importance of his discovery. Unable to accept the new sonorities, he declared his compositions to be "repugnant to the nature of the melodic line", thus denying his proper place in history. The fifth postulate would remain unresolved for many years.

The Greek Ideal

The failure of Saccheri, like that of his predecessors, was due to the dominant influence of Greek thought over musical theory. The Greeks conceived of a universe governed by divine ratios. The same ratios ruled the orbits of the planets, the harmonies of geometry, and the diagrams of music. Many examples of simple ratios were to be found. The Pythagorean theorem related the pitch of a hypotenuse to the lengths of the sides. If a length was doubled, the pitch was halved. Pythagorean tuning systems were based on the ratio 3/2, called the *Golden Interval* which was considered to be the interval most pleasing to the ear.

Ultimately, the Greeks related all ratios to the overtone series, a theme echoed for many centuries in the quotation "God created the overtone series; all else is the work of man". From the overtone series they derived the different tuning systems, or *modes*. Modal logicians believed that tuning had a strong effect on a person's temperament. Ratio and reality were inseparable; music, then was a revelation of the existing order of the universe.

Since there was only one music, there was no need to make it explicit. To a modern musician, Euclid's *common notations*, such as "the score is greater than the parts", sound curious. They allude to further notations which are equally obscure and fail utterly as self-contained definitions. Similarly, while composers following Euclid recognized the importance of proper voice-leading, they felt obliged to notate only those steps which might cause confusion.

This tendency towards abbreviation lead directly to a number of confusing notational practices, including *mathematica ficta*, figured bass, and ornamentation. The performer was required to fill in the missing steps in the appropriate manner. The great mathematician J. S. Bach was considered old-fashioned for his practice of writing out all ornamentation. The most subtle omission involved instrumentation. Composers frequently relied on the specific properties of their instruments. If a performer is to reconstruct the sound totally from scratch, much more information is needed.

Greek theory was not without dissenters. Critics raised fundamental questions of interpretation, some of which have been decided only in modern performance practice. Zeno of Alea questioned the interpretation of dotted notes. He argued that a whole note is impossible, since one must dot a·half note infinitely many times. The critic Epimenides made the notorious self-criticizing statement "all critics are liars". Epimenides' paradox in more modern form, "this song is unsingable," ultimately led to Gödel's "Universe Symphony", which is either uncriticizable or incomplete. The question of the universality of Euclid's harmony, however, remained unasked.

Noneuclid

The dissonance was finally resolved some 2000 years later by Noneuclid (1874–1951). Near the end of July 1921, Noneuclid told a pupil that "At one o'clock minus 12 minutes today, I made a discovery that will ensure the supremacy of German geometry for the next 100 years." He called the elements of his new system "at-one's" after the first two words of his famous quotation, and his system came to be known "at-onal" geometry.

Basically, 12-to-one geometry dispensed entirely with the notion of parallel intervals, which had been the cause of much worry, and declared that no intervals would be forbidden. Noneuclid took as his model the face of a clock. Melodic lines were to be identified with the hands. Since the hands always met at the center, there could be no parallel intervals. In Noneuclidean harmony, each atone receives a number from 12 to one, based on the numbers of a clock. Atone row, or series, is then constructed on this scale from 12 to one. Series may be added, subtracted, summed, or differentiated. The last operation (atone differentiation) is of particular interest to psychoacousticians.

Noneuclidean harmony upset the Greek ideal of simple ratio. Atonality requires the complete symmetry of equal-temperament, a tuning system in which intervals cannot be expressed in simple whole number ratios by flattening out the irregularities of the modes. Noneuclidean harmony severed its ties to temperament, and thus to any direct connection with reality.

For many years, the exact significance of the new harmonic system eluded the critics. Certainly one could invent a new set of axioms, and proceed to compose pieces in that style. But would the results be meaningful? Would the structure be audible? Tonality and atonality seemed totally contradictory. If one embraced the new system, what would be the logical status of Euclidean harmony? Would the efforts of Euclidean composers have been wasted?

Noneuclidean harmony forced musicians to broaden their attitudes towards the relation between music and the world. Noneuclid had shown that there existed harmonies in direct opposition to Euclidean harmony but nevertheless maintained internal consistency. This meant that harmony must be treated as an abstract game, quite independent of its descriptive power. Composers have now begun to shift their attention from writing pieces in existing styles to the invention of new styles themselves.

While all harmonies may be equally valid in a compositional sense, the question remains as to which system accurately describes acoustical reality. The final irony was revealed by recent work in psychoacoustics, which has discovered that the ear does not hear precise whole number ratios. Perceptual space is warped by many other factors, such as the volume of a sound, and the rate at which it is moving. So the Euclidean model is not *correct* even as a description of real sounds. In fact Euclidean harmony is just one point on a continuum of possibilities which includes all the varieties of atonality. For everday purposes Euclidean harmony suffices, since the intervals of equal-temperament are close approximations to those of the overtone series. The difference is only apparent in extreme cases, especially when modulations occur at a very fast rate. No wonder, then, that Noneuclidean harmonies were not realized earlier.

Recent Developments

With atonal geometry as their model, many other pathological compositional styles began to appear. Some people attempted to write pieces which could be played in more than one way, creating compositions of two or more functions. Others advocated the use of chance to express their compositions' indeterminate form. This trend met with considerable resistance. The acoustician Albert Einstein, who developed the revolutionary theory of relative pitch, was quoted as saying "God does not play dice".

In the 1870's, Georg Cantor began investigating the theory of supersonic pitches. Previously, composers had assumed that all pieces of unbounded speed were equally inaudible. Cantor showed otherwise, by using notions of 2 against 3 correspondence and uncountable rhythms. His research lead eventually to the ear-filling melodies, such as Piano's curve, sung by such birds as the orioles and the cardinals. While these sounds cover all frequencies, they are nondifferentiable. Cantor was also able to show that the violin has more notes than a piano, by the "glissando" method.

In 1976, the duo pianists Haken and Appel realized the long-unperformed "Map Quartet", which suggests that every national anthem can be sung with only four voices. Their realization used a computer program, which systematically enumerated the enormous number of permutations drawn from what was basically a very simple musical idea. In so doing, Haken and Appel revived basic questions concerning the nature of beauty. If no one can hear the structure of a composition can it claim to be beautiful? Who is the composer of a piece of computer music? The answers to these questions have yet to be found.

For further information on Noneuclid and discussions of truth in harmony, the reader is referred to "Gödel, Escher, Bach: an Eternal Golden Braid" by Douglas Hofstadter (Basic Books, 1979), which metaphorically weaves together the works of the music theoretician Kurt Gödel, the artist Maurits Cornelis Escher, and the mathematician Johann Sebastian Bach.

Magic
Cuboctahedrons

Charles W. Trigg
LOS ANGELES CITY COLLEGE

The surface of a cuboctahedron consists of eight equilateral triangles and six squares, with triangles and squares alternating around each of the twelve vertices, as indicated in the orthogonal projection of Figure 1.

The squares form a vertex-connected network with two squares at each vertex. They lie by twos in parallel planes. Between each parallel pair is a medial plane containing four vertices of a medial square. Thus, the vertices of the cuboctahedron lie on 9 squares, with three squares on each vertex.

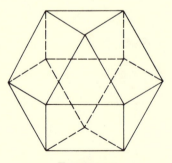

FIGURE 1

Orthogonal projection of cuboctahedron.

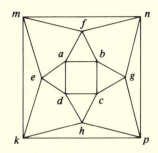

FIGURE 2

A Schlegel diagram of the cuboctahedron.

When twelve elements having a sum Σ are distributed on the vertices of a cuboctahedron in such a fashion that the perimeter sum of each surface square is θ, the cuboctahedron is said to be magic. Then, summing the perimeters, $6\theta = 2\Sigma$, so the magic constant $\theta = \Sigma/3$. It follows, from consideration of three parallel squares, that the perimeter sum of each medial square is $\Sigma - 2(\Sigma/3)$ or $\Sigma/3$, also. In the Schlegel diagram of Figure 2, the quartets on the vertices of the medial squares are e,f,g,h; a,c,p,m; and b,d,k,n.

The eight surface triangles lie by twos in parallel planes. When the perimeter sums of the three squares surrounding triangles afb and hpk are equated, and the equation is simplified, we have $a + f + b = h + p + k$; that is, on a magic cuboctahedron, the perimeter sums of opposite triangles are equal.

When the elements on the vertices are the first 12 positive integers, then $\Sigma = 78$ and $\theta = 26$. As a first step in identifying the magic cuboctahedrons, the 33 partitions of 26 into four distinct integers ≤ 12 are listed below. To achieve compactness, the integers 10, 11, and 12 are represented by X, Y, and Z, respectively.

12YZ	14XY	239Z	258Y	3472	349X	456Y
13XZ	159Y	248Z	267Y	356Z	358X	457X
149Z	168Y	257Z	259X	348Y	367X	4589
158Z	169X	23XY	268X	357Y	3689	4679
167Z	178X	249Y	2789			5678

With 1 assigned to fixed position a, two quartets having only 1 digit in common are needed for distribution in the positions a,b,c,d and a,e,m,f. Eleven quartet pairs meet this requirement, namely: 12YZ, 169X; 12YZ, 178X; 13XZ, 159Y; 13XZ, 168Y; 149Z, 168Y; 149Z, 178X; 14XY, 158Z; 14XY, 167Z; 158Z, 169X; 159Y, 167Z; and 159Y, 178X. With 1 fixed in position, the distribution of each quartet on the vertices of its quadrilateral can be completed in 3! ways. Thus there are $6^2(11)$ or 396 initial distributions to consider.

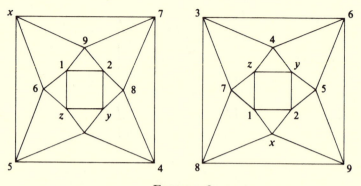

FIGURE 3

Complementary magic cuboctahedrons.

One distribution of the first quartet pair is shown in the diagram on the left in Figure 3. A quartet, $457X$, containing X and no other integers in common with the other quartets, was selected for distribution on the outer square. Then, a quartet containing 9 and 2 and one of 4, 5, and 7 was sought for placement on the *fngb* quadrilateral. The quartet 2789 was chosen. A quartet containing 8 and Y and one of 4 and 5 was needed next. When $348Y$ was placed on the quadrilateral *cgph*, the digit 5 was left for placement on the vertex k to complete the magic cuboctahedron.

In this manner, 40 magic cuboctahedrons have been identified. They exist as 20 reflective twins. One twin is converted into the other by reflection in the plane of vertices m, a, c, p. Only one of each twin pair has been recorded in Table 1. Distributions that can be rotated to coincide are not considered to be distinct.

Two integers will be said to be complementary if their sum is 13. Two magic cuboctahedrons will be said to be complementary if their corresponding elements are complementary. The two cuboctahedrons in Figure 3 are complementary.

In Table 1, the complementary cuboctahedron pairs are: A, B; C, D; E, F; G, H; and I, J. In each of the pairs, the a, b, c, d groups are permutations of the same quartet of integers. The same is true of their e, f, g, h and k, m, n, p groups, except that in the A, B and I, J pairs, the quartets are interchanged.

The ten cuboctahedrons, K through U, are self-complementary. Within the pairs N, P; R, S; and T, U, the a, b, c, d and k, m, n, p distributions are identical. Within the pairs N, P and R, S, the e, f, g, h distributions are reversed. In the set R, S, T, U, the a, b, c, d groups are permutations of the same integers, as the e, f, g, h and the k, m, n, p groups are also.

Since the corresponding triangles are opposite, the perimeter sums of the triangles bordering the outer square in Figure 2 are the same as the perimeter sums of the triangles bordering the inner square. Proceeding

Penta-gram	a	b	c	d	e	f	g	h	k	m	n	p	fmn	gnp	hkp	ekm
				Locations on Schlegel Diagram										*Triangle Sums*		
A	1	2	Y	Z	6	9	8	3	5	X	7	4	26	19	12	21
B	1	Z	Y	2	X	7	4	5	9	8	3	6	18	13	20	27
C	1	3	X	Z	5	Y	8	2	7	9	4	6	24	18	15	21
D	1	Z	X	3	Y	8	2	5	7	6	4	9	18	15	21	24
E	1	9	4	Z	7	8	6	5	2	X	3	Y	21	20	18	19
F	1	Z	4	9	8	6	5	7	2	Y	3	X	20	18	19	21
G	1	5	8	Z	4	X	9	3	7	Y	2	6	23	17	16	22
H	1	Z	8	5	X	9	3	4	7	6	2	Y	17	16	22	23
I	1	7	6	Z	4	Y	3	8	2	X	5	9	26	17	19	16
J	1	Z	6	7	5	9	2	X	4	Y	3	8	23	13	22	20
K	1	2	Z	Y	6	9	7	4	5	X	8	3	27	18	12	21
L	1	Y	Z	2	X	8	3	5	9	7	4	6	19	13	20	26
M	1	9	Z	4	8	X	5	3	Y	7	2	6	22	17	19	20
N	1	Y	4	X	6	7	5	8	2	Z	3	9	20	19	17	22
P	1	Y	4	X	8	5	7	6	2	Z	3	9	20	19	17	22
Q	1	8	Z	5	6	9	7	4	Y	X	2	3	21	12	18	27
R	1	9	Y	5	7	6	3	X	4	Z	8	2	26	13	16	23
S	1	9	Y	5	X	3	6	7	4	Z	8	2	23	16	13	26
T	1	Y	5	9	7	X	3	6	4	8	2	Z	20	17	22	19
U	1	Y	5	9	X	7	6	3	4	8	2	Z	17	20	19	22

TABLE 1

Magic cuboctahedrons, constant = 26.

clockwise from the topmost triangle, the perimeter sums of the bordering triangles are recorded in Table 1. In every case the four sums are distinct. The quartet of sums is the same in the pairs C, D; E, F; G, H; K, Q; L, M; and R, S, as well as in the group N, P, T, U. In C, D, E, and F, the four sums are in arithmetic progression. Alternate sums total 39 in C, D, E, F, G, H, K, L, M, and Q. In N, P, R, S, T, and U, there are consecutive sum pairs that total 39.

Cuboctahedrons L, M, R, and S are supermagic with 9 squares and 2 triangles having perimeter sums of 26.

There is no distribution of the first 12 integers such that the perimeter sums of the eight triangles are the same, since 2(78)/8 or 39/2 is not an integer.

Games, Graphs, and Galleries

Ross Honsberger
UNIVERSITY OF WATERLOO

The Bulging Railway

Let's begin by giving Martin a little test (due to Murray Klamkin) of his mathematical intuition. Imagine a flat stretch of straight railway track AB, 5000 feet long, which is immovably fastened down at each end (Figure 1). In the heat of the summer the track expands 2 feet, causing it to buckle. Assuming it bends symmetrically, how high do you think the bulge rises above the ground at the middle? An inch? 4 feet? A tenth of an inch?

Since the total length of the track is now 5002 feet, each half would be 2501 feet. While it is clear that the buckle would take the form of a gentle curve, we can get a rough idea of the situation by assuming it to be straight. On this basis we obtain an estimate of the height x by applying the theorem of Pythagoras:

$$x = \sqrt{2501^2 - 2500^2}$$
$$= \sqrt{(2501 - 2500)(2501 + 2500)}$$
$$= \sqrt{5001},$$

making it approximately 70 feet! This comes as a big surprise to many people. The actual value, taking into account the curve in the track, is closer to 67 feet. How close did you come, Martin?

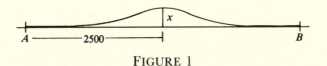

FIGURE 1

Arithmetic Mastermind

A year or so ago a mathematical game called *Mastermind* swept across the country. One player would secretly arrange 4 colored pegs in a row and the second player, by a combination of guessing and reasoning, would try to determine the colors and order of the pegs in as few steps as possible. Let us play a similar arithmetic game, also due to Murray Klamkin.

We begin by having you choose 5 positive integers less than 100, repetitions allowed:

$$(a_1, a_2, a_3, a_4, a_5).$$

It is up to me to determine these numbers as quickly as I can. I am permitted to enlist your help through the following kind of dialogue. I present to you a set of 5 numbers $(x_1, x_2, x_3, x_4, x_5)$ and you respond, not as the Code-maker does in *Mastermind*, by telling me how many of my x's are correct and how many are in the correct places, but by telling me only the single number S_1 obtained by multiplying the corresponding numbers in our selections and adding:

$$S_1 = a_1 x_1 + a_2 x_2 + a_3 x_3 + a_4 x_4 + a_5 x_5.$$

Armed with this value S_1, I then propose another slate $(y_1, y_2, y_3, y_4, y_5)$. If I am correct the game is over; otherwise you respond with the number S_2 given by

$$S_2 = a_1 y_1 + a_2 y_2 + a_3 y_3 + a_4 y_4 + a_5 y_5.$$

This procedure continues until I get them right.

It is clear that after 5 steps I am in possession of 5 equations in the 5 unknowns a_1, a_2, a_3, a_4, a_5 and can determine what they are. Thus the game never needs to go beyond 5 steps.

Of course, the idea is for me to determine your selections in as few steps as possible. Would you believe, Martin, that I can always do it in 4 steps? Would you believe 3 steps? 2 steps? The fact is, it can be done in just *one* step!

Since I know that each of your numbers is less than 100, I choose my first slate to be

$$(10^8, 10^6, 10^4, 10^2, 1).$$

This leads you to compute

$$S_1 = 100000000 a_5 + 1000000 a_4 + 10000 a_3 + 100 a_2 + a_1,$$

which displays your selections as 2-digit numbers from left to right. For example, if

$$(a_1, a_2, a_3, a_4, a_5) = (17, 68, 5, 42, 8),$$

then $S_1 = 1768054208$, and it's all over.

The First n Positive Integers

It is well known, and easy to establish, that

$$1 + 2 + \cdots + n = \frac{n(n+1)}{2}.$$

The following novel approach to this result is due to Donald Snow of Brigham Young University.

Let

$$S_1(n) = 1 + 2 + \cdots + n$$

and

$$S_2(n) = 1^2 + 2^2 + \cdots + n^2.$$

Then

$$
\begin{aligned}
S_2(n + m) &= 1^2 + 2^2 + \cdots + (n + m)^2 \\
&= (1^2 + 2^2 + \cdots + n^2) + (n + 1)^2 + (n + 2)^2 + \cdots \\
&\quad + (n + m)^2 \\
&= S_2(n) + (n^2 + 2n + 1^2) + (n^2 + 4n + 2^2) + \cdots \\
&\quad + (n^2 + 2mn + m^2) \\
&= S_2(n) + mn^2 + 2n(1 + 2 + \cdots + m) + (1^2 + 2^2 + \cdots \\
&\quad + m^2) \\
&= S_2(n) + mn^2 + 2nS_1(m) + S_2(m),
\end{aligned}
$$

that is,

$$S_2(n + m) = S_2(n) + S_2(m) + 2nS_1(m) + mn^2.$$

Interchanging n and m, we have

$$S_2(m + n) = S_2(m) + S_2(n) + 2mS_1(n) + nm^2.$$

Then the equality $S_2(n + m) = S_2(m + n)$ immediately simplifies to

$$2nS_1(m) + mn^2 = 2mS_1(n) + nm^2.$$

Since $S_1(1) = 1$, the substitution $m = 1$ gives

$$2n + n^2 = 2S_1(n) + n,$$

and out pops our formula $S_1(n) = \dfrac{n^2 + n}{2}$.

I never cease to be amazed at the remarkable relation

$$1^3 + 2^3 + 3^3 + \cdots + n^3 = (1 + 2 + 3 + \cdots + n)^2.$$

When he was visiting from Australia, Roger Eggleton pointed out the following geometric demonstration of this fact. Since the area of k squares of side k is $k \cdot k^2 = k^3$, let us arrange in the plane

> 1 square of side 1, 2 squares of side 2, 3 squares of side 3, $\cdots$, n squares of side n,

having a total area of $1^3 + 2^3 + 3^3 + \cdots + n^3$. Let us do our best to build up a layered square by adding the squares of a given size around the outside of the part that has been constructed (Figure 2). This succeeds pretty well, except for the "even" squares, $2, 4, 6, \ldots$, which both overlap (▨) and leave uncovered gaps ($\boxed{x}$).

A moment's reflection shows that each overlapping square corresponds to a blank square of the same size, yielding the conclusion that the total area is that of a square of side $(1 + 2 + \cdots + n)$.

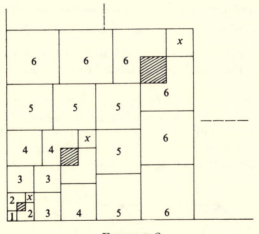

FIGURE 2

Herb Shank's Milkman's Route

In the middle of 1974, my colleague Herb Shank made an unusual discovery about graphs. I would like to describe his result without considering its proof. To avoid getting involved with the explanation of several technical terms in graph theory, let us confine ourselves to the familiar setting of geometry. In so doing, the theorem will not be presented in its full generality. In order to state the whole story, permit me to address

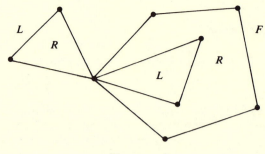

FIGURE 3

the following single sentence to those readers who are acquainted with simple graph theory:

> our story begins with any planar, Eulerian graph which has an odd number of spanning trees.

Suppose a configuration F is constructed in the plane by connecting polygons together at single vertices. Any number of polygons may be used in any arrangement you please. One polygon may even occur inside another. We need to observe only three things:

1 every polygon is to have an *odd* number of sides,

2 no two edges are permitted to cross,

3 no polygon is to be connected into the figure at more than one vertex (this avoids producing a "ring" of polygons).

Now, let the regions in F be colored, alternating red and a nice lime green so that regions which border along the same edge have different colors. The above conditions guarantee that such a coloring is always possible. In order to keep our example reasonably simple, we illustrate with two triangles and a pentagon (Figure 3). We use R for red and L for lime.

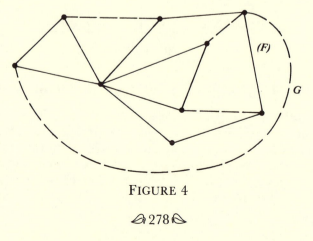

FIGURE 4

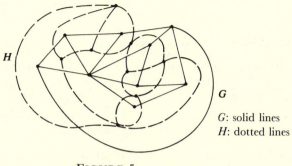

FIGURE 5

Next, we embellish the figure by inserting any additional edges we like between pairs of vertices of *F*, subject only to the condition that no two edges cross either themselves or edges which are already in *F* (this allows even for the introduction of multiple edges, that is, edges which connect vertices that are already joined by an edge). Let the resulting figure be called *G* (Figure 4).

From *G* we derive an independent figure *H*, called the *dual* graph, as follows:

1 place a vertex in each region of *G*, including the infinite outside region;

2 join the vertices of two neighbouring regions of *G* by drawing an edge from the one to the other so that it crosses through the interior of an edge in their common boundary; insert such an edge for each edge in their common boundary (Figure 5).

Finally, we assign the label *R* and *L* to each vertex of *H* according to the color of its region—red or lime (Figure 6).

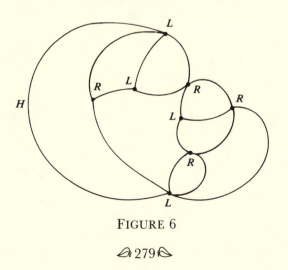

FIGURE 6

Now we have a labelled graph *H*, which, no matter what choices might have been exercised in its construction, always possesses the following remarkable property:

> Beginning at any vertex *V* as a starting point, and proceeding along any edge of that vertex, one will traverse each edge exactly twice, once in each direction, and end up at *V* again, by always making a hard right turn at each vertex labelled R and a hard left turn at each vertex labelled L.

Such a tour can be charted nicely by placing a numbered arrow in the appropriate direction on each edge as it is traversed. Thus, for example, we might obtain the following—Figure 7:

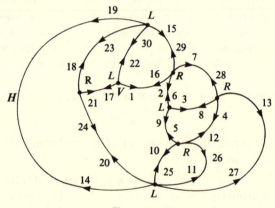

FIGURE 7

Who Stole The Apples?

I imagine that many of us have spent time during our holidays working on some of those brainteasers that call for "each man's occupation" or "who is married to whom?". At the age one is often introduced to such a pastime, a straightforward trial-and-error process of elimination is likely to be the only approach at one's disposal. However, certain problems of this sort have very ingenious solutions. For example, consider the following problem:

> Out of 6 boys, exactly two were known to have been stealing apples. But who? Harry said "Charlie and George". James said "Donald and Tom". Donald said "Tom and Charlie". George said "Harry and Charlie". Charlie said "Donald and James". Tom couldn't be found. Four of the boys interrogated had named one miscreant correctly, and one incorrectly. The fifth had lied outright. Who stole the apples?

The problem is not without its appealing twists—Tom couldn't be found; one of them lied outright. In my *Ingenuity in Mathematics* I gave a rather involved formal approach to the problem. Recently my good friend and colleague Scott Vanstone showed me the following brilliant solution which translates the whole thing into a simple problem concerning a graph. Making use of an observation of John Annulis (University of Arkansas at Monticello), the solution is virtually transparent.

Let a graph (Figure 8) be constructed with a vertex for each boy (C for Charlie, G for George and so on), and with an edge for each pair of boys named in the statements (edge CG reflects Harry's statement, etc.). Thus every edge but one has exactly one end at a thief, and the remaining edge, due to the outright liar, has neither end at a thief. That is to say, the two vertices which represent the thieves, have between them a total of 4 endpoints (one from each of four edges and none from the fifth edge). We should note also that the thieves themselves, are not joined by an edge. We see immediately that the culprits can only be Charlie and James (C and J).

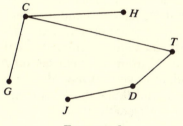

FIGURE 8

Chvátal's Art Gallery Theorem

Chapter 11 in my book *Mathematical Gems II* is entitled "Chvátal's Art Gallery Theorem". It concerns a certain problem of guarding the paintings in an art gallery. The way the rooms in museums and galleries snake around with all kinds of alcoves and corners, it is not an easy job to keep an eye on every bit of wall space. The question is to determine the minimum number of guards that are necessary to survey the entire building. The guards are to remain at fixed posts, but they are able to turn around on the spot. The walls of the gallery are assumed to be straight. Chvátal showed that if the gallery has n walls (that is, if the floor-plan is an n-gon as in Figure 9), then, whatever its zig-zag shape, the minimum number of guards needed is never more than $[n/3]$, the integer part of $n/3$. His proof is given in *Mathematical Gems II*. Now I would like to present a new proof of this theorem. Chvátal's analysis is certainly first-class, but the following proof by Steve Fisk of Bowdoin College is simply inspired.

First of all, the gallery is divided into triangles by drawing non-intersecting diagonals in its interior (Figure 10). Next, the vertices of the gallery are colored so that two vertices that are joined by an edge in the figure are given different colors. It is an easy application of mathematical

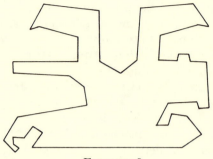

FIGURE 9

induction to prove that 3 colors suffice for this task. For $n = 3$, the entire configuration consists merely of a single triangle, and 3 colors obviously are sufficient. Let us suppose that 3 colors suffice for a triangulated n-sided gallery ($n \geq 3$) and that we have at hand a gallery G with $(n + 1)$ sides.

Clearly there is no shortage of diagonals (AB) which cut from G a single triangle (ABC). By the induction hypothesis, the triangulated n-sided configuration G', obtained from G by cutting off $\triangle ABC$, can be colored with 3 colors so that adjacent vertices have different colors. Since only 2 of the 3 colors are used at A and B, the third color can be used at C to extend this into a successful 3-coloring of the triangulated gallery G itself. Thus, by induction, 3 colors suffice in all cases.

Suppose, therefore, that the vertices of our gallery are properly 3-colored with the colors a, b, and c. Since there are only n vertices altogether, all three of the colors cannot each be used *more* than $n/3$ times. The color which is used least often, (say b), must occur only m times, where $m \leq n/3$. Since m is an integer, this is the same as saying $m \leq [n/3]$.

Since no two vertices of any triangle in the figure can be the same color, each triangle must have a vertex of each color a, b, and c. Clearly a watchman at the vertex colored b can survey the whole triangle from that

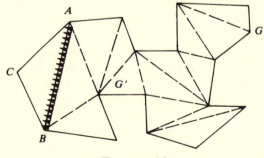

FIGURE 10

vertex, and m watchmen at the m vertices colored b can observe all the triangles, that is, the entire gallery.

The Track

Consider now the following engaging problem which, as far as I can determine, is due to the brilliant young Hungarian mathematician Laszlo Lovász.

Around a track T are n arbitrarily spaced depots $x_1, x_2, \ldots, x_n$, each containing a quantity of gasoline (Figure 11). The total amount of gas is exactly enough to take a car around T once. Prove that, no matter how the gas is distributed over the depots, there is some depot at which an empty car can gas up, proceed around T picking up the gas at the other depots as they are encountered, and get all the way around the track.

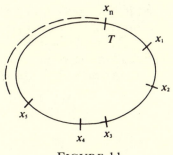

FIGURE 11

The following solution was contributed by Dean Hoffman (Auburn University), who learned of it from Lovász himself.

Suppose a practice run is made, starting from any point on the track, with a car that has lots of extra gas. Since the amount of gas required to make the run is exactly the quantity picked up along the way, the car will finish the lap with the same amount of gas it began with. Now, during the run we keep an eye on the gas gauge, and we note that when we arrived at depot x_i, the reading was lowest for the whole trip, say d gallons. During the run, then, we were always carrying around an extra d gallons of gas that was never needed. Clearly the same would have happened if this d gallons of gas had been left at home. Without this unnecessary gas, we would have come into depot x_i just as the tank ran dry, a condition that we achieve by starting the real trip at depot x_i. Beginning at x_i, the gas gauge will never be pressed to go below zero because it never went below d gallons during the practice run.

Lattice Cubes

The points (x, y, z) in 3-dimensional space which have all three coordinates integers are called *lattice points*. They occur at the vertices of the unit cubes into which space is divided by the planes $x = a$, $y = b$, $z = c$, where a, b, c are integers. Obviously there is any number of cubes, having their (8) vertices at lattice points, which sit in space with their faces parallel to the coordinate planes. The length of the edge of such a cube is clearly an integer.

There are also lots of cubes, having their vertices at lattice points, which do not occur in "standard" position but sit *obliquely* (relative to the grain of the lattice). Few of us, however, possess the intuition not to be surprised at the all-inclusive result:

> the length s of the edge of a cube C, having its vertices at lattice points, is always an integer.

Since it doesn't matter which point of the lattice we choose to take as the origin O of coordinates, let's use one of the vertices of C. Suppose the three edges from O go to the vertices $V_1(x_1, y_1, z_1)$, $V_2(x_2, y_2, z_2)$, $V_3(x_3, y_3, z_3)$. Having edge s, C has volume $V = s^3$. A well known formula from freshman mathematics gives the volume of a cube in terms of the edges at a vertex. For the appropriate choice of sign, we have

$$V = \pm \begin{vmatrix} x_1 & y_1 & z_1 \\ x_2 & y_2 & z_2 \\ x_3 & y_3 & z_3 \end{vmatrix}.$$

Because all of the entries here are integers, the volume V must also be an integer, that is,

$$s^3 \text{ is an integer.}$$

Now the length of the edge OV_1 is

$$s = |OV_1| = \sqrt{x_1^2 + y_1^2 + z_1^2} = \sqrt{\text{an integer.}}$$

Therefore we see s^2 is also an integer. Consequently, s itself must be rational:

$$s = \frac{s^3}{s^2} \quad \text{is rational.}$$

But, as we just observed, $s = \sqrt{\text{an integer}}$. And if the square root of an integer is not irrational, then it is not just rational, but integral. Thus s is an integer.

This result can be generalized in a straightforward manner to cubes in n-dimensional space, where n is odd.

Probing the Rotating Table

William T. Laaser
STANFORD UNIVERSITY

Lyle Ramshaw
STANFORD UNIVERSITY

ABSTRACT

Consider the following generalization of the Problem of the Rotating Table: the shape of the table is a regular n-gon; at each of the n corners of the table is a well containing a glass that is either upright or inverted; and the player has k hands. For each k and n, the question then arises whether or not there exists a procedure that is guaranteed to ring the bell after a finite number of moves. In this paper, we will see that such a procedure exists if and only if the parameters k and n satisfy the inequality $k \geq (1 - 1/p)n$, where p is the largest prime factor of n.

Section 1
Introduction

In his column in the February, 1979 issue of *Scientific American*, Martin Gardner presented the Problem of the Rotating Table as follows:

I open with a delightful new combinatorial problem of unknown origin. It was passed on to me by Robert Tappay of Toronto, who believes it comes from the U.S.S.R.

Imagine a square table that rotates about its center. At each corner is a deep well, and at the bottom of each well is a drinking glass that is either upright or inverted. You cannot see into the wells, but you can reach into them and feel whether a glass is turned up or down.

A move is defined as follows: Spin the table, and when it stops, reach each hand into a different well. You may adjust the orientation of the glasses any way you like, that is, you may leave them as they are or turn one glass or both.

Now, spin the table again and repeat the same procedure for your second move. When the table stops spinning, there is no way to distinguish its corners, and so you have only two choices: you may reach into any diagonal pair of wells or into any adjacent pair. The object is to get all four glasses turned in the same way, either all up or all down. When this task is accomplished, a bell rings.

At the start the glasses in the four wells are turned up or down at random. If they all happen to be turned in the same direction at this point, the bell will ring at once and the task will have been accomplished before any moves were made. Therefore it should be assumed that at the start the glasses are not all turned the same way.

Is there a procedure guaranteed to make the bell ring in a finite number of moves? Many people, after thinking briefly about this problem, conclude that there is no such procedure. It is a question of probability, they reason. With bad luck one might continue to make moves indefinitely. That is not the case, however. After no more than *n* correct moves one can be certain of ringing the bell. What is the minimum value of *n*, and what procedure is sure to make the bell ring in *n* or fewer moves?

Consider a table with only two corners and hence only two wells. In this case one move obviously suffices to make the bell ring. If there are three wells (at the corners of a triangular table), the following two moves suffice.

1 Reach into any pair of wells. If both glasses are turned the same way, invert both of them, and the bell will ring. If they are turned in different directions, invert the glass that is facing down. If the bell does not ring:

2 Spin the table and reach into any pair of wells. If both glasses are turned up, invert both, and the bell will ring. If they are turned in different directions, invert the glass turned down, and the bell will ring.

Although the problem can be solved in a finite number of moves when there are four wells and four glasses, it turns out that if there are five or more glasses (at the corners of tables with five or more sides), there are no procedures guaranteed to complete the task in n moves. Next month I shall give one solution for the problem with four glasses and discuss some generalizations of the problem developed by Ronald L. Graham and Persi Diaconis.

There are several fine points about the rules which deserve clarification, concerning the types of procedures that the player can legally follow. First, the player must announce exactly which wells she is planning to probe before she actually probes either of them. It is not legal for her to probe one well, and then to decide which other well to probe on the basis of what she finds in the first one. Secondly, once she has probed the wells and turned the glasses in some way, the bell will not ring to indicate success until after she has withdrawn her hands from the wells. That is, it is not legal for the player to try out several orientations of the glasses with her hands still in the wells, hoping to ring the bell on at least one of them.

The solution for the problem with four wells and four glasses given in Martin Gardner's column of March, 1979 makes the bell ring in no more than five moves. If the reader is not already familiar with the Problem of the Rotating Table, she is encouraged to reconstruct this solution.

The March, 1979 column also mentions two generalizations of the Rotating Table Problem, suggested by Ronald L. Graham and Persi Diaconis. They first suggested allowing the player more than two hands. For each k and n, we can imagine a k-handed player and a table in the shape of a regular n-gon, with n corners, wells, and glasses; the problem then is to discover a finite procedure guaranteed to ring the bell, or to prove that no such procedure exists. Graham and Diaconis also suggested generalizing the problem by replacing the glasses with objects that have more than two positions. And still other interesting generalizations are possible; for example, Scott Kim considered replacing the square table by a table with a richer group of symmetries, such as a cube.

This paper is devoted to the first of these generalizations. In particular, we define $f(n)$ for $n \geq 2$ to be the smallest integer k such that a finite procedure guaranteed to ring the bell exists when a k-handed player faces an n-cornered polygonal table, whose glasses have precisely two states. Our goal is to determine the values of the function f.

The rules that specify the Rotating Table Problem are sufficiently subtle that it is hard to be sure that they have been completely specified. Recall that we have already had to pause to clarify several fine points. Thus, it behooves us to state the problem more formally, in the hope that a more formal description will also be more precise. We will rephrase the problem as an asymmetric two-person game between the Player and the Table, involving the manipulation of circular strings.

Whenever we talk about polygons in this paper, we will be referring to regular polygons that are free to rotate about their centers, but that cannot be turned upside down. With this understanding, we define a *necklace of length n* to be a labelling of the vertices of an n-gon. A necklace of length n is like a string of length n, except that all circular shifts of the string correspond to the same necklace. We will specify particular necklaces either by drawing a picture, or by writing a string in angle brackets, where the string represents the labels of the necklace in clockwise order from an arbitrary starting point. For example, the necklace of length six in Figure 1 could be written $\langle 010011 \rangle$, or $\langle 100110 \rangle$, or in four other ways; but $\langle 101100 \rangle$ represents a different necklace.

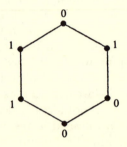

FIGURE 1

A necklace of length six.

At any point in time, the current state of the glasses in their wells corresponds to a necklace of length n over the alphabet $\{u, d\}$, where we will use u and d for "up" and "down" to denote upright and inverted glasses respectively. The first move of the Game of the Rotating Table is a move by the Table; in particular, the Table determines the initial state of the glasses by choosing a necklace over $\{u, d\}$. Call this state S_1.

Now it is time for the Player's first move, in which she must announce the circular arrangement of her probing hands around the table. We will call this arrangement a *probe pattern*. Formally, a probe pattern is represented as a necklace over the alphabet $\{h, g\}$, where h for "hand" denotes a position that the Player desires to probe, and g for "gap" denotes an unprobed position. Naturally, the Player must choose a probe pattern that has no more than k occurrences of the symbol h, since the Player has only k hands. Call this first probe pattern P_1.

Next, it is the Table's move again. The Table gets to choose one of the n possible ways in which the necklaces S_1 and P_1 can be superimposed. This choice models the spinning of the Rotating Table. After the Table has made this choice, the Player is told for each h in the probe pattern P_1

whether the corresponding symbol of the state S_1 is a u or a d. The Player may then order that any or all of the u's in probed positions be changed to d's, or vice versa. The result of these alterations defines the next state of the table, a necklace S_2 over the alphabet $\{u, d\}$. We will call a state necklace *monotone* if it consists entirely of u's or entirely of d's. If the state S_2 is monotone, a bell now rings and the Player has won the game in one probe.

At this point, the game enters a loop. Suppose that the non-monotone state S_i has just been determined. The Player then decides upon a probe pattern P_i; the Table determines how S_i and P_i are to correspond; and the Player is informed of the contents of S_i in those positions corresponding to h's in P_i. The Player then decides upon new contents for these positions, and her decisions determine the next state S_{i+1}. If S_{i+1} is monotone, the bell rings and the Player is deemed to have won in i probes. The Table's goal is to prevent the Player from winning; we will say that the Table wins if and only if the game continues forever. Recall that we have defined $f(n)$ to be the smallest integer k such that k hands allow the Player to win; a winning strategy for the Player exists if $k \geq f(n)$, while a winning strategy for the Table exists if $k < f(n)$.

The remainder of this paper is devoted to the proof of the following result:

THEOREM 1 For any integer $n \geq 2$, let $f(n)$ denote the smallest integer with the property that $f(n)$ hands are enough to enable the Player to win the Game of the Rotating Table. Let p denote the largest prime factor of n. Then, the value of $f(n)$ is given by the formula

$$f(n) = \left(1 - \frac{1}{p}\right)n.$$

Previous work established some special cases of this result. As announced in Martin Gardner's column of March, 1979, Ronald L. Graham and Persi Diaconis showed that $f(n) = n - 1$ if n is a prime number, while $f(n) \leq n - 2$ if n is composite. For all $a \geq 1$ and $b \geq 2$, Scott Kim showed the lower bound

$$f(ab) \geq a\left\lceil \frac{b}{2} \right\rceil,$$

where $\lceil x \rceil$ denotes x rounded up to the nearest integer. And James Boyce first conjectured the formula for $f(n)$ given in Theorem 1.

By the way, Theorem 1 has the curious corollary that, except for $n = 2$, the minimum number of hands needed is always an even number.

Section 2
The Lower Bound

In this section, we will demonstrate that the formula given in Theorem 1 constitutes a lower bound on the true value of $f(n)$. In particular, we will prove the following Lemma.

LEMMA 1   Let $n \geq 2$ be an integer, and let p be its largest prime factor. Then, if the Player is restricted to fewer than $(1 - 1/p)n$ hands, there is a winning strategy for the Table in the Game of the Rotating Table.

PROOF   We begin by noting that, if n factors in the form $n = lm$, an n-gon can be viewed as l distinct and interleaved m-gons. Because $6 = 2 \cdot 3$, for example, we can view a hexagon as a Star of David, as shown in Figure 2.

Let p be the largest prime factor of the integer $n \geq 2$, and determine l by the relation $n = lp$. In fact, we could take p to be any prime factor of n in this argument, but the largest prime factor gives us the strongest result. As noted above, the states and probe patterns of the Rotating Table Game can be viewed as composed of l interleaved p-gons. Our goal is to construct a winning strategy for the Table, assuming that the Player has fewer than $(1 - 1/p)n$ hands. The basic idea is to guarantee that, for all i, at least one of the p-gons that constitute the state S_i will be nonmonotone, that is, will contain both up and down glasses.

This condition is easy to establish initially, since the Table may choose the state S_1 arbitrarily. But we must show that the Table can maintain this condition as the Game progresses. Suppose that the state S_i contains a nonmonotone p-gon; choose one, and call it S_i^*. Let P_i denote the probe pattern which the Player decides to employ on the ith move. Since the Player is restricted to fewer than $(1 - 1/p)n = l(p - 1)$ hands, at least one of the l different p-gons of P_i must contain at least two gaps g.

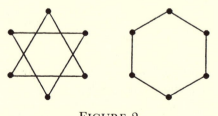

FIGURE 2

Viewing a hexagon as two interleaved triangles.

Choose such a p-gon, and call it P_i^*. Also choose two gaps which appear in P_i^*, and suppose that they appear at distance j around P_i^*, where we measure distance so that adjacent vertices are at distance 1. Now, consider touring S_i^* by walking around it in steps of j vertices each. Since p is prime, this tour will visit every vertex of S_i^* before returning to its starting place, no matter what the value of j. During this tour, we will certainly come across both u's and d's, since S_i^* was chosen to be nonmonotone. Hence, we will come across some d immediately after a u. This implies that S_i^* contains a u and a d which are at distance j.

We are now ready to advise the Table how to proceed. It should chose to align the necklaces S_i and P_i so that the p-gons S_i^* and P_i^* are superimposed, and furthermore, so that the u and d at distance j in S_i^* correspond to the two g's at distance j in P_i^*. Since the Player cannot alter the states of glasses which correspond to gaps, the resulting state S_{i+1} will contain at least one nonmonotone p-gon. By using this technique repeatedly, the Table can guarantee for all i that the state S_i will contain a nonmonotone p-gon. This constitutes a winning strategy for the Table.

Section 3
Perfect Powers

The lower bound result in Lemma 1 was the easy part. To complete the proof of Theorem 1, we are left with the more exacting job of constructing winning strategies for the Player that use the minimal number of hands. We begin this task with an insight that will allow us to use previously constructed strategies for smaller tables as subroutines when constructing strategies for larger tables.

We need a definition. Suppose that the size n of the table factors in the form $n = lm$. Recall that we can then view an n-gon as l distinct, interleaved m-gons. Suppose that some necklace T of length n has the property that all of the m-gons of which it is composed are monotone. Note that the necklace T must then consist of m copies of some string of length l, strung together end to end; that is, T must have the form $\langle X^m \rangle$ for some string X. We will distinguish such necklaces by calling them *perfect mth powers*. For example, an even-length necklace is called a perfect square when the labels at the two ends of every diagonal are the same, implying that the necklace has the form $\langle XX \rangle$ for some string X. Saying that a necklace of length n is a perfect nth power is another way of saying that it is monotone.

Suppose that we are constructing a strategy for the Player to use against a size n Table where n factors nontrivially as $n = lm$, and suppose that we get to a point where the current state S of the table is a perfect mth power. We can finish the strategy in an easy way as long as we have at least $mf(l)$ hands. The idea is to mimic an optimal Player strategy for a size l

table, with everything replicated m times; one might call this process raising a strategy to a power. Take the first probe pattern of the smaller strategy, and build a probe pattern for the larger strategy by stringing together m copies. Since the current state S of the table is a perfect mth power, each of our Player's m groups of $f(l)$ hands will feel the same sequence of ups and downs. We then instruct our Player to perform with each group of hands whatever adjustments of the table state the smaller strategy recommends. No matter what those adjustments are, the next state of the larger table will also be a perfect mth power, and thus, we will be able to continue mimicking the smaller strategy. This argument has demonstrated the truth of the following Lemma.

> **LEMMA 2** Suppose that the Rotating Table Game is being played with a table of size n, and that the current state S of the table is a perfect mth power for some integer $m \geq 2$ that properly divides n. Then, there exists a strategy that the Player may now begin to employ that is guaranteed to ring the bell in a finite number of probes, and that uses only $m f(n/m)$ hands.

Before we can apply Lemma 2, we first need to force the table into a state that is a perfect power. The next Lemma claims that there exist strategies that handle this part of the job without using too many hands.

> **LEMMA 3** If p denotes the largest prime factor of an integer $n \geq 2$, then there exists a strategy for the Player in the Rotating Table Game of size n with the following properties: it uses at most $(1 - 1/p)n$ hands, and, in a finite number of moves, it either rings the bell or forces the table into some state that is a perfect pth power.

We will postpone the proof of Lemma 3 until later. Instead, let us now show that Lemmas 1, 2, and 3 together allow us to complete the proof of Theorem 1.

> PROOF OF THEOREM 1 Theorem 1 claims that $f(n) = (1 - 1/p)n$, where p is the largest prime factor of n. From Lemma 1, we know that $f(n) \geq (1 - 1/p)n$. It remains to show that $f(n) \leq (1 - 1/p)n$. Our proof will be by induction on n; thus, we assume that the Theorem holds true for all tables smaller than n in size.

Our task is to demonstrate the existence of a winning strategy for the Player that uses only $(1 - 1/p)n$ hands. We begin by using the strategy from Lemma 3; this either rings the bell, or forces the table into a state S that is a perfect pth power. If n is prime, then p equals n, and any state that is a perfect pth power is also monotone. In this case, therefore, the strategy from Lemma 3 always rings the bell, and we are done.

On the other hand, if n is not prime, then p is a proper divisor of n, and hence Lemma 2 applies. The strategy from Lemma 2 will take the Player from the state S to a state that rings the bell. We only need to check that the Player will have enough hands to implement the Lemma 2 strategy. This boils down to checking that the inequality

$$\mathbf{1} \qquad\qquad \left(1 - \frac{1}{p}\right)n \geq pf(n/p)$$

holds. To see this, we will invoke our inductive hypothesis for the table size n/p. Let q be the largest prime factor of n/p, and hence the second largest prime factor of n. From the inductive hypothesis, we can conclude that

$$\mathbf{2} \qquad\qquad f(n/p) \leq \left(1 - \frac{1}{q}\right)\left(\frac{n}{p}\right).$$

Since $q \leq p$, Inequality **2** implies Inequality **1**, and thus our Player will have enough hands to use the Lemma 2 strategy.

The rest of this paper is devoted to the proof of Lemma 3. It turns out that this proof splits into two somewhat different cases. If the largest prime factor of n is at least 3, then a relatively simple strategy called the up-down strategy will work; this is discussed in Section 4. In Section 5, we will turn to the trickier up-flip strategy that is needed when n is a power of 2.

Section 4
The Up-Down Strategy

In this Section, we will give the proof of Lemma 3 for all table sizes n other than powers of 2. To develop some intuition for the general case, we will start by handling the particular case of $n = 12$. The largest prime factor of 12 is 3. Therefore, Lemma 3 for the case $n = 12$ makes the following claim: there exists a strategy that an 8-handed Player could employ against a 12-cornered Table that either rings the bell or forces the table into some state that is a perfect cube.

When $n = 12$, the table is a dodecagon. To understand the strategy that we are about to construct, that dodecagon should be viewed as composed of two disjoint, interleaved hexagons, and each of the hexagons should be viewed as two disjoint, interleaved triangles. Figure 3a shows a dodecagon with its vertices numbered like those of an ordinary, 12-hour clock face; in somewhat different language, Figure 3a could be described as a picture of the *clock necklace*

$$C = \langle (1)(2)(3)(4)(5)(6)(7)(8)(9)(10)(11)(12) \rangle.$$

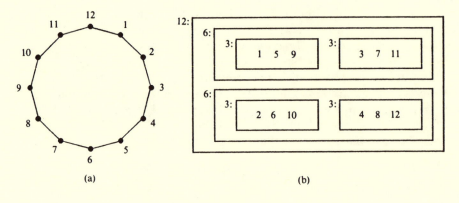

FIGURE 3

(a) The clock necklace, and (b) its structural diagram.

In Figure 3b, the clock necklace C has been redrawn to reflect the decomposition of the dodecagon into interleaved hexagons and triangles. Each rectangle in Figure 3b represents the regular subpolygon of the clock face whose vertices it encloses; at the upper left corner of each rectangle is a label that gives the number of enclosed vertices. We will call such pictures *structural diagrams*. We can draw a similar structural diagram for any necklace of length 12, in particular, for the states and probe patterns that arise as we construct our strategy.

Our strategy will use three different probe patterns, called P_1, P_2, and P_3. Figure 4 depicts the structural diagrams that these probe patterns should have. Pattern P_1 uses six hands to probe precisely one of the two hexagons; pattern P_2 uses six hands also, but probes precisely one of the triangles from each of the two hexagons; and P_3 uses eight hands to probe precisely two of the vertices of each of the four triangles. Furthermore, these insights about what vertices each pattern will probe are valid regardless of how the table rotates. We don't know which of the two hexagons P_1 will probe, but it will definitely probe one of them; the same is true for P_2 and P_3. It is this invariance under rotation that makes structural diagrams useful.

No matter what structural diagram we draw, there will always be at least one necklace that has that structural diagram, but there may be more than one. In the cases of P_1 and P_2 there is exactly one; we must choose $P_1 = \langle ghghghghghgh \rangle$ and $P_2 = \langle gghhgghhgghh \rangle$ if they are to have the structural diagrams in Figure 4. But we have some choice in the case of P_3. The two different necklaces $\langle gggghhhhhhhh \rangle$ and $\langle ghhghhghhghh \rangle$ both have the structural diagram that we require for P_3; and in fact, there are six other necklaces that we won't bother to mention that also have that structural diagram. It turns out that we can choose the P_i's to be any probe

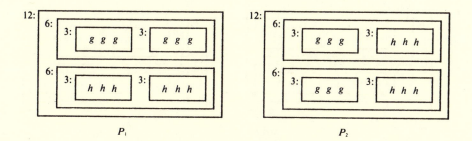

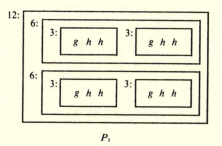

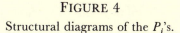

FIGURE 4

Structural diagrams of the P_i's.

patterns that have the correct structural diagrams; precisely which ones we choose doesn't matter.

Before exploring what these probe patterns can do, we pause to extend our notation for states. Recall that the exact state of the table is represented by a necklace of length n over the alphabet $\{u,d\}$. We can also use necklaces to represent partial information about the state of the table, by allowing the symbol e to appear as well: the symbol e for "either" will denote a position that contains either an up glass or a down glass. For example, the initial state of the table S_1 has the formula $S_1 = \langle e^n \rangle$, since the Table gets to choose this initial state arbitrarily.

For the first three probes of our strategy, we instruct the Player to probe with the patterns P_1, P_2, and P_3, in that order, turning up every glass that she feels, regardless of its initial state. The effects of these probes can be seen by examining Figure 5, which depicts the structural diagrams of the states S_1 through S_4. The probe with pattern P_1 forces one of the hexagons to be monotone up in state S_2. The six hands in pattern P_2 probe one triangle from each hexagon, and hence the second probe forces a third of the four triangles to be monotone up in state S_3. Pattern P_3 probes precisely two of the three vertices of every triangle, and hence the third probe turns up two of the three remaining e's. Thus, after the first three probes, the state S_4 of the table will be $S_4 = \langle eu^{11} \rangle$.

It might happen that, after one of these three probes, the remaining e's all happen to represent glasses that are actually up. If so, the bell will

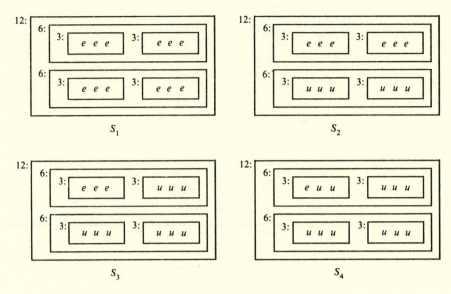

FIGURE 5

Structural diagrams of S_1 through S_4.

ring, and the Player will have won. We might as well ignore this case, therefore, and concentrate instead on the case where the final remaining e represents a down glass, so that the state S_4 is actually $S_4 = \langle du^{11} \rangle$.

We will refer to the portion of the strategy that we have discussed so far as the *first pass*. The net effect of the three probes in the first pass is to force the table into a state that is only one position away from being monotone up. We will refer to this lone down glass as the *exceptional glass*, and we will also use the term *exceptional* to refer to the triangle that contains it.

Where are we heading, anyway? We want to force the table into a state that is a perfect cube. We can do that in, at most, three more probes which we will call the probes of the *second pass*. These probes will also use the patterns P_1, P_2, and P_3, in that order. First, let us consider probing S_4 with P_1; recall that P_1 probes one of the two hexagons. If the table spins so that the Player feels the exceptional glass, the Player can merely turn that glass up, and the bell will ring. Hence, we might as well assume that the Player's hands probe the hexagon that is currently monotone up. We instruct the Player to turn this entire hexagon down; the structural diagram of the resulting state S_5 is shown in Figure 6.

On to the second probe of the second pass. Recall that P_2 probes one triangle from each hexagon. If one of the triangles that the Player feels is nonmonotone, it must be the exceptional triangle. In this case, we instruct the Player to turn the exceptional glass up, which makes every triangle in the resulting state monotone; this is equivalent to saying that the resulting state is a perfect cube. Since that is our goal, we can tell the Player to omit

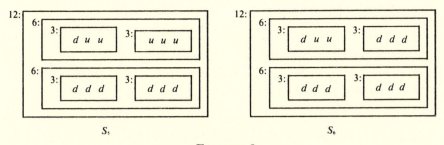

FIGURE 6

Structural diagrams of S_5 and S_6.

the rest of the second pass. On the other hand, it might happen that the Player feels only monotone triangles. One of them will be monotone down, and the other monotone up; we instruct the Player to leave the down triangle alone, and to turn the up triangle down. This results in the state S_6 whose structural diagram is shown in Figure 6.

We have now come to the third and final probe of the second pass, a probe using pattern P_3. Recall that P_3 probes two of the vertices of each of the four triangles. Currently, the exceptional triangle is in state $\langle duu \rangle$, while all of the nonexceptional triangles are monotone down. Note that the Player cannot get confused about which triangle is which. If the two vertices that she feels from a triangle are both down, then that triangle must be monotone down; we instruct the Player to leave those triangles alone. The Player also has two hands in the exceptional triangle. Those two hands feel either both of the up glasses, or one up glass and one down glass. In the former case, we turn both up glasses down; in the latter, we turn the down glass up. The net effect is to guarantee that the exceptional triangle becomes monotone in the next state S_7. But all of the other triangles in S_7 will also be monotone, in fact, monotone down. Thus, the Player has forced the table into a state that is a perfect cube, as demanded by Lemma 3.

We will call the strategy that we have just constructed the *up-down strategy*, since the first pass turns every glass but the exceptional one up, while the second pass turns every triangle but the exceptional one down. Of course, we have to prove Lemma 3 for all n, not just for $n = 12$. It turns out that we can handle all values of n other than powers of 2 by the up-down strategy. The only confusing thing is the notation. On the other hand, the up-down strategy fails for powers of 2, so we postpone our discussion of those cases until Section 5.

PROOF OF LEMMA 3 WHEN n IS NOT A POWER OF 2   Let n denote the size of the table, and let p denote the largest prime factor of n; since n is not a power of 2, we have $p \geq 3$. We are to construct a strategy for the Player with the following properties:

it either rings the bell or forces the table into a state S that is a perfect pth power, and it uses no more than $(1 - 1/p)n$ hands.

We begin by setting up some notation. Suppose that n has j prime factors, of which p is the largest. If we put the prime factors of n into increasing order, we can write

$$n = p_1 p_2 \cdots p_j \quad \text{for} \quad j \geq 1, \, p_1 \leq p_2 \leq \cdots \leq p_j = p.$$

From this factorization of n, we define some ancillary notations, l_i and r_i; the value of l_i is the product of the first up through but not including the ith prime factor of n, while r_i is the product of the ith through and including the jth. More formally, for $1 \leq i \leq j + 1$, we define the l's and r's by

$$l_i = \prod_{1 \leq k < i} p_k \quad \text{and} \quad r_i = \prod_{i \leq k \leq j} p_k.$$

Note that, for all i, we have $l_i r_i = n$.

In order to understand the general case of the up-down strategy, one has to view the n-gon in a fairly complicated way. If we start at the outside, we have the n-gon, or equivalently, the r_1-gon, composed of p_1 interleaved r_2-gons; each of the r_2-gons is composed of p_2 interleaved r_3-gons; and in general, each r_i-gon is composed of p_i interleaved r_{i+1}-gons. Finally, each r_j-gon is composed of p_j vertices; that is to say, r_j equals p_j. The p_j-gons are at the bottom of the hierarchy, and will play the role that the triangles played in the $n = 12$ case.

The previous paragraph demonstrates the meaning of the r_i's: they give the sizes of the various polygons into which we are breaking up the n-gon. The l_i's play the complimentary role of counting how many r_i-gons there are. In particular, for each i, the n-gon contains precisely l_i distinct r_i-gons.

We can now get a reasonable grip on what a structural diagram looks like in the general case. The rectangles are nested j levels deep, and, at each level of the nesting, the l_i distinct r_i-gons are each broken up into p_i distinct r_{i+1}-gons. Figure 7 is an attempt to show, in the general case, what the ith level of nesting of a structural diagram looks like.

The up-down strategy in the general case will use j different probe patterns, which we will call $P_1, P_2, \ldots, P_j$. As in the case of $n = 12$, it won't matter which probe patterns the P_i's are, as long as they have the correct structural diagrams. We can describe the structural diagram that P_i is required to have in terms of the ith level of nesting, shown in Figure 7. In particular, the pattern P_i should probe all but one of the r_{i+1}-gons that constitute each r_i-gon. We will abbreviate this condition by saying that such a pattern *resolves the ith level of nesting*.

We now choose the patterns P_i arbitrarily, subject to the condition that P_i must resolve the ith level of nesting. The next thing to check is that

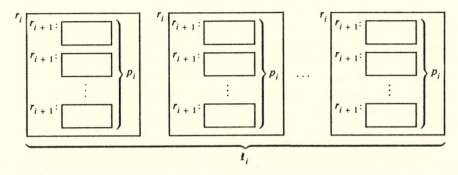

FIGURE 7

The ith level of nesting of a structural diagram in the general case.

our Player will have enough hands to use the probe patterns P_i that we have chosen. To see this, note that any pattern that resolves the ith level of nesting uses exactly

$$l_i(p_i - 1)r_{i+1} = \left(1 - \frac{1}{p_i}\right)n$$

hands. Since our Player is allowed $(1 - 1/p)n$ hands where $p = p_j \geq p_i$, she will have enough hands to use the patterns P_i.

We now advise our Player to execute the first pass, that is, to probe with each of the patterns from P_1 to P_j in order, turning up every glass that she feels. The pattern P_1 probes the entire table, an r_1-gon, except for a single r_2-gon; then, pattern P_2 probes all of each r_2-gon except for one of the constituent r_3-gons. This process continues, with pattern P_i further restricting the portion of the table that can still contain down glasses to a single r_{i+1}-gon. Therefore, after probing with P_j, the table can include, at most, one down glass. We might as well assume that there is exactly one down glass, or else our Player has already won. We will use the term *exceptional* to refer to this down glass, and to the p_j-gon that contains it. The first pass has effected the "up" portion of the up-down strategy: it has turned all but the exceptional glass up.

It follows from our choice of the patterns P_i that P_i probes all of the vertices of any r_{i+1}-gon of which it probes any of the vertices. This implies that P_i must in fact probe either all of or none of each r_k-gon for any $k > i$. This fact helps to explain our choice of terminology; in order to examine only part of an r_k-gon; that is, to resolve the kth level of nesting, one needs to use a pattern P_i where $i \geq k$. In particular, we note that all of the probe patterns P_i except for the last one, P_j, must probe either all of or none of each r_j-gon—that is to say—each p_j-gon. We could rephrase this observation by saying that the patterns P_i for $i < j$ are perfect p_jth powers.

On to the second pass. We advise our Player to run through the patterns $P_1, P_2, \ldots, P_j$ again, in order, with somewhat more complex instructions. Think of the table as composed of interleaved p_j-gons. When we start the second pass, all of the p_j-gons are monotone up except for the exceptional one, which includes a single down glass. And, until the last probe of the pass, we will be probing all of any p_j-gon that we probe any part of. We instruct our Player as follows, for all but the last probe. If you feel the exceptional p_j-gon, turn the exceptional glass from down to up. This puts the table into a state that is a perfect p_jth power, so you have satisfied the demands of Lemma 3; omit the rest of the second pass. If you feel a p_j-gon that is monotone up, it is nonexceptional, and it hasn't yet been reversed; turn all of the glasses in such a p_j-gon down. If you feel a p_j-gon that is monotone down, it has already been reversed; leave it alone.

We can show that only one p_j-gon can escape being probed during the first $j - 1$ probes of the second pass. The argument is essentially the same as the argument that we used to analyze the first pass. The probe with P_1 will hit all of the p_j-gons except for those lying in a particular r_2-gon; the probe with P_2 will hit all of the rest except for those lying in a particular r_3-gon; and so on. By the aftermath of the probe with P_{j-1}, all of the p_j-gons will have been probed except for those lying in a particular r_j-gon, and $r_j = p_j$. Furthermore, we can assume that the exceptional p_j-gon escapes being probed; if our Player was lucky enough to probe the exceptional p_j-gon, she was instructed to correct the exception and quit. Therefore, the first $j - 1$ probes in the second pass probe all of the non-exceptional p_j-gons, and turn them all down. By doing so, they accomplish the "down" phase of the up-down strategy.

Going into the last probe of the second pass, then, the table contains one exceptional p_j-gon which has $p_j - 1$ up glasses and one down glass, while all of the other p_j-gons are monotone down. For this last probe, we instruct our Player to use pattern P_j, which probes $p_j - 1$ of the vertices of each p_j-gon. And here we reach a subtle point. So far, we have not needed to use the assumption that $p_j = p \geq 3$, but now that assumption becomes crucial. Note that, because $p_j \geq 3$, the Player will be feeling at least two glasses from each p_j-gon, and hence she must feel at least one of the up glasses of the exceptional p_j-gon. This allows her to figure out which of her groups of $p_j - 1$ hands is probing the exceptional p_j-gon. We have to outlaw the case of $p_j = 2$ at this point precisely because this argument fails. If $p_j = 2$, the exceptional p_j-gon is a diagonal in the state $\langle du \rangle$; if the Player probes just one end of this diagonal, she might probe the down end, and then she would not be able to distinguish that down glass from all the other down glasses.

But for now, we are assuming that $p_j \geq 3$, so the Player will be able to locate the exceptional p_j-gon. We instruct our Player to leave all of the non-exceptional p_j-gons alone. For the exceptional p_j-gon, there are two cases. Either she will feel the lone down glass, in which case she should turn it up; or she will feel all $p_j - 1$ of the up glasses, in which case she should turn them all down. In either case, the exceptional p_j-gon will become monotone in the next state, and hence the demands of Lemma 3 will be satisfied. This completes the proof of Lemma 3 for every integer n other than powers of 2.

Section 5
The Up-Flip Strategy

Our remaining task is to construct the strategy implied by Lemma 3 when the size of the table is a power of 2. As we noted above, the up-down strategy comes to grief in this case, because the Player might not be able to locate the exceptional diagonal during the last probe of the second pass. In this Section, we will construct a strategy for the power of 2 case called the *up-flip strategy*, which is similar to the up-down strategy but involves a trickier second pass. *Flipping* a glass means turning it up if it is currently down, or down if it is currently up. During the second pass of the up-flip strategy, we will instruct our Player on each probe to flip every non-exceptional diagonal that she feels rather than to turn them all down. The repeated flips put the table into states that have lots of up and down glasses arranged in interesting ways, and it turns out that a careful Player can take advantage of the resulting structure. To prepare for the subtleties of the up-flip strategy, we need to study some facts about periods in circular strings.

For any integer k and necklace T, we will say that T is *k-periodic*, or that k is a *period* of T, if and only if each symbol in T is the same as the symbol k steps to its right around T. Putting it another way, a necklace is k-periodic when it looks exactly the same after being rotated k positions to the right. Every necklace is 0-periodic; if a necklace is l-periodic, then it must also be $(-l)$-periodic; and, if a necklace is both l-periodic and m-periodic, it must also be $(l + m)$-periodic. In fancier language, the set of all periods of a necklace T forms a subgroup of the integers under addition.

Since a necklace of length n is always n-periodic, every necklace has some positive period. We will refer to the smallest positive period of a necklace as its *basic period*. We then have the following result.

LEMMA 4 The periods of any necklace are exactly the multiples of its basic period.

PROOF Certainly any multiple of the basic period will be a period. To see the other implication, let k be the basic period of a necklace, let m be some other period, and let q denote the greatest common divisor of k and m. Recall that q can be written in the form $q = ak + bm$ for some integers a and b; hence, q must also be a period of the necklace. Since k, the basic period, is the smallest positive period of the necklace, we must have $q = k$. And since q divides m, we conclude that m must be a multiple of the basic period. Readers with algebraic background will recognize this as the proof that the integers form a principal ideal domain.

The notion of a period is pretty standard; in fact, it is closely connected to the idea of perfect powers. Note that, if a necklace of length n has basic period k, then k divides n, and the necklace is a perfect (n/k)-th power. But we need another concept as well, which is more exotic. This other notion only makes sense for necklaces over an alphabet with exactly two symbols; in our case, these two symbols will be u and d. We will call each of these symbols the *opposite* or *complement* of the other. Then, we will say that a necklace is k-*alternating* if and only if each symbol in the necklace is the opposite of the symbol k steps to its right. That is, rotating the necklace k steps to the right gives the same result as complementing every symbol and not rotating at all.

If a necklace is k-alternating, it must also be $2k$-periodic, because complementing something twice is the same as not complementing it. It can happen, however, that the necklace also has some periods smaller than $2k$. For example, consider the necklace $\langle 010101 \rangle$ of length six. This necklace is 3-alternating and 6-periodic; but it happens to be 2-periodic as well. Our next Lemma states that these small periods cannot arise if k is a power of 2. This result probably seems irrelevant at the moment; be patient.

LEMMA 5 If a necklace is 2^i-alternating, then its basic period is precisely 2^{i+1}; putting it another way, a 2^i-alternating necklace has no non-trivial periods.

PROOF Certainly 2^{i+1} will be a period; thus the basic period of the necklace must divide 2^{i+1}, so it must be a power of 2. But if the basic period were 2^l for some $l \leq i$, then 2^i would also have to be a period of the necklace. No necklace can be both 2^i-periodic and 2^i-alternating. Therefore, the basic period must be precisely 2^{i+1}.

With Lemma 5 out of the way, we are ready to start constructing the up-flip strategy.

PROOF OF LEMMA 3 WHEN n IS A POWER OF 2 Suppose that our Player is facing a table of size $n = 2^j$. We are to construct a strategy that uses no more than $n/2$ hands, and that either rings the bell or forces the table into a state that is a perfect square. If

$n = 2$, there is a trivial one probe strategy that suffices. The case of $n = 4$ is not that trivial, but there is a four probe strategy that does the job: just take the first four moves of the solution to the original Rotating Table Problem given in Martin Gardner's column of March, 1979. Therefore, we can restrict our consideration here to those cases where $n \geq 8$, or equivalently, where $j \geq 3$. It is somewhat surprising that the case of $n = 4$ has to be treated specially, but it seems to be necessary; the up-flip strategy that we are about to construct just doesn't work for a square table.

The first difference between the up-down and the up-flip strategies arises in the definition of the j probe patterns $P_1, P_2, \ldots, P_j$. For the up-down strategy, we were content to let these patterns be chosen arbitrarily, subject only to the condition that P_i resolve the ith level of nesting. Here, we will choose a particular sequence of patterns that possesses this property, but also has a simple structure. In detail, we will specify that the pattern P_i consist of 2^{j-i} blocks of 2^{i-1} consecutive hands separated by 2^{j-i} blocks of 2^{i-1} consecutive gaps. More formally, we can define the necklace P_i by the relation

1
$$P_i = \langle (g^{2^{i-1}} h^{2^{i-1}})^{2^{j-i}} \rangle.$$

The pattern P_1 is made up of alternating g's and h's, while the pattern P_j probes one semicircle and leaves the other unprobed. Consider the kth hand in each block of h's in P_i, for some k in the range $1 \leq k \leq 2^{i-1}$. These 2^{j-i} hands together probe one of the two 2^{j-i}-gons that constitute a 2^{j-i+1}-gon; furthermore, the other constituent 2^{j-i}-gon goes unprobed, since it falls under the kth gap in each block of g's in P_i. This demonstrates that the probe pattern P_i defined by Equation **1** actually does resolve the ith level of nesting.

The first pass of the up-flip strategy is the same as that of the up-down strategy: we instruct our Player to probe each of the patterns P_i, and to turn up every glass that she feels. By the same analysis as in the up-down case, we know that our Player will either win the entire Game at some point during this pass, or else, at the end of this pass, the table will have one down glass and $(n-1)$ up glasses. Call the lone down glass the *exceptional glass*, and the diagonal that contains it the *exceptional* diagonal. Since all of the fun is going to occur during the second pass in this argument, we will depart from our usual subscripting convention, and use the symbol S_1 to denote the state of the table at the beginning of the second pass. The first pass forces the table into the state

$$S_1 = \langle du^{2^{j-1}} \rangle.$$

The first probe of the second pass is also the same as it used to be. We instruct our Player to probe with pattern P_1. If she feels a down glass, it must be the exceptional glass; she can turn it up and win at once. So we can assume that she feels only up glasses; we instruct her to turn them all down. This forces the table into the state

2 $$S_2 = \langle dd(ud)^{2^{j-1}-1} \rangle.$$

Before our Player goes further in the second pass, we instruct her to find a piece of scrap paper upon which she can keep notes about the state of the table. She begins these notes with the state S_2 given in Equation **2**. With this paper at her side, we can give our Player a set of instructions that will guide her from the second probe up until, but not including, the last probe of the second pass. These instructions are designed to maintain the validity of certain conditions. Let S_i denote the state of the table immediately before the ith probe of the second pass for $2 \leq i \leq j$; we will maintain the following conditions:

3 $\Bigg\{$ The state S_i differs in exactly one position from a necklace T_i that is 2^{i-2}-alternating.

4 $\Bigg\{$ Just after the $(i-1)$st probe of the second pass, our Player can write the state S_i on her scrap paper.

Both of these conditions are satisfied for $i = 2$, since the state S_2 is given explicitly by Equation **2**, and it does differ in only the exceptional place from a 1-alternating necklace. Our task during the interior of the second pass is to arrange that these conditions continue to hold. Let us consider the ith probe of the second pass for $2 \leq i < j$, where we assume that the conditions have held so far. For this probe, we instruct our Player to use the probe pattern P_i. Since $i < j$, we can deduce by the same reasoning as in the up-down case that the pattern P_i is a perfect square; that is, that it probes both ends of any diagonal of which it probes either end. We first ask our Player to classify the diagonals that she finds.

By Condition **3**, the state S_i of the table that she is probing differs in only the exceptional place from a necklace T_i that is 2^{i-2}-alternating. The necklace T_i must also be 2^{i-1}-periodic, and thus, since $i < j$, it must be at least a perfect fourth power. We can conclude that all of the diagonals of S_i other than the exceptional diagonal must be monotone. Hence, our Player can determine at once if she is probing the exceptional diagonal. If she is, we instruct her to make it monotone, by either turning its down glass up or its up glass down; it doesn't matter. This will put the table into a state that is a perfect square, and thus satisfy the demands of Lemma 3; the rest of the second pass should be omitted.

Suppose instead that our Player does not probe the exceptional diagonal; that is to say, she probes only monotone diagonals. In the up-down strategy, we instructed her to turn down all of the monotone p_j-gons that she felt; but in this strategy, we instruct her to turn the monotone up diagonals down, and the monotone down diagonals up, that is, to flip each monotone diagonal that she feels. This explains the name of the up-flip strategy, but what is it good for? Since S_i was 2^{i-2}-alternating except for the exceptional glass, it was also 2^{i-1}-periodic except for the exceptional glass. The effect of flipping every glass that can be felt with P_i is to complement and leave alone alternating blocks of length 2^{i-1}. Therefore, this flipping will guarantee that the next state S_{i+1} is 2^{i-1}-alternating except for the exceptional glass. That is, the flipping guarantees the truth of Condition **3** at time $i + 1$.

But what about Condition **4**? Recall that we are assuming that our Player does not probe the exceptional diagonal, and hence, that she flips every glass that she feels. Since the state S_i is already written on her scrap paper, it is enough to show that she can tell without ambiguity which of the glasses in S_i she is probing, that is, where her hands are located around the table. Consider first what the situation would be if she were probing the necklace T_i with pattern P_i. The results of the probes by each block of 2^{i-1} hands would have the form $X\bar{X}$ for some string X of length 2^{i-2}, because the necklace T_i is 2^{i-2}-alternating. Now, suppose that the string $X\bar{X}$ actually appears in two different places around T_i, separated by a rotation through l positions. Since T_i is also 2^{i-1}-periodic, the entire necklace T_i must then match itself when rotated l steps; that is, l must be a period of T_i.

Finally, we can see the relevance of Lemma 5; from that result, we deduce that 2^{i-1} must actually be the basic period of T_i, and hence that l must be a multiple of 2^{i-1}. Considering the structure of P_i, this implies that there are only two different ways in which the necklaces P_i and T_i could have been aligned that would have produced the probe results $X\bar{X}$. In fact, T_i will consist precisely of 2^{j-i+1} copies of the string $X\bar{X}$, and the Players' hands must be probing either all of the odd or all of the even copies. Any other possibility would force T_i to have a nontrivial period, contrary to Lemma 5. But our Player is really probing S_i, not T_i. Since we are assuming that our Player does not feel the exceptional glass, one of the two possible orientations for her hands around S_i is eliminated; and this only leaves one possibility. Therefore, our Player will be able to tell from the results of the ith probe precisely where her hands are positioned around the necklace S_i. With this information, she can easily compute the necklace S_{i+1}, and write it on her scrap paper.

We have shown that our Player can maintain Conditions **3** and **4** throughout the interior of the second pass, the "flip" phase of the up-flip strategy. We turn now to the last probe of the second pass, where she will be probing one semicircle. First, we want to show that our Player will be able to tell whether or not she is feeling the exceptional glass. This time, she cannot distinguish the exceptional diagonal by looking for a diagonal that is non-monotone, since she is only probing one end of each diagonal. But from Condition **3** for $i = j$, we know that the state S_j of the table on input to the jth probe of the second pass will differ in only one position from a 2^{j-2}-alternating necklace T_j. Therefore, our Player can determine whether or not she is feeling the exceptional glass by checking to see whether or not the results of her probes have the form $X\bar{X}$ for some string X of length 2^{j-2}. In particular, if she misses the exceptional glass, then her probe results will have the form $X\bar{X}$; but if she is feeling the exceptional glass, her probe results will have the form $Y d Z \bar{Y} d \bar{Z}$, where one of the two d's is the exceptional glass.

Suppose that our Player does feel the exceptional glass. Her next task is to determine which of the d's is the exception, so that it can be corrected. And it is here that our assumption that $j \geq 3$ is critical. If we try and use the up-flip strategy for a square table, we will be faced at this point with trying to decide which of the d's in the block dd is the exceptional one and there is just no way to tell. But we have already handled the case of $n = 4$ by a special argument. Therefore, we can conclude that at least one of the two strings Y and Z is non-empty, and hence that our Player can actually feel some neighborhood of each of the two current candidates for the exceptional glass. Note that those neighborhoods will be complementary. Now, by Condition **4**, we know that our Player's piece of scrap paper has the exact necklace S_j written on it. From this necklace, it is easy to deduce what states the left and right neighbors of the exceptional glass are in. This observation allows our Players to determine which of the current candidates is actually the exceptional glass. We instruct her to flip just that glass. This forces the exceptional diagonal into the monotone state $\langle uu \rangle$, and hence satisfies the demands of Lemma 3.

There is only one more case to handle. We now suppose that, on the last probe of the second pass, our Player does not feel the exceptional glass; that is, the results of her probes take the form $X\bar{X}$. In this case, the necklace T_j from which the state S_j differs in only the exceptional place must be given by the formula $T_j = \langle X \bar{X} X \bar{X} \rangle$. Arguing from Lemma 5 again, we can show that the string $X\bar{X}$ will appear as a substring of T_j only in the two obvious places. Hence, it only appears once as a substring of S_j; this means that our Player will be able to determine uniquely the location of her hands around S_j. Since she is probing a semicircle, and she doesn't feel the exceptional glass, she must be probing the other vertex of the exceptional diagonal. We instruct our Player to look at her scrap paper, and figure out which of the up glasses that she can feel is the one diagonally opposite the exceptional glass. Turning just this up glass down will force the exceptional diagonal into the monotone state $\langle dd \rangle$, and thus satisfy the demands of Lemma 3.

We have completely determined the values of the function $f(n)$. But this does not mean that we can close the book on the Problem of the Rotating Table. On the contrary, we mentioned in the Introduction several other generalizations of the problem that are worthy of study, and even the variant that we have been considering might repay further analysis. For example, one might be interested in constructing short winning strategies for a k-handed Player, where k either equals or exceeds $f(n)$.

The strategy-building techniques that we have discussed can give us some preliminary feel for the lengths of one family of strategies. In particular, let $n = a_1 a_2 \ldots a_j$ be a factorization of n in which each factor a_i satisfies $a_i \geq 2$; suppose for convenience that the a_i have been numbered so that all of the 2's, if any, come first. By using the up-down strategy if $a_j > 2$, or the up-flip strategy if $a_j = 2$, we can reduce the strategy-building problem of size n to the problem of size n/a_j in at most $2j$ probes, using no more than

$$K = \max_{1 \leq i \leq j} \left(1 - \frac{1}{a_i} \right) n$$

hands. Therefore, we can construct a winning strategy for a size n table that uses only K hands and takes at most $j^2 + j$ probes all told. There is a trade-off here: either we can make the a_i's large, which makes j small and the resulting strategy short, but demands lots of hands; or we can make the a_i's small, which makes j large and the strategy long, but gets by with fewer hands. Furthermore, it is relatively easy to shorten these strategies at least a little. For example, it is often possible to save probes by carrying over information from one up-down stage of a strategy to the next. But are there any strategies that take substantially fewer probes in the worst case than those that we have constructed?

Note added in Proof: Ted Lewis and Stephen Willard determined $f(n)$ independently, and they have published their results in the article *The Rotating Table* in the May, 1980 issue of *Mathematics Magazine* (volume 53, number 3, pages 174–179). When n is not a power of 2, they employ a refinement of the up-down strategy in which carrying information over from one up-down stage to the next allows the Player to get by with roughly half as many probes. Furthermore, they discovered a close relative of the up-down strategy that handles the cases in which n is a power of 2: The first pass turns glasses up, and most of the second pass turns them down. Flipping is confined to the next to the last probe of the second pass, which flips a single diagonal.

Numbers and Coding Theory

Supernatural Numbers

Donald E. Knuth
STANFORD UNIVERSITY

"God," said Leopold Kronecker [10], "made the integers; everything else is the work of man." If Kronecker was right, it would be heresy to claim that any noninteger numbers are *supernatural* in the sense that they have miraculous powers. On the other hand, mathematicians generally refer to the nonnegative integers $\{0, 1, 2, \ldots\}$ as the set of *natural* numbers; therefore if any numbers are supernatural, they are also natural. The purpose of this essay is to discuss the representation of natural numbers that are "super" in the sense that they are extremely large; many superscripts are needed to express them in conventional notation.

It doesn't take a very big number to transcend the size of the known universe. For example, if we consider a cube that's 40 billion light years long on each edge, and if we pack that cube with tiny little 10^{-13} cm $\times 10^{-13}$ cm $\times 10^{-13}$ cm cubes (so that each tiny cube is much smaller than a proton or neutron), the total number of little cubes is less than 10^{125}. This quantity isn't really humungous; it has only 125 digits.

The great Archimedes seems to have been the first person to discuss the existence of extremely large numbers; his famous essay "The Sand-Reckoner" [1] concludes by demonstrating that fewer than 10^{63} grains of sand would be enough to fill the universe as defined in his day. Furthermore he introduced a system of nomenclature by which he could speak of numbers up to $10^{80,000,000,000,000,000}$.

The English language doesn't have any names for such large quantities, and it is instructive to consider how we could provide them systematically. In these inflationary days, we may soon need new words to express prices; during 1923, for example, Germany issued postage stamps worth 50 billion marks each, but these stamps were almost valueless.

When we stop to examine our conventional names for numbers, it is immediately apparent that these names are "Menschenwerk"; they could have been designed much better. For example, it would be better to forget about thousands entirely, and to make a *myriad* (10^4) the next unit after hundreds. After all, numbers like 1984 are conventionally read as "nineteen hundred eighty-four," not as "one thousand nine hundred eighty-four."* The use of myriads would provide us with decent names for everything up to 10^8. For example, the prime number 9999,9989 would be called "ninety-nine hundred ninety-nine myriad ninety-nine hundred eighty-nine."

The next unit we need after myriads is 10^8. Let's agree to call this a myllion (pronounced mile-yun); similar words like myllionaire will, of course, also be of use. After myllions come byllions; one byllion equals 10^{16}. It clearly is an improvement to have the unambiguous word byllion for this next step, since people have never been able to agree about what a "billion" is. (English and German people think it is a million million, while Americans and Frenchmen think it is a thousand million.) Note also the comforting fact that our national debt is only a small fraction of a byllion dollars. The nomenclature should now continue as follows:

10^{32}	tryllion	10^{2048}	nonyllion	10^{131072}	quindecyllion
10^{64}	quadryllion	10^{4096}	decyllion	10^{262144}	sexdecyllion
10^{128}	quintyllion	10^{8192}	undecyllion	10^{524288}	septendecyllion
10^{256}	sextyllion	10^{16384}	duodecyllion	$10^{1048576}$	octodecyllion
10^{512}	septyllion	10^{32768}	tredecyllion	$10^{2097152}$	novemdecyllion
10^{1024}	octyllion	10^{65536}	quattuor-decyllion	$10^{4194304}$	vigintyllion

For example, the total number of ways to shuffle a pack of cards is 52! = 8065::8175,1709;4387,8571:6606,3685;6403,7669;;7528,9505;4408, 8327:7824,0000;0000,0000, if we use punctuation marks to group the digits by fours, eights, sixteens, and so on. The English name for this number under the proposed system would be "eighty hundred sixty-five quadryllion eighty-one hundred seventy-five myriad seventeen hundred nine myllion forty-three hundred eighty-seven myriad eighty-five hundred seventy-one byllion sixty-six hundred six myriad thirty-six hundred eighty-five myllion sixty-four hundred three myriad seventy-six hundred sixty-nine tryllion seventy-five hundred twenty-eight myriad ninety-five hundred five myllion forty-four hundred eight myriad eighty-three hundred twenty-seven byllion seventy-eight hundred twenty-four myriad myllion." This number is inexpressible in our present American language, unless one resorts to something like eighty thousand six hundred fifty-eight vigintillion one hundred seventy-five nonillion ... eight hundred twenty-four trillion," since unabridged dictionaries don't give names for any units past one vigintillion ($=10^{63}$) except centillion ($=10^{303}$).

Of course we need to go past vigintyllions and even centyllions if we are to have names in our language for all of God's creation. The next unit presumably should be called "unvigintyllion," if we extrapolate the above pattern and go further into Latin-like nomenclature. Thus, from the formula

$$80{,}000{,}000{,}000{,}000{,}000 = 2^{56} + 2^{52} + 2^{51} + 2^{50} + 2^{45} + 2^{44} + 2^{42}$$
$$+ 2^{41} + 2^{40} + 2^{39} + 2^{36} + 2^{33} + 2^{32} + 2^{30}$$
$$+ 2^{29} + 2^{28} + 2^{27} + 2^{26} + 2^{25} + 2^{19},$$

a reader who understands the proposed system will know the new name of Archimedes' last number $10^{80{,}000{,}000{,}000{,}000{,}000}$. (The answer appears at the end of this article.)

Sooner or later we will run out of Latin names to give to the units, since Romans never did count very high. Even if Roman scholars had adopted Archimedes' scheme, we would have trouble naming a unit like

$$10^{2^{10^{80{,}000{,}000{,}000{,}000{,}003}}}$$

However, we can get around this problem by simply giving each basic unit $10^{2^{n+2}}$ the name "latin{the name of n with spaces deleted}yllion" for all large n. For example, $10^{2^{10000000000000002}}$ would be "latinbyllionyllion." In this way we obtain English names for all the natural numbers, no matter how super they are. For example, one of the names would be "latinlatinlatinbyllionyllionyllionyllion"; can the reader deduce its magnitude? (See the end of this article for the answer.)

At this point the reader may be thinking, "So what? Who cares about names for gigantic numbers, when ordinary decimal or binary notation gives a simpler representation that is much more suited to calculations?" Well, it's quite true that number names are more interesting to the linguistic parts of our brains than to those little grey cells we use for calculation; the mere fact that a great man like Archimedes wrote about some topic doesn't necessarily prove that it has any scientific interest. Our discussion has not been completely pointless, however, because it has prepared us to talk about a more important problem whose solution involves somewhat similar concepts.

The problem we shall discuss in the remainder of this article is this: How can arbitrarily large natural numbers be represented as sequences of 0's and 1's so that

1 The representation of each number is never a "prefix" of the representation of any other number.

In other words, if some number is represented by the sequence "01101", no other number will have a representation that begins with "01101". Furthermore the representation should satisfy a second condition,

2 The representation of each number should be "lexicographically smaller" than the representation of every larger number.

Lexicographically smaller means that it would appear earlier in a dictionary; for example, 01101 is lexicographically smaller than all sequences beginning with "1" or with "0111" or with "011010" or "011011".

The importance of this problem is that we often want to communicate a *sequence* of numbers to a computer. For example, such a sequence might stand for instructions telling the computer what to do. The critical problem is that the computer must know where one number stops and the next one begins; condition **1** ensures that this will be the case, since the computer can read the sequence from left to right and there will be no ambiguity. Condition **2** is not quite so critical, but it is a nice property to have: it implies that the representation is *order-preserving* in the sense that, whenever one sequence of numbers is lexicographically smaller than another, the representation of the first sequence will also be lexicographically smaller than the representation of the second.

Conditions **1** and **2** are easy to achieve if we need to represent only a few numbers. For example, if we happen to know that only the numbers 0 through 7 will ever be required, we can make all the representations of the same length: 000, 001, 010, 011, 100, 101, 110, 111. But we want to be able to encode *all* natural numbers, not only the small ones.

Perhaps the first correct solution that comes to mind is a "unary" representation, where we represent 0 as "0", 1 as "10", 2 as "110", and so forth; the representation of n is "1^n0", that is, n ones followed by a single zero. The zero acts as an end-marker or "comma" between successive numbers that the computer must read. This representation clearly satisfies **1** and **2**, but it is hardly a practical way to represent even medium-size numbers, since it requires $n + 1$ bits to represent n. We really want to have a representation that is as concise as possible, subject to conditions **1** and **2**.

Perhaps the next simplest solution is based on the ordinary binary notation for integers, namely $\{0, 1, 10, 11, 100, 101, \ldots\}$. This representation as it stands doesn't satisfy either **1** or **2**, but we can improve it by prefixing the binary notation by a suitable sequence indicating its length, as follows:

$0 \rightarrow 00$	$8 \rightarrow 1110000$	$64 \rightarrow 1111110000000$
$1 \rightarrow 01$	$9 \rightarrow 1110001$	$127 \rightarrow 1111110111111$
$2 \rightarrow 100$	$10 \rightarrow 1110010$	$128 \rightarrow 111111100000000$
$3 \rightarrow 101$	$15 \rightarrow 1110111$	$255 \rightarrow 111111101111111$
$4 \rightarrow 11000$	$16 \rightarrow 111100000$	$256 \rightarrow 1111111000000000$
$5 \rightarrow 11001$	$31 \rightarrow 111101111$	$511 \rightarrow 1111111011111111$
$6 \rightarrow 11010$	$32 \rightarrow 11111000000$	$512 \rightarrow 11111111100000000000$
$7 \rightarrow 11011$	$63 \rightarrow 11111011111$	$1000 \rightarrow 1111111110111101000$

In general for $n \geq 2$, if the binary representation of n is 1α where α is any sequence of 0's and 1's, the new representation of n is

$$1^{|\alpha|} 0\alpha$$

where $|\alpha|$ denotes the *length* of α. If m bits are needed to write n in binary notation, its new representation has $2m - 1$ bits. We have roughly doubled the number of bits needed to represent n, in order to indicate unambiguously where each representation ends.

This solution can be further improved, but before we pursue the investigation any further it is interesting to point out that our problem is essentially equivalent to a *guessing game*, where one person has thought of an arbitrary natural number and the other person tries to guess it. The rules of this game are something like Twenty Questions: the one who guesses is only allowed to ask "Is your number less than n?" for nonnegative integers n, and the other player answers "yes" or "no." Of course the game might last longer than twenty questions, because it is impossible to distinguish between more than 2^{20} numbers based on the answers to only 20 yes-no questions, while infinitely many secret numbers are allowed. But the general idea is to guess the number as quickly as possible.

The relation between this guessing game and the sequence representation is not difficult to see. Any strategy that can be used by the guesser to deduce all natural numbers leads to a solution to **1** and **2**: We simply let the representation of n be the sequence of answers that would be given when the secret number is n, with 0 standing for "yes" and 1 for "no." Conversely, given a solution to **1** and **2** we can construct a guessing strategy that corresponds to it under this rule, if we also allow the guesser to ask the stupid question "Is your number less than infinity?" (The answer will always be "yes"; some solutions to **1** and **2** have every representation beginning with 0.)

The unary solution that we discussed first corresponds to the strategy under which the guesser simply asks

Is your number less than 1?
Is your number less than 2?
Is your number less than 3?

and so on, until receiving the first "yes." Our second solution is more clever: it begins with the questions

> Is your number less than 2?
> Is your number less than 4?
> Is your number less than 8?

and so on, then it uses a "binary search" to pinpoint the secret number once the first "yes" answer has revealed its order of magnitude.

Under this equivalence between the guessing game and the representation problem, a good strategy for the guesser corresponds to a concise representation for numbers; so our search for good representations boils down to the same thing as the search for a good number-guessing scheme.

Incidentally, the number-guessing game is not simply frivolous, it has important practical applications. For example, if a polynomial $f(x)$ is known to have exactly one positive root, where $f(0)$ is negative and $f(x)$ is positive for all sufficiently large x, the root is less than n if and only if $f(n)$ is positive; thus a good strategy for number guessing leads to an efficient procedure for root location without evaluating derivatives of f.

We can improve on the second solution above by realizing that it is essentially using the unary strategy to represent the length of n's binary representation; the binary strategy can be substituted instead! In other words, the sequence $1^{|\alpha|}0$ used to encode the length of α can be replaced by a more concise representation of $|\alpha|$. By using this idea repeatedly, we obtain progressively shorter and shorter representations for large numbers. And we are eventually led to what might be called a *recursive* strategy, of the following form: First guess the number of binary digits of n, then use binary search to determine the exact value of n. To guess the number of digits of n, the same strategy is used, recursively; thus we first guess the number of digits in the number of digits of n, by guessing the number of digits in the number of digits of n, and so on. The first questions in the recursive strategy are:

> Is your number less than 1?
> Is your number less than 2?
> Is your number less than 4?
> Is your number less than 16?
> Is your number less than 65536?

and so on, until the answer is "yes." (Note that $2 = 2^1$, $4 = 2^2$, $16 = 2^4$, and $65536 = 2^{16}$; the next question would refer to 2^{65536}.) The recursive strategy proceeds in this way until finding an upper bound on the secret number; then it proceeds to unwind the recursion. If the secret number is really enormous, the guesser will soon be guessing really enormous values.

The recursive representation corresponding to this recursive guessing scheme can be defined very simply: Let $R(n)$ be the sequence of 0's and 1's that represents n. Then

$$R(0) = 0; \text{ and}$$
$$R((1\alpha)_2) = 1\ R(|\alpha|)\alpha,$$

where $(1\alpha)_2$ is the number whose binary representation is 1α and $|\alpha|$ is the length of the sequence α. Small integers are represented as follows:

$0 \to 0$	$8 \to 11101000$	$64 \to 11110010000000$
$1 \to 10$	$9 \to 11101001$	$127 \to 11110010111111$
$2 \to 1100$	$10 \to 11101010$	$128 \to 111100110000000$
$3 \to 1101$	$15 \to 11101111$	$255 \to 111100111111111$
$4 \to 1110000$	$16 \to 111100000000$	$256 \to 11110100000000000$
$5 \to 1110001$	$31 \to 111100001111$	$511 \to 11110100011111111$
$6 \to 1110010$	$32 \to 1111000100000$	$512 \to 111101001000000000$
$7 \to 1110011$	$63 \to 1111000111111$	$1000 \to 111101001111101000$

It is easy to see that this representation satisfies **1** and **2**. Computer scientists interested in the transformation of recursive methods to iterative methods will enjoy finding a simple nonrecursive procedure that evaluates n, given its representation; a solution appears at the end of this article.

The recursive representation of a large number n will be about half as long as its representation under the binary scheme. For example, the representation of $2^{65536} - 1$ in the binary scheme is the sequence

$$1^{65535}01^{65535}$$

of length 131071, while its representation in the recursive scheme is

$$1^501^{65554},$$

only 65560 bits long. On the other hand, it takes quite awhile for the recursive scheme to show any payoff when n is small; the binary scheme produces a representation shorter than the recursive one for all values of n between 2 and 127 inclusive. (The recursive scheme wins out only when $n = 0$ or when n is 512 or more.) Thus the binary scheme will be preferable in many applications.

We can improve the recursive scheme's performance for small n by observing that all sequences except for the one representing 0 begin with 1; if we only want to represent strictly positive numbers, we can safely drop the initial "1". Furthermore the natural numbers are in one-to-one correspondence with the strictly positive integers, so there is a representation $Q(n)$ such that

$$1Q(n) = R(n + 1).$$

Under this modified recursive scheme we have

$$0 \to 0$$
$$1 \to 100$$
$$2 \to 101$$
$$3 \to 110000$$
$$4 \to 110001$$

and so on; the binary scheme beats this one only for $n = 1, 3, 4, 5, 6, 7$, and for $15 \leq n \leq 63$. The same transformation can be applied to the Q scheme, for the same reasons, yielding a P scheme where

$$1P(n) = Q(n + 1),$$

so that we have

$$0 \to 00$$
$$1 \to 01$$
$$2 \to 10000$$
$$3 \to 10001$$
$$4 \to 10010$$

and so on; now the binary scheme wins only when $n = 2, 3, 6, 7, 62, 63$, and when $14 \leq n \leq 31$.

In some sense we would like to find the *best possible* strategy, but it is hard to say exactly what that means. The binary strategy looks a lot better than the unary one, but even the unary strategy beats the binary one when $n = 0$. We will see below that the situation is inevitable: No strategy for the guessing game dominates another—in the sense that the first needs no more questions than the second does to determine n, for all n, and the first sometimes needs fewer questions—unless, of course, the dominated strategy asks stupid questions whose answer "yes" or "no" is already deducible from the previous answers. If one nonstupid strategy is better than another for some n, it will necessarily be worse for others. You win some, you lose some.

In order to analyze just how good these schemes are, we ought to be more quantitative. (Let the reader beware: The rest of this article involves mathematical technicalities that are elementary but sometimes a bit subtle.) If $n \geq 1$, let us write λn for the unique natural number such that $2^{\lambda n} \leq n < 2^{1 + \lambda n}$. Thus if $n = (1\alpha)_2$ in binary notation, the number λn is $|\alpha|$, the length of α. It is convenient to define $\lambda 0 = 0$, so that λn is a natural number whenever n is a natural number. We shall write $\lambda \lambda n$ for $\lambda(\lambda n)$, and so on; furthermore $\lambda^m n$ will stand for the m-fold repetition of the λ function, so that $\lambda^0 n = n$ and $\lambda^3 n = \lambda \lambda \lambda n$, etc. Finally, we also write $\lambda^* n$, to indicate the least integer m such that $\lambda^m n = 0$.

It is easy to express the length of the representation of n in terms of such functions. Let $c(n)$ be this length, that is, the "cost" of representing n. In the guessing game, $c(n)$ is the number of questions needed to determine a given secret number. The unary guessing strategy has a rather large cost,

$$c_U(n) = n + 1,$$

while the binary strategy reduces this to

$$c_B(n) = \begin{cases} 2, & \text{if } n = 0 \text{ or } n = 1; \\ 2\lambda n + 1, & \text{if } n > 1. \end{cases}$$

The cost of the recursive strategy is

$$c_R(n) = \lambda n + \lambda\lambda n + \lambda\lambda\lambda n + \cdots + \lambda^* n + 1,$$

where the infinite series implied by the "$\cdots$" is really finite, since $\lambda^m n = 0$ whenever $m \geq \lambda^* n$. Finally, the modified recursive strategies have costs

$$c_Q(n) = c_R(n+1) - 1, \quad c_P(n) = c_R(n+2) - 2.$$

These formulas verify our earlier remark that the recursive strategies cost about half as much as the binary scheme when n is large.

Let us say that a representation scheme is *irredundant* if the corresponding guessing game never asks any stupid questions (questions that have answers already known or that could be deduced from those known). This means that if α is any sequence of 0's and 1's such that α is not the representation of any n, but such that α does occur as a prefix of some representation of an integer, then both $\alpha 0$ and $\alpha 1$ occur as prefixes of representations. All reasonable schemes will be irredundant; for if, say, $\alpha 0$ occurs as a prefix but $\alpha 1$ doesn't, we can shorten the representation of some integers without violating **1** or **2** merely by deleting the 0 following α whenever α appears as a prefix.

The cost functions of irredundant representations satisfy an important arithmetical relation:

FACT 1   Let $c(n)$ be the cost of representing n in an irredundant representation scheme. Then

$$\frac{1}{2^{c(0)}} + \frac{1}{2^{c(1)}} + \frac{1}{2^{c(2)}} + \cdots = 1.$$

PROOF   In fact, if α is the representation of n, we have

$$\frac{1}{2^{c(0)}} + \frac{1}{2^{c(1)}} + \cdots + \frac{1}{2^{c(n-1)}} = (.\alpha)_2$$

in binary notation, for all n. This formula clearly holds when $n = 0$, since there can be no 1's in an irredundant representation of 0. Let β be the representation of $n + 1$; we need to show that

$$(.\alpha)_2 + \frac{1}{2^{|\alpha|}} = (.\beta)_2.$$

No number is represented by a sequence of 1's only, since the sequence 1^m is followed in lexicographic order only by sequences of which it is a prefix. Thus $(.\alpha)_2 + 2^{-|\alpha|}$ is less than 1, and in fact $(.\alpha)_2 + 2^{-|\alpha|} = (.a_1 a_2 a_3 \ldots)_2$ is the smallest binary number greater than $(.\alpha)_2$ not having α as a prefix. If

$(.\beta)_2 = (.b_1 b_2 b_3 ...)_2$ is unequal to $(.a_1 a_2 a_3 ...)_2$, let j be minimal such that $b_j \neq a_j$ and let k be maximal such that $a_k = 1$. We have $(.b_1 b_2 b_3 ...)_2 > (.a_1 a_2 a_3 ...)_2$, hence $b_j = 1$ and $a_j = 0$. If $k < j$, the sequence $b_1 ... b_{j-1}$ is a redundant prefix, because $b_1 ... b_{j-1}1$ occurs as a prefix of the representation of $n + 1$ but $b_1 ... b_{j-1}0$ never occurs as a prefix. If $k > j$, the sequence $a_1 ... a_{k-1}$ is a redundant prefix, because $a_1 ... a_{k-1}0$ occurs as a prefix of the representation of n but $a_1 ... a_{k-1}1$ never occurs as a prefix. (Note that k is the position of the rightmost 0 in α.)

It is easy to verify that 1^k occurs as a prefix for all k. Now if 1^k is a prefix of the representation of n, the sum $2^{-c(0)} + 2^{-c(1)} + \cdots + 2^{-c(n-1)}$ lies between $(.1^k)_2 = 1 - 2^{-k}$ and 1, hence the infinite sum as $n \to \infty$ converges to 1. *Q.E.D.*

(This proof uses both properties **1** and **2**, and indeed the result would be false if only property **1** were assumed. For example, the following procedure generates representations of length $n + k$ for each integer n in an irredundant fashion, for any fixed $k \geq 2$: Represent 0 by 0^k, then for $n = 1, 2, 3, ...$, find a sequence α such that (a) α has occurred as a prefix of a representation for some number $< n$ but not as a representation; (b) $\alpha 0$ and $\alpha 1$ have not both occurred as prefixes; and (c) α is as short as possible satisfying (a) and (b). Then the representation of n can be any sequence of length $n + k$ containing $\alpha 0$ or $\alpha 1$ as a prefix, whichever hasn't previously occurred. For example, one such representation scheme for $k = 2$ begins

$$
\begin{array}{lll}
0 \to 00 & 3 \to 11000 & 6 \to 11100000 \\
1 \to 100 & 4 \to 011000 & 7 \to 010100000 \\
2 \to 0100 & 5 \to 1010000 & 8 \to 0111000000.
\end{array}
$$

On the other hand, it is not difficult to prove the inequality

$$
\frac{1}{2^{c(0)}} + \frac{1}{2^{c(1)}} + \frac{1}{2^{c(2)}} + \cdots \leq 1
$$

for all representations which satisfy property **1**; this relation is known as "Kraft's inequality" [9].)

There is also a converse to Fact 1:

FACT 2 Let $c(0), c(1), c(2), ...$ be a nondecreasing sequence of positive integers such that

$$
\frac{1}{2^{c(0)}} + \frac{1}{2^{c(1)}} + \frac{1}{2^{c(2)}} + \cdots = 1
$$

Then there is an irredundant representation scheme with cost function $c(n)$.

PROOF   By letting the representation α of n be defined by the formulas $2^{-c(0)} + 2^{-c(1)} + \cdots + 2^{-c(n-1)} = (.\,\alpha)_2$ and $|\alpha| = c(n)$, we obtain a scheme with the desired properties. Q.E.D.

If X is an irredundant representation scheme whose cost function c_X is not monotonic, we can permute the integers to obtain another representation scheme with the same costs sorted into nondecreasing order. Facts 1 and 2 now show that there exists an irredundant scheme having these sorted costs. Representation schemes with monotonic cost functions can be called *standard*, since most applications prefer to have $c_X(n) \leq c_X(n+1)$ for all n.

Let us conclude our investigations by trying to find the best possible scheme, putting an emphasis on asymptotic questions (that is, on the efficient representation of very large integers). Clearly the binary method is more efficient than the unary method, and we also prefer the recursive method to the binary method because of its superior performance on large numbers. These examples suggest the following definition: "A representation scheme X with cost function $c_X(n)$ *dominates* another scheme Y with cost function $c_Y(n)$ if $c_X(n) \leq c_Y(n)$ for all large n and $c_X(n) < c_Y(n)$ for infinitely many n." If X dominates Y and Y dominates Z by this definition, then X dominates Z.

Clearly the recursive methods P, Q, R all dominate the binary method B, and B dominates the unary method U. When we try to compare the three recursive schemes with each other, however, it turns out that none of them is dominant. Method P is best nearly all the time; in particular, whenever $\lambda n = \lambda(n+2)$ we have $c_P(n) = c_Q(n) - 1 = c_R(n) - 2$. But there are infinitely many n where Q is better than both P and R (namely when $n = 2^{2^k} - 2$ and $k \geq 1$), and there are infinitely many n where R is better than both P and Q (namely when $n = 2^{2^{2^k}} - 1$). These facts suggest that it might be impossible to dominate method R, and in that sense we might conclude that method R is "optimal." However, any such hopes are quashed by

FACT 3   If X is any standard representation scheme, there is another standard representation scheme Y that dominates X and satisfies $c_Y(n) \leq c_X(n)$ for all but one value of n.

PROOF   The general idea of the proof is to choose a sequence of length c and replace it by infinitely many sequences of respective lengths $c+1, c+2, c+3$, etc., because $2^{-c} = 2^{-c-1} + 2^{-c-2} + 2^{-c-3} + \cdots$.

Proceeding more formally, we may assume that X is irredundant. For $k > 1$ let a_k be the number of n's such that $c_X(n) = k$, and let j be minimal such that $a_j > 0$. Let $b_j = a_j - 1$ and let

$b_k = a_k + 1$ for all $k > j$. Then there is a unique nondecreasing function $c_Y(n)$ having exactly b_k values of n with $c_Y(n) = k$, and this function satisfies $2^{-c_Y(0)} + 2^{-c_Y(1)} + \cdots = 2^{-c_X(0)} + 2^{-c_X(1)} + \cdots$. By Fact 1 and Fact 2, there is an irredundant representation scheme with cost function c_Y. Furthermore it is easy to see that $c_Y(n) \leq c_X(n)$ except for the single value of n such that $c_X(n) = j$ and $c_X(n+1) > j$; and $c_Y(n) < c_X(n)$ whenever $j + 1 < c_X(n-1) < c_X(n)$. Q.E.D.

It is hopeless to find a strictly optimum scheme, because repeated application of the construction in the proof of Fact 3 will give an infinite family of better and better schemes, each dominating its predecessors. However, it turns out that no scheme can be much better than our recursive scheme R, in spite of all the apparent improvement guaranteed by Fact 3.

FACT 4 Let Λn be the function defined by

$$\Lambda n = \lambda n + \lambda\lambda n + \lambda\lambda\lambda n + \cdots.$$

Every cost function $c(n)$ corresponding to a representation scheme satisfies

$$c(n) > \Lambda n + \lambda\lambda^* n$$

for infinitely many n.

PROOF Let $d(n) = \Lambda n + \lambda\lambda^* n$; we shall show that the sum $\sum_{n \geq 0} 2^{-d(n)}$ diverges. This suffices to complete the proof, for if $c(n) \leq d(n)$ for all $n \geq m$ we have the impossible inequality $1 = \sum_{n \geq 0} 2^{-c(n)} \geq \sum_{0 \leq n < m}(2^{-c(n)} - 2^{-d(n)}) + \sum_{n \geq 0} 2^{-d(n)}$. In general if $f(n) = \Lambda n + g(\lambda^* n)$ for any function g, we have

$$\sum_{n \geq 0} \frac{1}{2^{f(n)}} = \sum_{n \geq 0} \frac{1}{2^{g(n)}},$$

since the left-hand side can be written

$$\sum_{m \geq 0} \frac{1}{2^{g(m)}} \sum_{\lambda^* n = m} \frac{1}{2^{\Lambda n}}$$

and for $m \geq 1$ we have

$$\sum_{\lambda^* n = m} \frac{1}{2^{\Lambda n}} = \sum_{\lambda^* \lambda n = m-1} \frac{1}{2^{\lambda n + \Lambda \lambda n}} = \sum_{\lambda^* k = m-1} \frac{1}{2^{k + \Lambda k}} \sum_{\lambda n = k} 1$$

$$= \sum_{\lambda^* k = m-1} \frac{1}{2^{\Lambda k}}.$$

Thus the sum $\sum_{n \geq 0} 2^{-d(n)}$ diverges if and only if the sum $\sum_{n \geq 0} 2^{-\lambda n}$ diverges, and it does. *Q.E.D.*

Using the same proof technique, we can show in fact that

$$c(n) > \Lambda n + \Lambda \lambda^* n + \Lambda \lambda^* \lambda^* n + \cdots + \Lambda (\lambda^*)^m n + \lambda (\lambda^*)^{m+1} n$$

infinitely often, for any fixed representation scheme and any fixed m.

Let us close by showing how to improve scheme R so that we obtain "ultimate" schemes that come very close to this lower bound. The sequence $R(n)$ begins with $1^{\lambda^* n} 0$, a sequence of length $\lambda^* n + 1$ that serves to identify $\lambda^* n$, followed by a sequence of length Λn that characterizes n once $\lambda^* n$ is known. In this sense, the R method starts by using a unary approach to guess $\lambda^* n$; and we know that we can do better. Let us therefore use the R method to determine $\lambda^* n$, then continue to determine n as before, calling this the RR method. The new cost function will be

$$c_{RR}(n) = \Lambda n + c_R(\lambda^* n) = \Lambda n + \Lambda \lambda^* n + \lambda^* \lambda^* n + 1,$$

and the RR representation of n will begin with the sequence $1^{\lambda^* \lambda^* n} 0$.

But wait, let's start by guessing $\lambda^* \lambda^* n$; then we obtain an RRR method whose cost function is

$$
\begin{aligned}
c_{RRR}(n) &= \Lambda n + \Lambda \lambda^* n + c_R(\lambda^* \lambda^* n) \\
&= \Lambda n + \Lambda \lambda^* n + \Lambda \lambda^* \lambda^* n + \lambda^* \lambda^* \lambda^* n + 1.
\end{aligned}
$$

The R^{m+1} method has a cost function equal to

$$c(n) = \Lambda n + \Lambda \lambda^* n + \cdots + \Lambda (\lambda^*)^m n + (\lambda^*)^{m+1} n + 1,$$

and this upper bound is almost the same as our lower bound.

Acknowledgements

Philip Davis [4] has written a delightful introduction to the elementary facts about large numbers. The problem of representation schemes was originally treated by Levenshteĭn [11], who presented Method R and proved that $c(n) > \Lambda n - \lambda^* n$ for infinitely many n; he also discussed representations with more than two symbols. The representation problem has also been studied independently by Elias [5] and by Even and Rodeh [6]; the guessing game for unbounded search was suggested by Bentley and Yao [2]. Methods P and Q were proposed by Jim Boyce and David Fuchs during a recent conversation with the author.

Numbers much larger than those considered here are discussed in [8]. Using the notation of that article, we have $\lambda^*(2 \uparrow \uparrow m) = m + 1$ and $\lambda^*(2 \uparrow \uparrow \uparrow m) = (2 \uparrow \uparrow \uparrow (m - 1)) + 1$, suggesting that the upper and lower bounds we have derived are really not so close together after all; perhaps the introduction of functions $\lambda^{**} n$ and $\lambda^{***} n$, etc., will lead to further clarification of "optimum" representation schemes when we imagine ourselves dealing with supersupernatural numbers.

An ancient Chinese mathematician named Hsü Yo (c. 200 A.D.) discussed a nomenclature for large numbers containing the units *wan* $= 10^4$, $i = 10^8$, *chao* $= 10^{16}$, and *ching* $= 10^{32}$, in his interesting book *Shu Shu Chi I*; see [12, p. 87]. I am indebted to Tung Yun-Mei for this reference.

There is a beautiful connection between the representation problem discussed here and the information-theoretic concept of algorithmic complexity originated independently by L. A. Levin in Russia and G. J. Chaitin in Argentina. A certain function $l(n)$, called $KP(n)$ by Levin (cf. [7]) and $H(n)$ by Chaitin [3], has the following two properties: (a) There is a representation scheme for natural numbers that satisfies condition **1** and has cost $l(n)$. (b) For every representation scheme satisfying condition **1** and having cost $c(n)$, there is a constant C such that $l(n) \leq c(n) + C$. Intuitively, $l(n)$ represents the length of the "simplest" description of an algorithm to compute n. This function $l(n)$ is not computable, but it is semicomputable from above, in the sense that, if $l(n)$ is actually less than a given number m, we can prove this fact in a finite amount of time. If $c(n)$ is the cost function for any representation scheme satisfying conditions **1** and **2** then there is a constant C such that $\lambda n + l(\lambda n) \leq c(n) + C$. Furthermore there is a constant C_0 such that $|\lambda n + l(\lambda n) - \max_{1 \leq k \leq n} l(k)| \leq C_0$. Thus the constructions we have given provide bounds on $\max_{1 \leq k \leq n} l(k)$. I am indebted to Péter Gács for these references.

References

1 Archimedes. 1956. The Sand Reckoner. In *The World of Mathematics*, vol. 1, ed. James R. Newman, pp. 420–31. New York: Simon and Shuster.

2 Bentley, J. L. and Yao, A. C. 1976. An almost optimal algorithm for unbounded searching. *Inf. Proc. Letters* 5: 82–87.

3 Chaitin, Gregory J. 1975. A theory of program size formally identical to information theory. *Journal of the ACM* 22: 329–340.

4 Davis, Philip J. 1966. *The Lore of Large Numbers*. New Mathematical Library, vol. 6, New York: Random House.

5 Elias, P. 1975. Universal codeword sets and representations of the integers. *IEEE Trans. Inform. Theory* IT-21: 194–203.

6 Even, S. and Rodeh, M. 1978. Economical encoding of commas between strings. *Comm. of the ACM* 21: 315–317.

7 Gács, Péter. 1974. On the symmetry of algorithmic information. *Soviet Math. Doklady* 15, 5: 1477–1480, 15, 6: v.

8 Knuth, D. E. 1976. Mathematics and computer science: coping with finiteness. *Science* 194: 1235–1242.

9 Kraft, L. G. 1949. A device for quantizing, grouping and coding amplitude modulated pulses. M.S. thesis, Electrical Eng. Dept., Mass. Inst. of Technology.

10 Kronecker, L. 1893. Remark in a lecture at the Berlin scientific congress, 1886: "Die ganzen Zahlen hat der liebe Gott gemacht, alles andere ist Menschenwerk." Quoted by H. Weber in *Math. Annalen* 43: 15.

11 Levenshteĭn, V. E. 1968. On the redundancy and delay of decodable coding of natural numbers. *Systems Theory Research* 20: 149–155.

12 Needham, Joseph. 1959. *Science and Civilisation in China* 3 Cambridge: Cambridge University Press.

Solutions

1 Archimedes' last number: One septendecyllion trevigintyllion quattuorvigintyllion quinvigintyllion sexvigintyllion septenvigintyllion octovigintyllion trigintyllion untrigintyllion quattuortrigintyllion septentrigintyllion octotrigintyllion novemtrigintyllion quadragintyllion duoquadragintyllion trequadragintyllion octoquadragintyllion novemquadragintyllion quinquagintyllion quattuorquinquagintyllion. (His name for this number was much shorter, but the new system beats his in lots of other cases.)

Incidentally, American names for large numbers are not based on good Latin (sexdecillion should be "sedecillion" and novemdecillion should be "undevigintillion"); so it is not clear how much further the American system can be extrapolated.

2 One latinlatinlatinbyllionyllionyllionyllion is

$$10^{2^{(10^{2^{(10^{2^{10000000000000002}+2)}}+2)}}}$$

We might call this one umptyllion, for short.

3 Iterative decoding procedure for Method R: Let the input sequence be $s_1 s_2 s_3 \ldots$, where each s_i is 0 or 1. Let k be the number of input bits already read, let l be the length of a number to be input, let m be the stack depth in a recursive implementation, and let n be the answer returned. The following recursive method corresponds directly to the definition, with k initially zero, using only global variables.

PROCEDURE R

Set $k \leftarrow k + 1$.
If $s_k = 0$, set $n \leftarrow 0$ and exit from R.
Otherwise, call R (recursively).
Then set $l \leftarrow n$, $n \leftarrow 1$.
While $l > 0$, repeatedly set $k \leftarrow k + 1$, $n \leftarrow 2n + s_k$, $l \leftarrow l - 1$.
Exit from R.

By mechanical transformations this recursive program can be converted into the following purely iterative procedure:

Set $k \leftarrow 1$ and $m \leftarrow 0$.

While $s_k = 1$, repeatedly set $m \leftarrow m + 1$ and $k \leftarrow k + 1$.
Set $n \leftarrow 0$.
While $m > 0$, repeatedly do the following operations:
Set $m \leftarrow m - 1$, $l \leftarrow n$, $n \leftarrow 1$.
While $l > 0$, repeatedly set $k \leftarrow k + 1$, $n \leftarrow 2n + s_k$, $l \leftarrow l - 1$.

The corresponding encoding procedure can be treated similarly.

---

* We might also consider changing "nineteen" to "onety-nine," etc.; but that might make our nomenclature *too* logical.

The Graph Theorists Who Count—and What They Count

Ron Read
University of Waterloo

What is a graph? To the majority of people the word "graph" conjures up a picture either of the kind of businessman's chart shown in Figure 1, or of a smooth curve, like Figure 2, which displays the property of some mathematical function. But to a large and growing number of mathematicians this same word suggests something completely different. These are the people who are concerned, in one way or another, with the branch of mathematics known as graph theory. When they think of a graph they

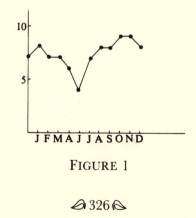

FIGURE 1

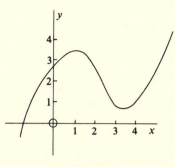

FIGURE 2

have in mind a diagram consisting of a number of points, or *nodes*, and lines or curves joining some of these nodes to others. These lines are often called *edges*; an example is given in Figure 3. (Most of the other figures in this article depict graphs of one sort or another.) These objects are what the man in the street would tend to refer to as networks (a word that graph theorists use in a somewhat special sense).

Graph theory—the study of these graphs and their properties—started as a branch of topology. Indeed, it was once referred to as the "slums of topology," but it has since become respectable (if it ever was otherwise) and is now of considerable importance, finding application to many problems throughout mathematics and the sciences.

Why should a subject which, at first sight, appears to be concerned with nothing more substantial than paper and pencil doodling be of any great use? We can answer this question by looking around us and seeing how many things of importance in everyday life are essentially graphs, or can be represented by graphs. Examples that come to mind are communication systems of various kinds; telephone networks, where telephone lines link certain cities with certain other cities; road, railroad or airline

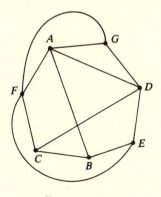

FIGURE 3

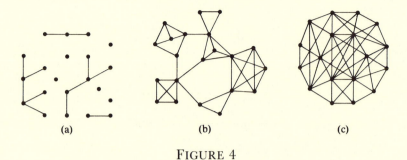

(a) (b) (c)

FIGURE 4

networks, and so on. In these examples we have actual physical objects that are themselves graph-like in nature, with lines, roads, etc. being the *edges* of the graph which join certain pairs of nodes—in this case, cities.

We can apply graph theory also to something that does not in itself resemble a graph, by constructing a graph which summarizes some of its important properties. Consider, for example, a group of people at a party. We might construct a graph by representing each person by a node, and joining two nodes by an edge if the two persons in question are acquainted. The result will be a graph something like those shown in Figure 4. We can tell a lot from such a graph; we can tell, for example, that the party of Figure 4a will be deadly dull; the party of Figure 4b will perhaps be good, but "cliquey"; and the party in Figure 4c will probably be wild!

We may note here that it may not be possible to draw a graph on a plane surface without making some of the edges cross each other; the graph of Figure 3 is a case in point. It is for this reason that we have drawn the nodes as small circles, so that, for example, the point where the edges AB and CD cross, will not be mistaken for a node of the graph. Some graphs can be drawn in the plane without any such accidental crossings. When they are drawn in this fashion they are called *plane graphs*, and we shall meet them again later on.

As an example of a practical problem in graph theory, consider a saboteur, or an enemy agent, who wishes to disrupt a complicated telephone network. He would like to know the minimum number of telephone lines to cut, or the minimum number of exchanges to blow up, in order to sever all communication between certain key points. This is a problem concerning the "connectivity" of graphs. We may wish to extend the concept of a graph by attaching a number to each edge, in which case we obtain what is known as a *network*. An example is given in Figure 5. These numbers could be the distances between points, or the cost of transporting commodities from one depot to another, or any of a host of other possibilities. So wide-spread are structures and situations that can be represented by networks, that a large part of the important and growing field of Operations Research is concerned with networks and their properties.

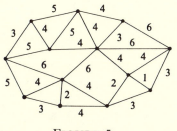

FIGURE 5

For this reason almost all research in graph theory contains the seeds of possible practical application, and is at least potentially useful. There are some exceptions, however. Tucked away in a dark corner of the graph theory edifice, with eyes carefully shielded from the light of the real world, are some graph theorists busy solving problems which are not even remotely practical. These are the devotees of graphical enumeration, a subject which is concerned with answers to questions of the form "How many different graphs are there of such and such a kind?" These graphical enumerators spend their time either in counting graphs, or in devising better and better ways of counting graphs. A strange occupation, but one that is not without interest.

Who are these "Graph theorists who count" and what sort of things do they count? By the end of this article I hope to have given at least a rough answer to these questions.

Cayley and the Enumeration of Trees

The pioneer of graphical enumeration was Sir Arthur Cayley, who, from 1857 onwards, stated and solved many problems concerning the enumeration of those graphs called *trees*. Graph theorists define a tree to be a connected graph (one that is all in one piece) which has no circuits. To understand what a circuit is, think of a graph as a network of roads; a circuit is then a walk on these roads that comes back to its starting node and does not pass through any junction (node) more than once. If the condition about returning to the starting node is dropped, we have what graph theorists call a "path" between two nodes. Another way of defining a tree is that it is a graph in which there is a unique path between any two given nodes. The graph of Figure 3 is not a tree because it has the circuit *ABCDEFGA*, among others; Figure 6 shows a graph that is a tree, and one can easily see why these graphs are so called.

In a tree the number of edges is always one less than the number of nodes (Problem No. 1; Prove this.) Therefore the basic question of tree enumeration is "How many trees are there with n nodes?" This is one of the questions which Cayley solved.

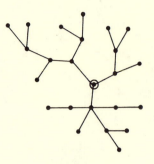

FIGURE 6

In graphical enumeration problems it is very important to be clear as to what exactly is being counted. When we say "How many different graphs of such and such a kind are there?," we need to know in particular what we mean by two graphs being the *same* or being *different*, and this may vary from one problem to another. Usually we regard two graphs as the same (or to use the technical term, *isomorphic*) if it is possible to label the nodes of each with the integers $1, 2, \ldots, p$ so that for any two labels, the nodes with those labels are either joined in both graphs or not joined in either.

Thus the two graphs of Figure 7, which at first sight seem to be very different, are in fact isomorphic. The figure shows one way of labelling the two graphs so as to show this. (Problem No. 2: Figure 8 shows four graphs on 10 nodes. Three of them are isomorphic, while the other is different from them. Which is the odd man out?)

Thus when we ask "How many different graphs are there?" we must say what we mean by "different." Frequently it will mean "non-isomorphic" but not necessarily; we may have something else in mind.

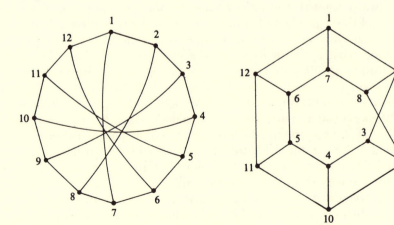

FIGURE 7

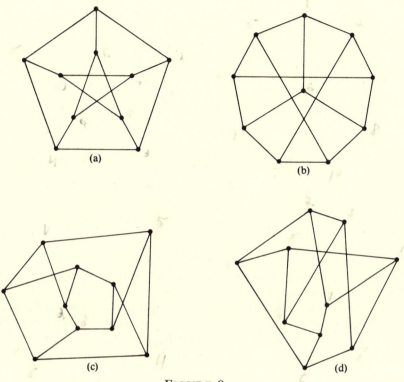

FIGURE 8

Counting Binary Trees

We shall now look in detail at a fairly simple problem, just to get a taste of the kinds of methods that are used in graphical enumeration. We shall enumerate binary trees. A *binary tree* is, first, a rooted tree. This means that one of its nodes, called the root, is distinguished in some way from the others. In drawing a diagram of a rooted tree, the root can be indicated by circling it (as in Figure 6) or by drawing the tree so that the root is at the bottom of the page. Next, at each node of the binary tree, there are, at most, two edges going upwards, that is, away from the root. If there are two, then one goes to the left and one to the right. If there is only one, it may go to the left or to the right; these two possibilities are considered to be different. If there is none, then the tree continues no further from that node, and the node in question is an *endnode* or *leaf* of the tree. Figure 9 shows all the binary trees on 4 nodes. Note that because the distinction between left and right is important—as part of the definition of binary tree—all these trees are different. Had we not made this distinction, the first two trees would have been isomorphic, as would many other pairs of trees in the figure.

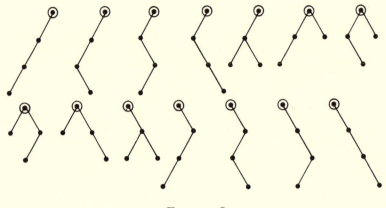

FIGURE 9

To start out the enumeration of binary trees we note that from the root of any given tree having, let us say $n + 1$ nodes, there will be two upward branches—one to the left and one to the right. These are defined as two *subtrees*—a left subtree and a right subtree, as shown in Figure 10. Let B_n denote the number of binary trees with n nodes, and let us pretend that we know the values of the numbers $B_1, B_2 \ldots B_n$. To find the number B_{n+1} we consider how we could construct binary trees on $n + 1$ nodes. This amounts to choosing two subtrees (having n nodes between them) as the left and right subtrees, and joining their roots to a new node—the root of the tree being constructed. In how many ways can we do this? If the left and right subtrees have l nodes and r nodes respectively, then we can make our choices in $B_l \cdot B_r$ ways; for any one of the B_l trees on l nodes can be taken with any one of the B_r trees on r nodes.

If we repeat this with different values of l and r (remembering that $l + r = n$) and add up the results, we shall obtain the number of binary trees with $n + 1$ nodes. But hold on! We have assumed that there are necessarily two edges going upwards from the root, and therefore exactly two subtrees every time; but there may be only one. Luckily we can avoid this difficulty by means of a little trick. If the tree being counted has only one subtree, we shall pretend that it really has two but that the other is a sort of "phantom tree"—one that has no nodes at all!

By means of this stratagem we can say that *every* binary tree gives rise to exactly two subtrees. Since there is only one such "phantom" tree we take $B_0 = 1$, and allow 0 as a possible value for l and r. In this way we arrive at the following result

1 $\qquad B_{n+1} = B_0 B_n + B_1 B_{n-1} + B_2 B_{n-2} + \cdots + B_{n-1} B_1 + B_n B_0$

which enables us to calculate the number B_{n+1} if the numbers of trees up to B_n are already known.

The reader may care to verify that these numbers begin as follows:

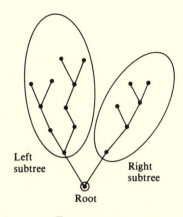

Left
subtree

Right
subtree

Root

FIGURE 10

n	0	1	2	3	4	5	6	7	8	9	10
B	1	1	2	5	14	42	132	429	1430	4862	16796

These numbers, the B_n, are the well-known Catalan numbers, which crop up in a large variety of combinatorial problems. Besides being the number of binary trees on n nodes, B_n is also the number of ways of dissecting an n-gon into triangles, the number of ways of joining in pairs $2n$ points on the circumference of a circle by nonintersecting chords, and the number of ways of properly nesting n pairs of parentheses. An interesting account of these numbers and their properties was given by Martin Gardner in the June 1976 issue of *Scientific American*.

It is a common and useful practice in enumeration to give the answer to a problem in the form of a "generating function." To derive a generating function from a sequence such as the Catalan numbers $\{B_n\}$ we make these numbers the coefficients in an infinite power series. The generating function for the Catalan numbers is thus

$$B(x) = B_0 + B_1 x + B_2 x^2 + B_3 x^3 + B_4 x^4 + \cdots.$$

Now the right-hand side of **1** is easily seen to be the coefficient of x^n in the product of $B(x)$ with itself, and therefore the coefficient of x^{n+1} in $x[B(x)]^2$. The left-hand side of **1** is the coefficient of x^{n+1} in $B(x)$. Hence we derive the following result

2 $$B(x) = 1 + x[B(x)]^2$$

where the "1" is added to the right-hand side to take account of the phantom tree, to which equation **1** does not apply.

Equation **2** is a quadratic equation in $B(x)$. Solving it we find that

$$B(x) = \frac{1 - \sqrt{1 - 4x}}{2x}$$

and from this, by a tedious but not difficult bit of algebraic manipulation, we can get an explicit expression for the general Catalan number. It is

$$B_n = \frac{(2n)!}{(n+1)!\, n!} \ .$$

The equation above illustrates a common technique for enumerating rooted trees. Remove the root and look at the subtrees that we get. There are many things that may happen when we do this, for there are many different kinds of trees waiting to be counted. We may get more than two subtrees—perhaps a variable number even; the subtrees may be ordered (as with the distinction between left and right that we just made) or they may not. Each subtree will, however, be rooted (at the node that was joined to the original root), and provided they are of the same type as the original tree, we may expect to be able to derive some kind of recurrence for the required numbers. Almost all tree enumeration has followed this kind of pattern.

Cayley enumerated, in particular, what we can call "ordinary" rooted trees, for which there are no restrictions on how many edges meet at a node, and where there is no importance attached to the order of these edges or how the tree is depicted on the plane page. He showed that, if T_n is the number of such trees on n nodes, then the following rather curious equation holds

$$T_1 x + T_2 x^2 + T_3 x^3 + \cdots = x(1-x)^{-T_1}\,(1-x^2)^{-T_2}\,(1-x^3)^{-T_3}\cdots .$$

As before, if we know the values of T_n up to some given value of n, then, by plugging these values into the right hand side of this equation we can extract the next number in the sequence from the left-hand side.

Cayley also enumerated unrooted trees. This is a more difficult problem because with no root at which to pull the graph apart (so to speak) it is not as easy to express the problem in terms of smaller problems, as we did above. Cayley also enumerated the numbers of different chemical compounds of certain types, thus opening up one of the few regions of enumerative graph theory that is possibly of practical use. The formulae for these compounds (the alkanes—or paraffins—and some of their derivatives) are essentially trees of rather special types.

The Enumeration of Graphs

For graphs there is no relation between the numbers of nodes and edges as there is for trees; a graph with p nodes can have any number of edges from 0 in the case of the "empty graph" up to $p(p-1)/2$ for the "complete" graph in which every node is joined to every other node. Thus the question to ask about graphs is "How many different graphs are there with p nodes and q edges" where p and q are given numbers.

This is a problem which Cayley did not attempt to solve. Had he tried, he would probably not have succeeded, since the tool for its solution had not then been forged. This tool is a theorem which G. Pólya published in 1938 (see [9]) though it is implicit in some of his earlier papers. Pólya's theorem is fundamental to the solution of many combinatorial problems, and although we cannot discuss it in detail, it will be useful to have a general idea of its scope and its usefulness in solving graphical enumeration problems.

Pólya's Theorem

Quite a number of combinatorial problems can be formulated in terms of the idea of "putting things in boxes." Let us suppose that we have a number of distinguishable boxes, and that in each box we may put an object—usually referred to as a *figure*. Moreover, suppose that each figure has associated with it, a nonnegative integer that is its *value*. For each box there is a collection of figures, one of which is to be placed in the box. (It frequently happens that we have the same range of choices for each box, but this is not required). In each box we put exactly one of the figures appropriate to the box, and in this way we obtain a *configuration*—the boxes and the figures that they contain. The value of a configuration is then defined to be the sum of the values of the figures that have been placed in the boxes.

The first question we may ask is "Given all the relevant information concerning which figures may be placed in which boxes, in how many ways can we produce configurations having a given value?" As an example of a problem of this type, suppose we have the task of placing 21 candles on a rectangular birthday cake, the top having been divided into six sections as shown in Figure 11. We shall suppose that we may place any number of candles in each of the sections and that the precise location of the candles within a section is not important; all that matters is how many candles are there. In this example a figure consists of a certain number (possibly zero) of candles; and the value of a figure is the number of candles. The problem then is to find the number of configurations that have the value 21.

It turns out that this is not a particularly difficult problem; so we shall immediately make it harder. We now suppose that the boxes are not

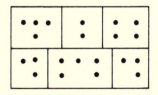

FIGURE 11

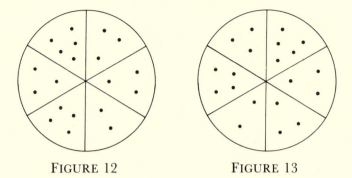

FIGURE 12   FIGURE 13

entirely distinguishable, and that certain rearrangements of the boxes will not be distinguishable from others.

For example, if, instead of the rectangular birthday cake, we have a circular one, divided into six identical sectors, then it would not be realistic to regard the two configurations of Figures 12 and 13 as different, since a clockwise rotation through 60 degrees will convert the first into the second. Two configurations that differ only in this sort of way will be regarded, for the purposes of this problem, as being the *same* configuration. In the general setting we suppose that we have a certain set of permutations (which will form a group) and that any two configurations which can be obtained, one from the other, by permuting the boxes (by some permutation in this group) will be regarded as *equivalent*. When we ask for the number of different configurations we are therefore asking for the number of *nonequivalent* configurations. For the circular birthday cake the group consists of the permutations of the boxes that correspond to rotations of the cake through multiples of 60 degrees. Note that the cake shown in Figure 14 is different from those in Figure 13, since no rotation through any angle will make it look the same as them.

In this article we shall not go into Pólya's theorem in detail; indeed we shall not even state it. We merely note that it provides a way of solving

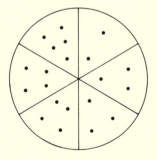

FIGURE 14

any problem of the general form just outlined. Given the data for such a problem, (namely the information on what figures may be placed in the boxes and their values) and given the group of permutations which determines whether two configurations are equivalent or not, we can apply Pólya's theorem and find the answer to the problem. Or rather the *answers*; for Pólya's theorem gives us a generating function which contains the number of configurations for any given value.

How can we reformulate the graph enumeration problem in terms of putting things in boxes? We do it as follows. Consider graphs on p nodes, labelled $1, 2, 3, \ldots, p$. There are $p(p-1)/2$ pairs of nodes, and each of these pairs is a possible site for an edge of the graph. Think of these pairs as the "boxes" for the problem, and allow just two figures for each box, namely the figure "no-edge" with value 0, and the figure "edge" with value 1. The result of putting one of these figures in each box corresponds, in an obvious way, to a graph—nodes i and j are joined if the box (i,j) contains the figure "edge"—and the value of a configuration (graph) is just the number of edges in it. Graphs constructed in this way will, however, be labelled graphs, since each node has its distinctive label from the set $\{1, 2, \ldots, p\}$. This is not what we want. If two graphs are isomorphic we do not want to count them as distinct just because they happen to be differently labelled. So we must allow for the fact that since the nodes are not all distinct, neither are the pairs of nodes—the boxes. Permuting the labels on the nodes automatically permutes the boxes among themselves. Thus if we relabel the nodes 1 and 2 so as to become nodes 5 and 9, the $(1,2)$-box becomes the $(5,9)$-box, and so on. Every permutation of the nodes gives rise to a permutation of the boxes, and it is the set of these permutations of the boxes that forms the group appropriate to this problem. By using Pólya's theorem the answer can then be obtained.

It was in this way that Frank Harary of the University of Michigan, now one of the most well-known people in the graph theory world, enumerated graphs in 1955. The generating function which comes out as the solution to this problem is quite complicated—fearsome even—but can be used to compute the numbers of graphs, especially with the aid of an electronic computer. C. A. King and E. M. Palmer have calculated the total numbers of graphs up to $p = 24$. The number of nonisomorphic graphs on 24 nodes is the following more-than-astronomical number:

195704906302078447922174862416726256004122075267063365754368.

(How's that for a really useful piece of information!)

The same technique will succeed in enumerating many variations on graphs, of which the most important is that of "directed graphs," or (as now usually abbreviated) *digraphs*, which are like graphs except that each edge is given a direction (indicated by an arrow) going one way or the other, or possibly both ways. Figure 15 shows a typical digraph. The enumeration problem for digraphs differs only in details from that for

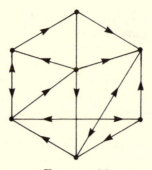

FIGURE 15

ordinary graphs and Pólya's theorem again gives the answer. This problem was also solved by Harary, in the same paper [5] in which he enumerated graphs.

Although Pólya's theorem is very powerful, it will not handle every problem of graphical enumeration. A problem that is proof against it is the enumeration of self-complementary graphs. The "complement" of a graph G is defined to be the graph G with the same set of nodes as G, having all the edges that are not in G and none of the edges that are in G. Figure 16 shows a graph (black edges) and its complement (grey edges) superimposed on the same set of nodes so as to display how a graph and its complement have, between them, all the $p(p-1)/2$ edges that are possible in a graph with p nodes.

Now it may happen that the complement of a graph is isomorphic to the original graph. The simplest example of such a "self-complementary" graph has 4 nodes and 3 edges and is shown in Figure 17. With 5 nodes there are two different self-complementary graphs, shown in Figure 18. These small examples are easily found by trial and error, but by the time we get up to 8 nodes trial and error is no longer a reliable way of finding out how many self-complementary graphs there are. (Problem No. 3: How many self-complementary graphs are there with 6 and 7 nodes?) What we

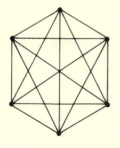

FIGURE 16

FIGURE 17

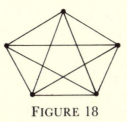

FIGURE 18

need is a theoretical result giving the number of these graphs for any value of p.

This problem requires something more powerful than Pólya's theorem because it has an added complication. A graph and its complement differ to the extent that we have interchanged the two figures "edge" and "no-edge" in going from one to the other. Thus we have a different kind of general problem—one in which we allow permutations of the *figures* as well as permutations of the boxes to decide whether two configurations are equivalent or not. We therefore have *two* permutation groups—one for the boxes and one for the figures—and everything becomes much more complicated. A method of coping with this more general type of problem was first given by the Dutch mathematician N. G. deBruijn in a paper [1] published in 1959, and it was not long before the enumeration of self-complementary graphs was achieved (see [10]). Complementation for digraphs is defined analogously to the corresponding graph concept. Figure 19 shows a digraph that is isomorphic to its complement, and the enumeration of such self-complementary digraphs came out along with that of self-complementary graphs.

One very curious fact that emerged from these results is that the number of self-complementary digraphs on an even number of nodes is always the same as the number of self-complementary graphs on twice the number of nodes! Such a direct relation suggests (intuitively, but very strongly) that there ought to be some simple connection between the

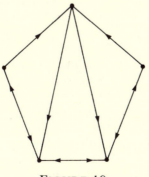

FIGURE 19

graphs and the digraphs; that there should be some easy way of "seeing" that the numbers must be the same, and why. Maybe there is; but if so, no one has yet discovered it.

As time went on, more powerful enumeration theorems were found—theorems that generalized those of Pólya and de Bruijn, or which struck out in quite different directions. Among the most important of these is the "Power Group Enumeration Theorem" of Harary and Palmer; but it is too complex to be described here. The reader can find further information in the book *Graphical Enumeration* [7] by these two authors—a comprehensive treatise on this specialized branch of graph theory.

The Strange Case of
J. H. Redfield

We should say something now about an unusual character who appeared on the stage of graphical enumeration. Perhaps it would be more correct to say "who reappeared" since ... but we shall see. In the early 1960's Frank Harary drew the attention of his fellow enumerators to a remarkable paper [11] called to his attention by one of his grad students. It was entitled "The theory of group-reduced distributions" and had been published by a hitherto unknown mathematician, J. H. Redfield, in 1927. In this paper (as Harary discovered, and as the rest of us verified with astonishment) Redfield had given a theorem equivalent to Pólya's theorem; he had carried out the enumeration of graphs (at least in part—and it was clear that he could have gone further); he had anticipated several other "recent" discoveries in enumeration theory; and, to crown it all, he had proved a powerful new theorem. I remember well this "rediscovery" of Redfield's paper since one of the results that he had anticipated was a theorem (concerning the enumeration of graphs obtained by superposing several graphs on the same set of nodes) which had been the mainstay of my own Ph.D. thesis a few years earlier! It seemed that for 30 years or so, graphical enumerators, and others, had been painfully solving problems and proving theorems that Redfield had known about and had published way back in 1927! And yet his paper had completely escaped notice. How was this possible?

It was not that it had been published in an obscure journal; the *American Journal of Mathematics*, in which it appeared, is a well-known and highly respected publication. Perhaps it was rather that several factors that normally succeed in drawing attention to a paper, just happened to be absent. Redfield was not a known worker in this field—in fact this was his only mathematical paper; the title of the paper gave no inkling of the nature of its contents; and Redfield's notation, including some rather bizarre symbols, was such that even the printed page did not give an immediate indication of what the paper was about. Above all, perhaps, was a contributing factor best described by saying that "the time was not ripe."

Redfield was not primarily a mathematician—his field was, or had been, linguistics. Nevertheless his one and only mathematical paper is a remarkable piece of work, and had it been recognized earlier for what it was, the course of enumerative graph theory would undoubtedly have been very different. Certainly, since its rediscovery, Redfield's paper has inspired several new advances. In particular R. W. Robinson of the University of Newcastle in Australia has adapted and improved some of Redfield's ideas, and hence solved several problems that were previously considered intractable.

The Enumeration of Plane Graphs

We have already seen what is meant by a plane graph; it is one that has been drawn in the plane without any edges crossing. Let us look at some enumeration problems pertaining to them. A natural question to ask is "How many plane graphs are there with p nodes and q edges?," but this is much too difficult. Let us consider "cubical plane graphs," by which we mean graphs in which exactly 3 edges come together at each node. In a cubical graph the numbers p and q of nodes and edges are related by $2q = 3p$ (Question 4: Why is this?), so that a single number determines the size of the graph—the number of nodes (which must be even). We might expect that this would make the problem easier, but it seems to be just as hard. This is typical of plane graphs. The enumeration problems are usually fiendishly difficult, however, when a problem *can* be solved, it often happens that the solution is simpler than graphs are in general.

We can see this very well if we further restrict the problem, and ask "How many plane cubical graphs are there with a fixed Hamiltonian circuit?" A Hamiltonian circuit in a graph is a circuit—a "Grand Tour" one might say—that visits every node of the graph. By being "fixed" we mean that the Hamiltonian circuit has been drawn once and for all in the plane, and only the remaining edges may be altered in position to form the several graphs that we are counting. We are therefore looking at objects like the one in Figure 20, in which the edges not in the Hamiltonian circuit are drawn in grey. Now the grey edges lie either entirely inside or entirely outside the circuit; for if they crossed it the graph would not be plane. Further, each node is an end of exactly one thick edge. We can therefore rephrase the enumeration problem for those graphs as follows: "Given a polygon with $2n$ vertices, in how many ways can we join some pairs of vertices by nonintersecting chords lying inside the polygon, and join the remaining vertices in pairs by non-intersecting 'chords' (for want of a better word) lying outside the polygon?"

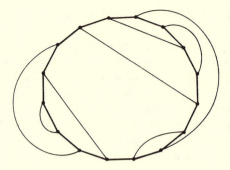

FIGURE 20

This problem was solved in a most elegant manner by W. T. Tutte, Distinguished Professor at the University of Waterloo, who has made an extensive study of plane graphs, and who is responsible for the bulk of the work that has been done on their enumeration. Let us briefly look at his result and its proof. (For full details see [12].)

First note that if we want all the chords to be on the inside of the polygon we have a simpler problem. In fact, it is one of the many problems for which the answer is the sequence of Catalan numbers, as already mentioned. Thus if we stipulate that a certain set of $2r$ nodes shall be the ends of the "internal" chords, then the number of ways of joining up these nodes in pairs is the Catalan number

$$\frac{(2r)!}{r!(r+1)!}.$$

A similar result holds for the remaining $2s$ nodes (where $r + s = n$) and the "external" chords.

Now any way of disposing the internal chords can be taken with any way of disposing the external chords, so we must multiply these two numbers together. But we must also multiply by the number of ways of choosing the two sets of nodes, and this, being just the number of ways of choosing the $2r$ nodes in the first set from the total of $2n$ nodes, is the binomial coefficient

$$\binom{2n}{2r} = \frac{(2n)!}{(2r)!(2s)!}.$$

Adding together these products for the various values of r and s we obtain

$$\sum_{r+s=n} \frac{(2n)!}{(2r)!(2s)!} \cdot \frac{(2r)!}{r!(r+1)!} \cdot \frac{(2s)!}{s!(s+1)!};$$

that is,

$$\sum_{r+s=n} \frac{(2n)!}{r!(r+1)!\,s!(s+1)!}.$$

A few simple manipulations now reduce this expression to the final, very simple, formula

$$\frac{(2n)!\,(2n+2)!}{n!(n+1)!^2(n+2)!}$$

for the required number of graphs.

When $n = 2$ this formula gives the value 10. It may seem strange that there should be so many possibilities for graphs on only 4 nodes, but we can see from Figure 21 why this is so. The circuit is a square, and since we are not allowed to rotate it, graphs 1 and 2 in the figure are counted as different. Moreover, although graphs 3 and 5 are isomorphic (even taking account of the colours of the edges) they are regarded as different here because they are drawn differently in the plane.

Note also that we have allowed the possibility of two edges between two nodes. Such "double edges" are sometimes allowed, sometimes not, according to the nature of the problem. It would seem to be more reasonable to stipulate here that double edges are not to occur; but if we do that, we come across an example of the capriciousness of enumeration problems. For this apparently minor change gives us a much more difficult problem—one that, as far as I know, is still unsolved.

There are many unsolved problems in enumerative graph theory. Some of these hover on the borders of possibility, seeming to need only the right trick or the right generalization of existing theorems and techniques to force them into submission. Others show all the signs of being stubbornly recalcitrant, and resistant to all known forms of attack. Either way they offer a fascinating challenge to one's mathematical ingenuity, and the prospect of many hours of good clean fun—that is if you like that sort of thing!

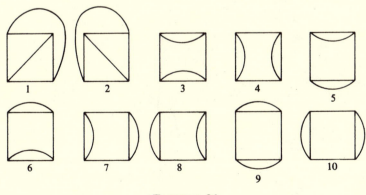

FIGURE 21

Solutions to the Problems

PROBLEM 1   The number of edges in a tree is one less than the number of nodes.

We can draw a tree by starting with a single node and adding edges one by one in such a way that each new edge joins a new node to a node already drawn. At each stage we increase by 1 the number of nodes and the number of edges. Hence, since the result is true when we first start (with 0 edges and 1 node) it is true when we finish drawing the tree.

PROBLEM 2   The graphs in Figure 9.

Graph (c) is the odd man out. The fact that the others are isomorphic is seen from the labelling given in Figure 21. Graph (c) contains a circuit of 4 nodes ($ABCD$), whereas the other graphs have no circuit of less than 5 nodes. This shows that (c) cannot be the same as the other graphs.

PROBLEM 3   Self-complementary graphs on 6 or 7 nodes.

There are none. A graph and its complement have between them all the $p(p-1)/2$ possible edges. Moreover, since they are isomorphic they each have the same number of edges. It follows that $p(p-1)/2$ must be an even number, and this is

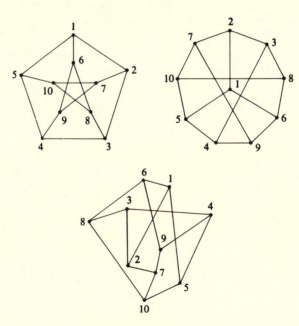

FIGURE 22

not the case for $p = 6$ or 7. More generally, self-complementary graphs can exist only when $p(p - 1)/4$ is an integer, and this happens only if p leaves remainder 0 or 1 when divided by 4.

PROBLEM 4   The equation $2q = 3p$ for cubical graphs.

Since each edge has 2 "ends" there are $2q$ ends in the graph as a whole. Each node accounts for exactly 3 ends, from which the result follows.

Bibliography

This bibliography includes some papers of general interest besides those specifically referred to in the text.

1   deBruijn, N. G. Generalization of Pólya's fundamental theorem in enumerative combinatorial analysis. *Indag. Math.* 21: 59–69.

2   Cayley, A. 1857. On the theory of the analytical forms called trees. *Phil. Mag.* 13: 172–176.

3   Cayley, A. 1874. On the mathematical theory of isomers. *Phil. Mag.* 47: 444–446.

4   Cayley, A. 1875. On the analytical forms called trees, with applications to the theory of chemical compounds. *Rep. Brit. Ass.* 257–305.

5   Harary, F. 1955. The number of linear, directed, rooted and connected graphs. *Trans. Amer. Math. Soc.* 78: 445–463.

6   Harary, F. 1969. *Graph Theory*. Reading, Massachusetts: Addison-Wesley.

7   Harary, F. and Palmer, E. M. 1973. Graphical enumeration. New York: Academic Press.

8   King, C. A. and Palmer, E. M. Calculation of the number of graphs of order $p = 1(1)24$. (unpublished.)

9   Pólya, G. 1937. Kombinatorische Anzahlbestimmungen für Gruppen, Graphen und chemische Verbindungen. *Acta Math.* 68: 145–254.

10   Read, R. C. 1963. On the number of self-complementary graphs and digraphs. *J. London Math. Soc.* 38: 99–104.

11   Redfield, J. H. 1927. The theory of group-reduced distributions. *Amer. J. Math.* 49: 433–455.

12   Tutte, W. T. 1976. Hamiltonian circuits. Colloquio Internazionale sulle Teorie Combinatorie. *Atti de Convegni Lincei.* 17: 194–199.

Error-Correcting Codes and Cryptography

N. J. A. Sloane
BELL LABORATORIES

This paper is intended to serve as an introduction to the exciting developments in secret codes that have taken place in the last ten years. David Kahn's interesting book *The Codebreakers* appeared in 1967 [29], which unfortunately was just before IBM described its Lucifer encryption scheme [11], [20], [51] and triggered the developments that I am going to describe.

The mathematical articles published in cryptography before 1967 were pretty dull (some say they were deliberately made dull, to discourage research in this area). But by the end of this paper I think you will agree that the new developments are very exciting indeed. These seem to me to be some of the most fascinating ideas in communication theory to come along in a long time. (Since most of this is not my own work I don't have to apologize for being so enthusiastic about it.) In order to avoid interlacing the text with superlatives let me say here once and for all that much of what follows is based on six classic, brilliant, elegant papers:

Shannon (1949, [49]),
Feistel, Notz and Smith (1975, [13]),

Wyner (1975, [56]),
Diffie and Hellman (1976, [5]),
Rivest, Shamir and Adelman (1978, [44]),
Merkle and Hellman (1978, [38]).

There will be nothing new here, except perhaps my way of looking at things. On the other hand I am going to mention so many different kinds of encryption schemes that most readers will find something that is unfamiliar.

By the way, I should like to stress that I am an unclassified cryptographer: I know nothing about the classified work on this subject. Fortunately those readers who do are under oath not to talk about it, and will not be able to correct me if what I say is inaccurate. All the encryption schemes I am going to describe have been published in the open literature. Nevertheless the opinions expressed here are my own, not of the authors I have referenced nor of the Bell System.

There are five parts to the article. The common theme is the problem of transmitting information down a wire (see Figure 1). Let us assume that

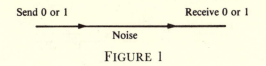

Send 0 or 1   Receive 0 or 1

Noise

FIGURE 1

we can send dots and dashes down the wire, or, to be more mathematical, 0's and 1's. When we send a 0 usually a 0 comes out at the far end, but sometimes (because of noise) a 1 is received. Conversely a 1 is usually received as a 1, occasionally as a 0. This wire is sometimes called a *binary symmetric channel*, and is described by the diagram shown in Figure 2.

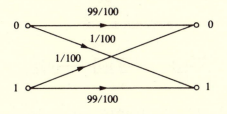

FIGURE 2

A binary symmetric channel. When a 0 is transmitted, 99 times out of a hundred a 0 is received, but 1 time in a hundred a 1 is received—similarly if a 1 is transmitted. Thus the probability of error on this channel is 1/100.

The data to be transmitted has already been converted into a string of 0's and 1's (perhaps it is being produced by a computer). Our problem is the following.

> SEND AS MUCH INFORMATION AS POSSIBLE DOWN THE WIRE, AS QUICKLY AND AS RELIABLY AS POSSIBLE. IT SHOULD ALSO BE PROTECTED AGAINST EAVES-DROPPERS.

I will describe five different kinds of solutions to this problem, under progressively more and more difficult conditions (see Figure 3). The first case (see Section 1) deals with the situation when the channel is simply that shown in Figure 1, and makes use of error-correcting codes. In the next part we have the same noisy channel, but now a second and suspicious-looking wire has appeared (see also Figure 10 below): there is an

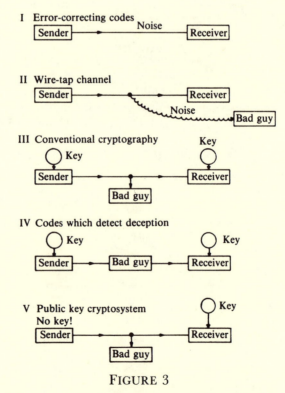

FIGURE 3

The five different communication systems considered in this paper.

eavesdropper! To begin with we assume the eavesdropper is working with poor quality equipment, and we will show how to defeat him easily (Section 2). The third part deals with the case when we are up against a well-equipped spy with the best equipment. This is the situation which conventional cryptography deals with, by using a secret key which is known to the sender and the receiver but not to the bad guy (Section 3). The fourth part is an extreme situation. Now we actually give the message to the bad guy to transmit for us. Yet we can still defeat him (Section 4). The final part describes the so-called public-key cryptosystems: these are ways of modifying the set-up of Section 3 so that the sender does not need to know the key (Section 5). In all cases we are able to defeat the bad guy and send our messages without him understanding them (or modifying them, in Section 4).

Section 1
Error-Correcting Codes

The problem is to send information down the channel shown in Figures 1 and 2 as quickly and as reliably as possible. The coding theorist's solution is to allow not just any old sequence of 0's and 1's to be sent, but only certain special sequences called *codewords*. The list of all the allowable codewords is called the *code*; it is known to both sender and receiver, who are cooperating in the attempt to eliminate transmission errors.

The messages are represented by the codewords. When a particular message is to be sent, the corresponding codeword is transmitted. At the receiver a (possibly distorted) version of the codeword arrives, and the decoder must decide which codeword was sent and hence what the message was. Figure 4 shows a simple example of a code: there are just two codewords, 00000 and 11111, of length 5, corresponding say to the messages Yes and No. Suppose the message is Yes, the codeword 00000 is transmitted, and (because of noise) 01000 is received. The decoder knows that errors are not very likely (see Figure 2) and so should decide that 00000 was transmitted and that the message is Yes. In general the decoder

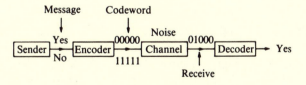

FIGURE 4

A simple error-correcting code in use. There are two codewords, 00000 and 11111. The sequence 01000 is received and is correctly decoded as Yes.

should choose that codeword which in some sense is "closest" to the received sequence.

To make this precise, let us define a distance between sequences. The *Hamming distance*, abbreviated $dist(u,v)$, between two sequences $u = (u_1, u_2, \ldots, u_n)$ and $v = (v_1, v_2, \ldots, v_n)$ of length n is the number of places where they differ. For example

$$dist(01000, 00000) = 1,$$

$$dist(01000, 11111) = 4.$$

We can now give the decoder more precise instructions: he should decode a sequence as that codeword which is closest to it in Hamming distance. This will be the most likely codeword to have been transmitted. Figure 4 illustrates this procedure, and we see that in this example the decoder was able to correct one error. In fact this code can even correct any pair of errors. For example if 00000 is sent and 01100 is received (the second and third digits are in error), 01100 is still closer to the true codeword than to 11111. On the other hand it is obvious that this code cannot correct three errors (if 11100 is received we would decode it incorrectly as 11111). So this is a double-error-correcting code.

The reason for this is plain: the codewords themselves have Hamming distance 5 apart. Precisely the same argument shows that for any code, the most important parameter is the minimum Hamming distance between codewords:

$$d = \min_{u \neq v} dist(u,v);$$

taken over all distinct codewords u and v. For then if not more than $e = [(d-1)/2]$ errors occur, the received vector will be closer to the true codeword than to any other, and the decoder will make the correct decision. (Here $[x]$ denotes the largest integer not exceeding x.)

To sum up, a code with minimum distance d between codewords is an $e = [(d-1)/2]$ error-correcting code. We wish therefore to find codes in which d is large (to correct many errors) and the length n is small (to reduce transmission time).

The following examples of codes are more interesting than the one in Figure 4.

1   The even weight code E_n, consists of all binary words of length n containing an even number of 1's—that is, having even *weight*. Figure 5 shows E_4. E_n contains 2^{n-1} codewords and has minimum distance $d = 2$ between codewords. This code is of no use for correcting errors, although it is able to detect if a single error has occurred. We shall meet it again in Section 2.

0	0	0	0
0	0	1	1
0	1	0	1
1	0	0	1
0	1	1	0
1	0	1	0
1	1	0	0
1	1	1	1

FIGURE 5

The code E_4, consisting of all eight vectors of length 4 that contain an even number of 1's.

2 The Hamming code of length 7. This is one of the many codes with a geometric construction. A *projective plane of order* p (a geometric object that will reappear in Section 4) is a collection of $p^2 + p + 1$ points and $p^2 + p + 1$ lines, arranged so that there are $p + 1$ points on each line and $p + 1$ lines through each point. Figure 6 shows a projective plane of order 2.

In order to describe a projective plane to someone else, it is not necessary to draw a picture. Instead one can just specify

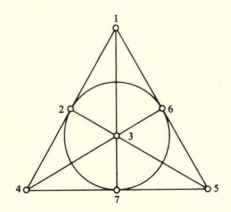

FIGURE 6

A projective plane of order 2. There are seven points, labeled $1, 2, \ldots, 7$, and seven lines (one of which is curved). Each line contains three points and each point lies on three lines.

Points

	1	2	3	4	5	6	7
	1	1	0	1	0	0	0
	0	1	1	0	1	0	0
	0	0	1	1	0	1	0
Lines	0	0	0	1	1	0	1
	1	0	0	0	1	1	0
	0	1	0	0	0	1	1
	1	0	1	0	0	0	1

FIGURE 7

A matrix of 0's and 1's which describes the projective plane of Figure 6. For example the first row of the matrix specifies that there is a line containing the points 1, 2 and 4.

which points lie on each of the lines (this could be done over the telephone). For example Figure 6 contains the lines

$$1,2,4$$
$$2,3,5$$
$$3,4,6$$
$$4,5,7$$
$$1,5,6$$
$$2,6,7$$
$$1,3,7$$

Alternatively we could describe the picture by a matrix, in which the rows represent lines and the columns represent the points (see Figure 7).

The Hamming code we want consists of the complements of the rows of this matrix (obtained by interchanging 0's and 1's), together with the zero codeword. It is shown in Figure 8, has minimum distance $d = 4$, and is a single-error-correcting code.

All the codes we have described have a special property which makes them easier to encode and decode: they are *linear* codes. That is, the componentwise sum of two codewords, taken mod 2, is always a codeword. For example the mod 2 sum of the second and third codewords in Figure 5,

$$0011$$

and

$$0101$$

$$
\begin{array}{ccccccc}
0 & 0 & 0 & 0 & 0 & 0 & 0 \\
0 & 0 & 1 & 0 & 1 & 1 & 1 \\
1 & 0 & 0 & 1 & 0 & 1 & 1 \\
1 & 1 & 0 & 0 & 1 & 0 & 1 \\
1 & 1 & 1 & 0 & 0 & 1 & 0 \\
0 & 1 & 1 & 1 & 0 & 0 & 1 \\
1 & 0 & 1 & 1 & 1 & 0 & 0 \\
0 & 1 & 0 & 1 & 1 & 1 & 0
\end{array}
$$

FIGURE 8

The Hamming code of length 7, containing eight codewords. This is obtained from the complements of the rows of Figure 7.

is

$$0110,$$

which is indeed in the code.

A linear code can be concisely defined by giving a so-called *parity-check matrix* for it. This is a matrix H with the property that a vector $u = (u_1, \cdots, u_n)$ is in the code if and only if

$$Hu^T \equiv 0 \ (mod \ 2),$$

where the T denotes transpose. It would require too much of a digression to explain why this is called a parity-check matrix—see [35, Chapter 1]. A parity-check matrix for the code of Figure 5 is

$$[1111],$$

since (u_1, u_2, u_3, u_4) is in the code if and only if

$$u_1 + u_2 + u_3 + u_4 \equiv 0 \quad (mod \ 2).$$

For the code of Figure 8 we could use

$$
\begin{bmatrix}
1101000 \\
0110100 \\
0011010 \\
0001101
\end{bmatrix}
$$

as a parity-check matrix.

In the thirty years or so that error-correcting codes have been in existence a large number of good examples have been discovered, and there is by now an extensive body of theory. In fact, F. J. MacWilliams and I have just written a 760-page book on the subject [35]. Many of these codes are defined by an ingeniously chosen parity-check matrix. One particularly powerful family are the Goppa codes. Unfortunately their description requires some knowledge of finite fields, so the following paragraph is optional reading.

3 Goppa codes of length $n = 2^m$. Choose an irreducible polynomial $G(z)$ of degree t over the Galois field $GF(2^m)$. Construct the parity check matrix

$$H = \begin{bmatrix} \dfrac{1}{G(\alpha_1)} & \cdots & \dfrac{1}{G(\alpha_n)} \\[2mm] \dfrac{\alpha_1}{G(\alpha_1)} & \cdots & \dfrac{\alpha_n}{G(\alpha_n)} \\[2mm] \cdots & \cdots & \cdots \\[2mm] \dfrac{\alpha_1^{t-1}}{G(\alpha_1)} & \cdots & \dfrac{\alpha_n^{t-1}}{G(\alpha_n)} \end{bmatrix} \tag{1}$$

where $\alpha_1, \alpha_2, \ldots, \alpha_n$ are the elements of $GF(2^m)$. The codewords are all the binary vectors u such that $Hu^T = 0$. The properties of this code are summarized in Figure 9.

Goppa codes (like BCH, Reed-Muller, and other codes we have not mentioned) have an efficient decoding algorithm—there is an algebraic process for taking the received sequence and finding the closest codeword. This is certainly *not* the case for most codes, a fact which underlies one of the public-key cryptosystems described in Section 5e. For further information about error-correcting codes the reader is referred to [35].

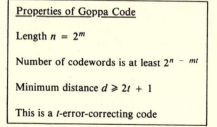

Properties of Goppa Code

Length $n = 2^m$

Number of codewords is at least 2^{n-mt}

Minimum distance $d \geqslant 2t + 1$

This is a t-error-correcting code

FIGURE 9

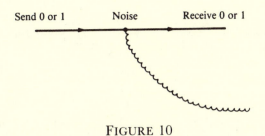

Send 0 or 1   Noise   Receive 0 or 1

FIGURE 10

A wire-tap appears!

Section 2
The Wire-Tap Channel

The communication channel now changes from that shown in Figure 1 to this (Figure 10).

The second wire is a tap leading to the Bad Guy, who is listening to everything that is said. The problem has changed: the goal is now to send information down the channel as quickly and as reliably as possible, and simultaneously to minimize what the wire-tapper learns.

In this section we consider the case of a low-budget wire-tapper, using imperfect equipment. That is, we assume that there is noise on the wire tap itself—the eavesdropper is listening over a binary symmetric channel like that in Figure 2. The main reference for this section is Wyner [56]; see also [1], [31], [32], [55]. The simplest case is when there is no noise on the direct wire—see Figure 11.

There is a beautiful and surprisingly simple solution to this problem. The key idea is the following.

> ENCRYPT 0 AS A LONG, RANDOM-LY CHOSEN STRING OF 0'S AND 1'S WITH AN *EVEN* NUMBER OF 1'S.
>
> ENCRYPT 1 AS A LONG, RANDOM-LY CHOSEN STRING OF 0'S AND 1'S WITH AN *ODD* NUMBER OF 1'S.

The length of the string is made great enough so that there are likely to be several errors introduced by the wire leading to the eavesdropper. For example, we might encrypt 0 as

$$Y = 101110100000011111010100$$

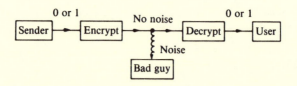

FIGURE 11

The simplest version of the wire-tap channel. There is no noise on the main wire, but the wire tap is a binary symmetric channel (see Figure 2) with a certain error probability p_0.

(a sequence of length 24 containing an even number of 1's), but by the time this reaches the eavesdropper this sequence might have been changed to

$$Z = 100110100110011111010100.$$

Even though the eavesdropper knows the encryption rule (but not of course which sequence Y has been chosen) it is clear that he is completely baffled. He knows that Z is a corrupted version of a sequence Y, but has no way of telling whether Y contained an even or an odd number of 1's.

On the other hand the legitimate receiver sees Y exactly as it was transmitted, and has merely to count the 1's in Y. He can do this very simply by forming the sum (mod 2) of all the bits ($=$ binary digits) in Y.

I think you will agree this is a wonderfully clever idea. Of course it has one obvious disadvantage—it is very slow. The rate of transmission of information is almost zero, since we have to send all of Y to communicate one bit to the receiver.

But this is easily fixed. First let me describe the scheme in another way. Let F^n denote the set of all binary vectors of length n. We divide F^n into two subsets: the subset E_n (see Section 1) consisting of all vectors with an even number of 1's, and the remaining vectors (call them D_n) which contain an odd number of 1's. Thus

$$F^n = E_n \cup D_n. \tag{2}$$

The language of group theory is helpful here. F^n is a group (under componentwise addition) and E_n is a subgroup (remember we observed in Section 1 that it was a *linear* code). Also D_n is a *translate* or *coset* of E_n:

$$D_n = 100 \cdots 0 + E_n,$$

and Equation **2** is an illustration of the principle that it is always possible to partition a group into translates of a subgroup. Our encryption scheme reads as follows. Partition $F^n = E_n \cup D_n$. To send a 0, transmit a random vector from E_n. To send a 1, transmit a random vector from D_n.

We can now state the general solution to the wire-tap problem. Choose a good linear error-correcting code C_1 containing 2^{n-k} codewords of length n (see Section 1). Partition F^n into 2^k cosets of C_1, say

$$F = C_1 \cup C_2 \cup C_3 \cup \cdots \cup C_{2^k}.$$

Number the possible messages to be sent from 1 to 2^k. Then:

> ### ENCRYPT THE i-th MESSAGE AS A RANDOMLY CHOSEN VECTOR FROM C_i.

It is easy for the legitimate receiver to discover which message was sent, since it is simple to decide to which coset of C_1 a vector belongs (one just computes the syndrome of the vector—see [35, p. 16]). But the wire-tapper is still baffled. The *transmission rate* of this scheme is k/n (n bits are transmitted in order to specify one of 2^k messages), and Wyner [56] was able to prove the following result.

THEOREM 1 If p_0 is the probability of error on the wire-tap, then it is possible to transmit at any rate below

$$-p_0 \log_2 p_0 - (1 - p_0) \log_2 (1 - p_0)$$

while keeping the eavesdropper in ignorance of what is being said.

The way this theorem is proved is by introducing a function H called *entropy* or *uncertainty* ([16, Chapter 1], [36, Chapter 2]). It is defined in such a way that if X is the message, Y is the encrypted message, and Z is the distorted version of Y heard by the eavesdropper, then $H(X)$ measures the uncertainty the eavesdropper had about X *before* hearing Z, and $H(X|Z)$ is a measure of his uncertainty about X *after* hearing Z. Of course

$$H(X|Z) \leqslant H(X).$$

To prove the theorem one shows that in fact

$$H(X|Z) \approx H(X),$$

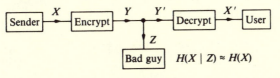

FIGURE 12

Perfect secrecy is assured if the uncertainty $H(X|Z)$ that the bad guy has about the message X *after* learning Z is essentially equal to his uncertainty $H(X)$ about X *before* installing the wire-tap.

or in other words that the eavesdropper has essentially learned nothing by installing the wire-tap (Figure 12). He might just as well have stayed home.

For the proof itself the reader is referred to [56]. Wyner also analyzes the more general situation when the direct wire from sender to receiver is noisy.

Section 3
Conventional Cryptography

In the previous section we easily defeated a wire-tapper who was using inferior equipment. But what if we are up against a professional, whose equipment is so good that he overhears what is being said without distortion? This is the situation to which conventional cryptographic methods apply. To begin, we consider encryption schemes which make use of a key that is known by the sender and receiver but not by the bad guy (see Figure 13). Section 5 will describe some recently discovered schemes in which only the receiver needs to know the key.

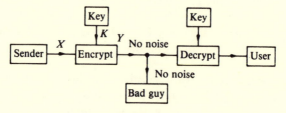

FIGURE 13

A conventional encryption scheme which makes use of a key known by the sender and receiver but not by the bad guy.

3A   The One-Time Pad

The simplest and most secure of all schemes is the one-time pad (Figure 14). (Several names are associated with this invention, especially that of G. S. Vernam, who at the time was working in American Telephone and Telegraph's research department [29, Ch. 13], [54].)

A long random string of 0's and 1's is formed, perhaps by tossing a coin many times. This is the key, which is punched onto paper tape and a copy is given to both sender and receiver. The sender adds this string bit-by-bit to the message to produce the encrypted text. For example if the message is

$$\cdots 0100001101$$

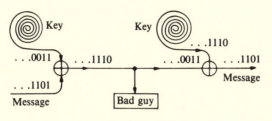

FIGURE 14

The one-time pad. The key is added mod 2 to the message. The key
sequence is as long as the message sequence and is used only once.

and the key string is

$$\cdots 1101110011$$

then the encrypted message is their sum:

$$\cdots 1001111110$$

(The sum is formed mod 2, bit-by-bit, with no carries.) At the receiving
end the key string is again added mod 2 to the encrypted message, giving
the original message back again:

Encrypted message $\quad \cdots 1001111110$

plus key string $\quad\quad \cdots 1101110011$

equals original message $\cdots 0100001101$.

The key string is used once and then destroyed. This is a perfect,
unbreakable cipher. For let X denote the message sequence, K the key
string, and

$$Y = X + K \tag{3}$$

the encrypted message. The bad guy knows Y; but since all the different
key strings K are possible and equally likely, so are all the possible
messages. Again he has learned essentially nothing by installing his wire-
tap (see Figure 12).

The only disadvantage of this scheme is that it requires as much key
as there is data to be sent. Nevertheless it is widely used for important
messages.

However for more routine matters it is desirable to have a scheme
which uses a smaller amount of key. The art of designing a good encryption
scheme is to find a way of expanding the key, that is, of taking a small
amount of key and using it as the *seed* to produce a much longer key string.
A great many ways of doing this are known (see for example [8], [15], [29]),
but here we shall only mention some techniques using shift registers.

3B Linear Feedback Shift Registers

The first, simplest and weakest method is to use the output from a linear feedback shift register as the key string (Figure 15).

If the shift register has m stages it is possible, by appropriate choice of the feedback connections, to obtain an output sequence of period $2^m - 1$. (See for example [22] or [34].) It is worth mentioning that these output sequences are codewords in a certain Reed-Muller error-correcting code [35, p. 406]. If m is large (for example, 100) this period is astronomical ($2^{100} - 1 \approx 10^{30}$).

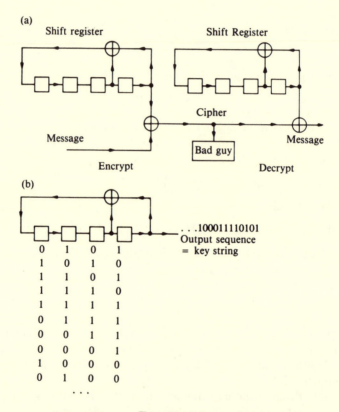

FIGURE 15

(a) The output from a linear feedback shift register is used as the key string. (b) An example of a four-stage shift register showing the successive contents of the register and the output sequence. The initial contents of the register (0101) is a secret four-digit key, and produces the output sequence

$$\cdots 111010110010001111 0101$$

of period 15 which is added bit-by-bit to the message.

In spite of this the cipher is still very easy to break. For to test the strength of a cipher it must be assumed that the bad guy knows the encryption algorithm (in this case the shift register), and has obtained a number of pairs

(message X, corresponding encrypted message Y)

—that is, pairs of matching plain text and encrypted text. He is trying to determine the key string K. In the case of a linear feedback shift register, this is simple. Once m successive matching bits of the X and Y sequences are known, their difference gives the contents of the shift register and hence (by letting the shift register run) the complete key string (cf. [18], [39]). In spite of its weakness this encryption scheme is very popular, perhaps because the large periods produce an illusion of strength.

3C   Nonlinear Shift Registers

However as soon as we allow nonlinear components to be used in the feedback circuit (Figure 16) the situation changes completely. Nonlinear shift register sequences can be made as secure as we wish.

The number of possible feedback functions f that can be used is 2^{2^m}, which when m is large is a very large number indeed. We wish to find a subset of the possible f's which

1   efficiently scramble the message, that is, such that the output sequence resembles the coin-tossing sequence of Section 3a,

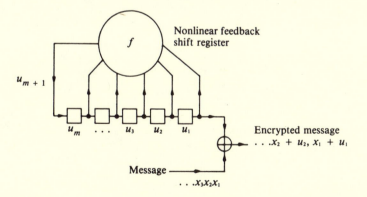

FIGURE 16

Encrypting by means of a nonlinear shift register. Only the encryption circuit is shown, although there is an identical circuit for decryption. There are m stages in the shift register, initially containing a secret m-digit key $u_m, u_{m-1}, \ldots, u_2, u_1$. The $(m+1) - st$ digit u_{m+1} which enters from the left is a complicated nonlinear function of $u_1, \ldots, u_m$, say $u_{m+1} = f(u_1, \ldots, u_m)$. Then $u_{m+2} = f(u_2, \ldots, u_{m+1})$, and so on.

2 are simple to evaluate, and

3 are easily changed (so that by changing the key we can change to a different f).

One way of doing this is to build complicated functions by combining simple components. Encryption schemes of this type are called *product ciphers*; they are analogous to the classical algorithm for mixing dough:

> ROLL IT
> FOLD IT
> ROLL IT
> FOLD IT
> ROLL IT
> . . .

In other words, perform a sequence of non-commuting operations on the dough! For example we might combine two digital operations such as a permutation and a simple nonlinear function. A useful component for constructing nonlinear functions is the two-state read-only memory device shown in Figure 17.

(a)

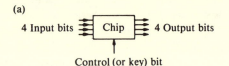

4 Input bits — Chip — 4 Output bits

Control (or key) bit

(b)

Truth table

Input	Output	
	Key = 0	Key = 1
0 0 0 0	1 0 1 0	0 1 1 1
0 0 0 1	0 0 1 1	0 1 0 0
0 0 1 0	1 0 0 1	1 1 1 1
0 0 1 1	0 0 0 0	1 0 1 0
. . . .	. . . .	. . . .
1 1 1 1	0 1 0 0	1 0 0 1

FIGURE 17

(a) A two-state read-only memory or 2-ROM with four input bits, four output bits, and one control or key bit. (Widely available as an inexpensive chip). (b) The output bits are specified as a function of the input bits and the key bit by means of a truth table. (Of course vastly more complicated components may be used: compare the circuit shown on page 79 of the July 1979 *Scientific American*.)

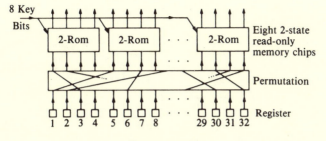

FIGURE 18

Two non-commuting operations applied to 32 bits of data. The first is a permutation, while the second is a nonlinear function formed from eight chips of the type shown in Figure 17, and controlled by eight bits of key.

To illustrate how these operations might be combined in an encryption scheme, let us consider a shift register with 32 stages (bottom of Figure 18). The basic operations are a permutation of the 32 bits (middle of Figure 18) and a nonlinear function built out of eight different two-state devices of the kind shown in Figure 17 (top of Figure 18).

These operations might be successively applied say 15 times, the resulting 32 bits fed back into the register (Figure 19), and simultaneously used as 32 bits of the key string and added to 32 message bits (Figure 20).

An alternative mode of operation is to first add the 32 message bits to the contents of the shift register, then apply the fifteen permutations and nonlinear functions, and finally take the resulting 32 bits as the encrypted text.

Product ciphers and their analogy with mixing dough were perhaps first described by Shannon [49]. The implementations using permutations and nonlinear circuits were developed by several people at IBM—see for example the papers by Feistel [11], [12], Feistel, Notz and Smith [13], [14], Girsdansky [20], [21], and Smith [51]. The circuits shown in Figures 17–20 are just simple examples. The reader will have no difficulty in inventing other, more complicated, circuits of his own.

Two actual encryption devices of this type that have received a lot of attention are the IBM Lucifer Cipher and the National Bureau of Standards Data Encryption Standard. The Lucifer cipher ([11], [20], [51]) uses a 128-bit key (rather than the 152 bits required in Figure 20), and encrypts the data in blocks of 128 bits (instead of the 32-bit blocks in Figure 20). The Data Encryption Standard ([3], [4], [40], [41]) uses a 56-bit key and encrypts the data in blocks of 64 bits.

Unfortunately not much seems to be known about the security of these ciphers, although some preliminary studies have been published ([2], [24]). If the key size is small, the bad guy can break the cipher by simply testing all possible keys. The Data Encryption Standard itself has been

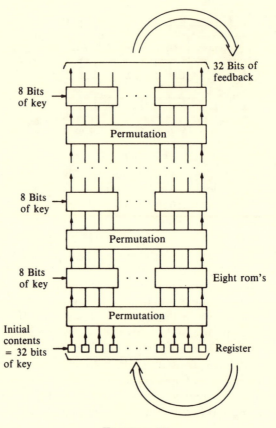

FIGURE 19

The operations shown in Figure 18 are applied fifteen times in succession to the contents of the shift register, and the final 32 bits are then fed back into the shift register. Each read-only memory requires one bit of key, and the initial contents of the register are specified by 32 bits of key, so the whole circuit is specified by a total of $8 \times 15 + 32 = 152$ bits of key.

criticized on these grounds, since the number of different keys is only $2^{56} \approx 10^{17}$ (see [6], [7], [27], [41], [53], [57]). Standards by which to judge the security of this type of cipher are badly needed.

In all the encryption schemes described in this section it has been assumed that the channel is free from errors. If channel errors do occur their effect is disastrous, for good ciphers have the property that altering one bit in the encrypted text changes about half the bits in the decrypted message. To avoid this an error-correcting code (see Section 1) should be used after the encrypting circuit, as shown in Figure 21.

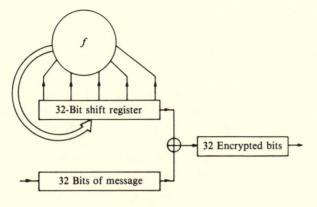

<figure>FIGURE 20</figure>

Shows how the circuit in the previous figure might be incorporated into a communication system. A fixed 152-bit key is used to specify the shift register and the function f, as in Figure 19. The data to be transmitted is divided into blocks of 32 bits each. To each block of data we add the 32 bits in the shift register to produce 32 bits of encrypted text. Then the fifteen permutations and nonlinear functions are applied, giving the next 32 bits in the shift register; these are added to the next 32 bits of data, and so on.

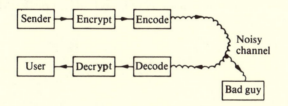

<figure>FIGURE 21</figure>

If the channel is noisy, an error-correcting code should be applied after encryption, since even one channel error changes about half the bits in the decrypted message.

Section 4
Codes Which Detect Deception

The channel continues to deteriorate: now it is actually controlled by the bad guy (see Figure 3, part 4). He has promised to faithfully transmit our messages, but we do not completely trust him. We therefore wish to *sign* or *authenticate* our messages in such a way that he is unable to replace the true message by a false one without being detected. E. N. Gilbert, F. J. MacWilliams and I studied this problem in a paper published in 1974 [19].

There are a number of situations where this problem arises. It was originally presented to us by G. J. Simmons of Sandia Corporation in connection with monitoring the production of certain materials in the interests of strategic arms limitation. However it is simpler to describe the problem in terms of a gambling casino.

The casino is managed by the Bad Guy, who is cheating the owner (the Good Guy) by reporting the daily takings from the slot machines to be less than they actually are and keeping the difference for himself. To prevent this, the owner proposes to install in each slot machine a secret key K and a device which takes as its input the day's takings X and the key K, and produces as output a signature or authenticator

$$Z = \Phi(X, K).$$

The device punches X and Z onto paper tape. The bad guy will then mail the tape to the owner, who will read X, recalculate Z from X and K, and check that this value of Z agrees with that punched on the tape. If it does, he will assume that X is correct (Figure 22).

On the other hand the bad guy knows X, Z and Φ (that is, he knows how the device works) but not K; and he wishes to replace X and Z by another pair X' and Z'. If he can do this in such a way that

$$Z' = \Phi(X', K)$$

then the substitution will not be detected, and he can pocket the difference $X'-X$.

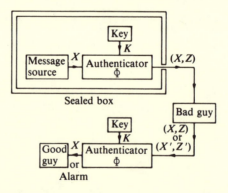

FIGURE 22

The good guy is planning to install a sealed box inside the slot machine. This will record the day's takings X on paper tape, together with a signature Z which is a function of X and the key K. Thus $Z = \Phi(X, K)$. The bad guy, who knows everything except K, wishes to replace (X, Z) by another pair (X', Z'). If Z' satisfies $Z' = \Phi(X', K)$ he will not be caught.

Figure 23 illustrates the kind of analysis the bad guy will perform. The figure shows all possible messages $X_1, X_2, X_3, \ldots$ and the authenticators $Z_1, Z_2, \ldots$ corresponding to the different keys. Suppose the bad guy sees that the message is X_1 and the authenticator is Z_2. By looking at the figure he sees that the key is one of 2, 3, 4, 5 or 6. He wishes to replace X_1 by X_2 and must decide which authenticator (Z_4, Z_5 or Z_6) to use. His chance of escaping detection is greatest if he chooses Z_5, for then he will not be caught if the key turns out to be 2, 5 or 6—he has 3 chances in 5 of succeeding. But if he were to choose Z_4 or Z_6 his chances would only be 1 in 5.

This argument suggests that a good way to design an authenticator system is to ensure that there are a large number of possible keys corresponding to each message–authenticator pair. Even so it is difficult to make the bad guy's probability of success very small. In [19] we proved the following result.

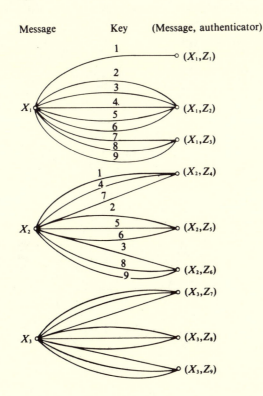

FIGURE 23

Shows the correspondence between messages, keys and authenticators.

THEOREM 2   Suppose there are M possible messages $X_1, \ldots, X_M$ and N possible keys $K_1, \quad , K_N$. If the message and the key are picked at random, then with optimal play the bad guy can guarantee that his probability of success is at least $1/\sqrt{N}$.

The proof uses techniques from information theory and will not be given here. It is possible that if the system is badly designed then the bad guy's probability of success may be greater than $1/\sqrt{N}$. However we also showed how to design the system so that his probability of success is *exactly* $1/\sqrt{N}$, which by the theorem is best possible. (For simplicity we assume that N is a perfect square.)

The construction makes use of a projective plane of order $p = \sqrt{N}$, as defined in Section 1 (see Figure 24).

Recall that there are $p^2 + p + 1$ points and $p^2 + p + 1$ lines in the projective plane. We arbitrarily choose one of the lines and call it the *equator*. The messages X will be represented by the $p + 1$ points on the equator. The keys K are represented by the remaining p^2 points. Finally the authenticator $Z = \Phi(X, K)$ is the unique line joining X and K. The device will punch the coordinates of X and the equation to the line Z onto the paper tape.

Now consider this from the bad guy's point of view. He knows X and Z, but all he can say about the key is that it is one of the remaining p points on Z. If he wants to replace X by X' he cannot do better than to pick one of these p keys at random. For the possible keys (the points on Z) are in one-to-one correspondence with the authenticators (the lines), given that one

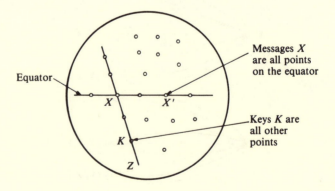

Messages X
are all points
on the equator

Equator

X X'

Keys K are
all other
points

K

Z

FIGURE 24

Optimal authentication system based on a projective plane of order p. The authenticator $Z = \Phi(X, K)$ is (the equation to) the line through X and K.

end of the line is fixed at X'. Thus his chance of success is $1/p = 1/\sqrt{N}$, as claimed.

This construction is mainly of theoretical interest, since it requires a great deal of key, even more than the one-time pad described in Section 3a. Nevertheless it is nice to see a scheme that can be proved to be secure (in contrast to those described in Sections 3c and 5).

It is worth pointing out that this is one situation where the one-time pad is useless. For in that case (see Equation **3**) the authenticator would simply be the sum of the message and the key, and the bad guy could find the key immediately (since he knows the message).

In [19] we also described some other authentication schemes (based on projective spaces and on random codes) which require a smaller amount of key. All of our analysis (and in particular the proof of Theorem 2) allows the bad guy to do as much computing as he needs. Of course in practice he is restricted by the size of his computer. Because of this limitation the conventional encryption schemes described in Section 3c— the Lucifer cipher, for example—can also be used as authenticators. For they produce an encrypted message Y which is a complicated function of the message X and the key K (see Figure 25), and this function is specifically designed to make it hard to find K given X, Y and Φ. This is a much more practical solution than our construction using projective planes, but suffers from the disadvantage that one does not know precisely how secure it is.

Another solution to the authentication problem can be obtained by using trapdoor functions—see Section 5f.

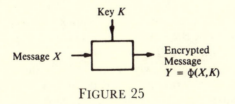

FIGURE 25

Section 5
Public-Key Cryptosystems

5A   ONE-WAY FUNCTIONS

The predecessor of the schemes to be described in this section is the simple and elegant *one-way function*. This is a function f which maps binary strings of some fixed length (say 100) onto binary strings of some other fixed length (say 120),

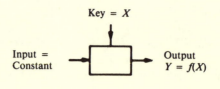

Key = X

Input =
Constant

Output
$Y = f(X)$

FIGURE 26

A conventional encryption scheme (such as the Lucifer cipher) operated as a one-way function. The usual input is set equal to an arbitrary constant, and the argument X is used as the key. The cipher is explicitly designed so that it is essentially impossible to find X given the output Y.

$$f : F^{100} \rightarrow F^{120},$$

and which has no known inverse! In other words, if you are told that

$$f(X) = 1010001 \cdots 11011110$$

there is no way that you can find X, even knowing how f is computed. This is not quite accurate, since in theory one could find X by testing each possible X in turn to see if $f(X)$ is equal to the given string. Of course this is essentially impossible, since there are too many different X's to try. So we should say that f is a one-way function if it is easy to compute $f(X)$ for any X, while it is infeasible to compute $f^{-1}(Y)$ for almost all Y in the range of f.

One-way functions certainly exist. For example we could operate one of the conventional encryption schemes of Section 3c as shown in Figure 26.

Such functions have a very pretty application to computers ([9], [42]). It is customary to allow the users of a computing system to choose their own passwords. These identify the users when they log on. However it is dangerous to store these passwords on a file in the computer since it is difficult to keep such files private. Instead one can use a one-way function f and store the values f(password) on the file. Since f cannot be inverted, a bad guy cannot obtain the passwords from this file. On the other hand when someone logs onto the computer he types in his password P, and the computer calculates $f(P)$ and compares it with the entry on the file. This is a simple and (almost) foolproof scheme. (To make it more secure, users must be encouraged not to choose passwords, such as short English names, which are easy to guess...)

5B TRAPDOOR FUNCTIONS

Of course a one-way function can't be used for encryption, for although encrypting X into $f(X)$ is secure, no one (including the legitimate receiver) can recover X. As Diffie and Hellman pointed out in [5], the way around this is to use what they call a *trapdoor function*. This is a function

$$E: F^{100} \to F^{120} \text{ (say)},$$

which has an inverse

$$D: F^{120} \to F^{100},$$

and both E and D are easy to compute. However the inverse cannot be discovered by studying E : unless you are told what the inverse is, E appears to be a one-way function. Then E can be used for encrypting and D for decrypting, since

$$D(E(X)) = X, \text{ for all } X \text{ in } F^{100}.$$

Furthermore the trapdoor property implies that someone can know how to encrypt without knowing how to decrypt. Examples of trapdoor functions will be given in Section 5c–5e.

Once we know how to find trapdoor functions we can set up what Diffie and Hellman call a *public-key cryptosystem*. Suppose there are a group of people who wish to be able to talk to each other in privacy. Each person i chooses a trapdoor function E_i with an inverse D_i. The functions $E_1, E_2, \ldots$ are listed in a public directory, while each person keeps his inverse function D_i secret. When j wishes to send a message X to i he simply transmits

$$Y = E_i(X)$$

over a public wire. Since only i knows the inverse function D_i, only i can compute

$$D_i(Y) = D_i(E_i(X)) = X$$

and read the message. Thus we have a communication network which ensures privacy without the use of keys.

5C   TRAPDOOR FUNCTIONS BASED ON PRIME NUMBERS

Rivest, Shamir and Adelman were the first to find a really satisfactory construction of trapdoor functions. This appeared in a paper [44] published in 1978, although Martin Gardner had already outlined their method in his August 1977 column [17]. In view of the extensive publicity this method has received, only a brief description will be given here.

Choose two large prime numbers p and q (each for example about 50 digits long), and a number s which is relatively prime to both $p-1$ and $q-1$. Calculate $r = pq$, and find t such that

$$st \equiv 1 \ mod \ (p-1) \cdot (q-1). \tag{4}$$

Then the encryption scheme is as follows.

THE RIVEST-SHAMIR-ADELMAN
ENCRYPTION SCHEME

PUBLISH r AND s.
KEEP p, q, t SECRET.

ENCRYPT:

$$E(x) \equiv x^s (mod\ r) = y.$$

DECRYPT:

$$D(y) \equiv y^t\ (mod\ r) = x.$$

Then $D(y) = x^{st} = x$ follows from Equation **4** and some elementary number theory. E is (it is believed) a trapdoor function because the only known way to find D, given r and s, is to factor r and so find p, q and hence t. But r is a 100-digit number and at present it appears to be essentially impossible to factor such large numbers. So the strength of this scheme rests on the fact that while it is relatively easy to find large primes [51], it is thought to be impossible to factor very large numbers [25], [46].

5D   TRAPDOOR FUNCTIONS BASED ON KNAPSACK PROBLEMS

Shortly after Rivest, Adleman and Shamir found their trapdoor functions, Merkle and Hellman [38] discovered another very simple family based on what are called knapsack problems.

The prototype of a knapsack problem is this: given a knapsack and a pile of objects that you would like to take on a hike, can you find a subset of the objects that *exactly* fill the knapsack? A simple numerical problem of the same type is:

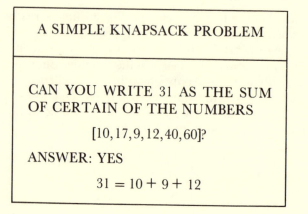

A SIMPLE KNAPSACK PROBLEM

CAN YOU WRITE 31 AS THE SUM
OF CERTAIN OF THE NUMBERS

$$[10, 17, 9, 12, 40, 60]?$$

ANSWER: YES

$$31 = 10 + 9 + 12$$

Thus 31 can be represented as the sum of the first, third and fourth numbers on the list, a fact which we indicate by writing

$$31 \leftrightarrow (1,0,1,1,0,0).$$

Conversely given $(1,0,1,1,0,0)$ we recover 31 by taking the sum of the first, third and fourth numbers.

In this way we can use this list of numbers as the basis for a primitive encryption scheme, as shown in Figure 27.

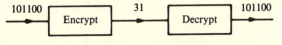

FIGURE 27

A primitive encryption scheme in which both the sender and receiver know the list of numbers $[10,17,9,12,40,60]$.

Some numerical knapsack problems are very easy to solve, for example if the numbers in the list are powers of two (or just increase very rapidly).

A VERY EASY KNAPSACK PROBLEM

EXPRESS 19 AS A SUM OF DISTINCT NUMBERS FROM THE LIST

$$[1,2,4,8,16,32].$$

ANSWER: $19 = 1 + 2 + 16$, OR

$$19 \leftrightarrow (1,1,0,0,1,0).$$

For this just amounts to finding the binary expansion of the number.

On the other hand some are extremely difficult to solve. Suppose for example the list contains 100 numbers $a_1, \cdots, a_{100}$, each about 40 digits long, and y is a 42-digit number.

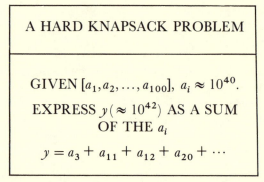

> **A HARD KNAPSACK PROBLEM**
>
> GIVEN $[a_1, a_2, \ldots, a_{100}]$, $a_i \approx 10^{40}$.
>
> EXPRESS $y(\approx 10^{42})$ AS A SUM
> OF THE a_i
>
> $$y = a_3 + a_{11} + a_{12} + a_{20} + \cdots$$

If the numbers on the list are chosen at random then there appears to be no possible way of finding out which subset of them add up to y. For there are $2^{100} \approx 10^{30}$ different subsets to be considered. If we used this list of numbers in the communication system shown in Figure 27 it would certainly be secure, but unfortunately even the legitimate receiver would not be able to recover the message.

What we need is a trapdoor knapsack problem which is easy for the legitimate receiver to solve, but which looks to anyone else like a hard problem. Merkle and Hellman [38] described several methods of doing this. The following variant of one of their constructions was discovered independently by Graham [23] and Shamir (see [48]).

We first form an easy knapsack problem by choosing numbers $a_1, \ldots, a_{100}$, say, which have powers of two embedded in their decimal expansions (Figure 28 shows a smaller example).

Figure 29 shows how this easy knapsack problem *could* be used for encryption. Of course this is very easy to break and is not how we will actually use it.

$$a_1 = \quad 8 \; 3 \; 0 \; 1 \; 0 \; 5 \; 1$$

$$a_2 = \quad\;\; 2 \; 0 \; 2 \; 0 \; 6 \; 1$$

$$a_3 = \quad 7 \; 8 \; 0 \; 4 \; 0 \; 9 \; 0$$

$$a_4 = \quad 3 \; 5 \; 0 \; 8 \; 0 \; 4 \; 9$$

$$a_5 = \quad 2 \; 4 \; 1 \; 6 \; 0 \; 1 \; 3$$

$$a_6 = \quad 3 \; 3 \; \underline{3 \; 2} \; 0 \; 7 \; 8$$

Powers of 2   Column of zeros

FIGURE 28

Six numbers $a_1, \ldots, a_6$, with embedded powers of two, to be used in a trapdoor knapsack problem.

Instead we choose two large numbers r and s such that there exists a number t with

$$st \equiv 1 \ (mod \ r).$$

Then we scramble the a's by multiplying them by s and reducing $mod \ r$. The resulting numbers

$$[b_1, b_2, \ldots, b_{100}], \quad b_i \equiv sa_i \ (mod \ r)$$

will appear to be random numbers in the range 0 to $r - 1$, and look like a very hard knapsack problem to anyone who does not know r, s and t. It is this hard knapsack problem that will actually be used for encrypting. We keep $a_1, \ldots, a_{100}$, r, s and t secret, and publish $b_1, \ldots, b_{100}$. The encryption scheme works like this.

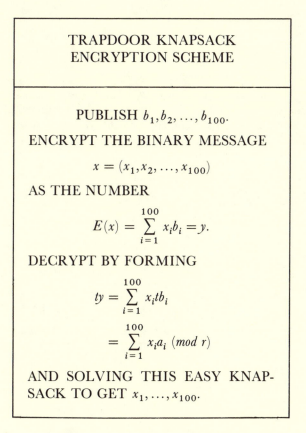

TRAPDOOR KNAPSACK ENCRYPTION SCHEME

PUBLISH $b_1, b_2, \ldots, b_{100}$.

ENCRYPT THE BINARY MESSAGE

$$x = (x_1, x_2, \ldots, x_{100})$$

AS THE NUMBER

$$E(x) = \sum_{i=1}^{100} x_i b_i = y.$$

DECRYPT BY FORMING

$$ty = \sum_{i=1}^{100} x_i t b_i$$

$$= \sum_{i=1}^{100} x_i a_i \ (mod \ r)$$

AND SOLVING THIS EASY KNAP-SACK TO GET $x_1, \ldots, x_{100}$.

Figure 30 shows an example. Anyone who does not know r, s and t is faced with solving an extremely hard knapsack problem, but the legitimate receiver recovers the message immediately.

| | Easy | |
Message	Knapsack	Encrypt
1	8 3 0 1 0 5 1	8 3 0 1 0 5 1
0	2 0 2 0 6 1	
1	7 8 0 4 0 9 0	7 8 0 4 0 9 0
0	3 5 0 8 0 4 9	
1	2 4 1 6 0 1 3	2 4 1 6 0 1 3
1	3 3 3 2 0 7 8	3 3 3 2 0 7 8
		2 1 8 5 3 2 3 2

53 = 101011 in binary
which is the message!

FIGURE 29

Shows how the easy knapsack numbers from Figure 28 could be used in an encryption scheme. The encrypted message is formed, just as in Figure 27, by adding the numbers from the list corresponding to the 1's in the message. Decryption is immediate: the binary expansion of the central pair of digits *is* the message!

| | Easy | Hard | |
Message	knapsack	knapsack	Encrypt
1	8 3 0 1 0 5 1	5 1 5 8 6 2 9 6 7	5 1 5 8 6 2 9 6 7
0	2 0 2 0 6 1	6 7 9 2 7 3 3 9 7	
1	7 8 0 4 0 9 0	5 1 3 4 3 8 3 0 5	5 1 3 4 3 8 3 0 5
0	3 5 0 8 0 4 9	4 0 2 1 6 1 7 8 3	
1	2 4 1 6 0 1 3	4 2 7 5 3 1 7 9 1	4 2 7 5 3 1 7 9 1
1	3 3 3 2 0 7 8	3 7 6 0 5 2 6 4 1	3 7 6 0 5 2 6 4 1
			1 8 3 2 8 8 5 7 0 4

Multiply by
$s = 324358647$
and reduce mod
$r = 786053315$

Cipher

FIGURE 30

The easy knapsack problem of Figure 28 is made into a hard problem by multiplying the numbers by $s = 324358647$ and reducing them mod $r = 786053315$. To decrypt we use the number $t = 326072163$, which satisfies $st \equiv 1 \pmod{r}$. The encrypted message 1832885704 is multiplied by t mod r, producing 21853232. The central digits, 53, when expanded in binary (see Figure 29) give the message.

5E   TRAPDOOR FUNCTIONS BASED ON GOPPA CODES

McEliece [37] has constructed a family of trapdoor functions using Goppa codes (see Example 3 of Section 1). We fix two numbers $n = 2^m$ and t, choose an irreducible polynomial $G(z)$ over $GF(2^m)$ of degree t, and form the parity-check matrix H of the corresponding t-error-correcting Goppa code (see Equation **1** above). From H we calculate a *generator matrix* for the code: this is a $k \times n$ matrix M, where $k = n - mt$, such that if $x = (x_1, \ldots, x_k)$ is the message vector, the corresponding codeword $c = (c_1, \ldots, c_n)$ is given by the matrix product

$$c = xM,$$

calculated mod 2. (It is easy to find M from H—see [35, Ch. 1].) The key idea is to scramble M by choosing a random invertible $k \times k$ binary matrix S and a random $n \times n$ permutation matrix P, and forming the new generator matrix

$$M' = SMP.$$

Then M' is published while M, S and P are kept secret. Here is the encryption scheme:

**ENCRYPTION SCHEME
USING GOPPA CODES**

ENCRYPT k-BIT MESSAGE x AS

$$y = xM' + z$$

WHERE z IS A RANDOM VECTOR CONTAINING t ONES, CHOSEN BY THE SENDER.

TO DECRYPT, CALCULATE

$$yP^{-1} = (xS)M + (zP^{-1}),$$

DECODE AS USUAL TO OBTAIN xS HENCE x.

The sender obscures the codeword xM' by changing t randomly chosen bits. Since the Goppa code is capable of correcting t errors, the legitimate receiver can remove this distortion, for example by using the error-correcting procedure described in [35, Chapter 12]. But an eavesdropper, who does not know M, S or P, must try and decode the code defined by the generator matrix M'. This is a very large random-looking linear code, and such codes are believed to be extremely difficult, if not impossible, to decode.

As a numerical example we might take $n = 1024 = 2^{10}$ and $t = 50$. There are about 10^{149} possible choices for $G(z)$, and an even larger number of ways of choosing S and P. The dimension of the code is at least $k = 1024 - 10 \cdot 50 = 524$, and the eavesdropper is faced with the problem of decoding an apparently random 50-error-correcting code of length 1024. For further details see [37].

5F SIGNED MAIL

Some of the public-key cryptosystems make it possible to send signed mail, that is, for i to be able to send j a message encrypted in such a way that only i could have sent it. This property is of course essential when these encryption schemes are used for transferring money.

For this to be possible, the trapdoor functions E_i and D_i (see Section 5b) should satisfy

$$E_i(D_i(X)) = X \text{ for all } X, \tag{5}$$

as well as

$$D_i(E_i(X)) = X \text{ for all } X.$$

The prime number schemes in Section 5c certainly satisfy this condition. Then i simply encrypts the message X as

$$Z = E_j(D_i(X))$$

and sends it to j, who can recover X by calculating

$$E_i(D_j(Z)) = E_i(D_i(X)) = X.$$

Only i could have sent this message, since only i knows D_i. Thus Z is a signed version of X.

Trapdoor functions which satisfy Equation **5** can also be used to solve the authentication problem described in Section 4. The message X can simply be accompanied by the authenticator $D(X)$. Everyone involved knows E and can verify that X is genuine by checking that

$$E(D(X)) = X.$$

But since D itself is secret, the bad guy cannot find the authenticator $D(X')$ needed to match his false message X'.

For more about signal mail see [30], [44] and [45]; the paper by Shamir, Rivest and Adleman [47] in this volume contains a nice application of these ideas to "mental poker."

5G CONCLUSION

Other public key cryptosystems have also been proposed ([26], [33]). But in every one of these schemes there is an important open question:

> HOW SECURE ARE THEY?

They appear secure, but up to now nothing has been proved about their strength, and until that happens there is always the possibility that someone will invent a clever way of breaking them. Some methods of attack on both the prime number schemes and the knapsack schemes have been proposed, but do not appear to pose a serious threat (see [28], [43], [48], [50]).

I hope the reader has enjoyed this introduction to an exciting and rapidly developing field. We have seen that, even under the most severe conditions, ingenious schemes are available which enable us to communicate in privacy.

References

1   Carleial, A. B. and Hellman, M. E. 1977. A note on Wyner's wiretap channel. *IEEE Trans. Info. Theory* IT-23: 387–390.

2   Coppersmith, D. and Grossman, E. 1975. Generators for certain alternating groups with applications to cryptography. *SIAM J. Applied Math.* 29: 624–627.

3   Data Encryption Standard, Federal Information Processing Standard Publication No. 46, National Bureau of Standards, U.S. Dept. of Commerce, January 1977.

4   Davis, R. M. 1978. The Data Encryption Standard in perspective. *IEEE Communications Society Magazine*, 16 (November): 5–9.

5   Diffie, W. and Hellman, M. E. 1976. New directions in cryptography, *IEEE Trans. Info. Theory* IT-22: 644–654.

6   ———. 1976. A critique of the proposed Data Encryption Standard. *Comm. ACM* 19: 164–165.

7   ———. 1977. Exhaustive analysis of the NBS data encryption standard. *Computer* 10: (June) 74–84.

8   ———. 1979. Privacy and authentication: an introduction to cryptography. *Proc. IEEE* 67: 397–427.

9   Evans, A. Jr., Kantrowitz, W., Weiss, E. 1974. A user authentication scheme not requiring secrecy in the computer. *Comm. ACM* 17: 437–442.

10 Fåk, V. 1979. Repeated use of codes which detect deception. *IEEE Trans. Info. Theory* IT-25: 233–234.

11 Feistel, H. 1970. Cryptographic coding for data-bank privacy. *Report RC-2827*, Yorktown Heights, N.Y.: IBM Watson Research Center.

12 ———. 1973. Cryptography and computer privacy. *Scientific American* 228 (May): 15–23.

13 Feistel, H., Notz, W. A. and Smith, J. L. 1971. Cryptographic techniques for machine to machine data communications. *Report RC-3663*. Yorktown Heights, N.Y.: IBM Watson Research Center.

14 ———. 1975. Some cryptographic techniques for machine-to-machine data communications. *Proc. IEEE* 63: 1545–1554.

15 Gaines, H. F. 1956. *Cryptoanalysis*. New York: Dover.

16 Gallager, R. 1968. *Information Theory and Reliable Communication*. New York: Wiley.

17 Gardner, M. 1977. A new kind of cipher that would take millions of years to break, *Scientific American* 237 (August): 120–124.

18 Geffe, P. R. 1967. An open letter to communication engineers. *Proc. IEEE* 55: 2173.

19 Gilbert, E. N., MacWilliams, F. J. and Sloane, N. J. A. 1974. Codes which detect deception. *Bell Syst. Tech. J.* 53: 405–424. For a sequel to this paper see reference [10].

20 Girsdansky, M. B. 1971. Data privacy—Cryptology and the computer at IBM Research. *IBM Research Reports* 7: (No. 4), 12 pages.

21 ———. 1972. Cryptology, the computer and data privacy. *Computers and Automation* 21 (April): 12–19.

22 Golomb, S. W. ed., 1964. *Digital Communications with Space Applications*, Englewood Cliffs, N.J.: Prentice-Hall.

23 Graham, R. L. Personal communication.

24 Grossman, E. K. and Tuckerman, B. 1977. Analysis of a Feistel-like cipher weakened by having no rotating key. *Report RC*-6375. Yorktown Heights: N.Y.: IBM Watson Research Center.

25 Guy, R. K. 1975. How to factor a number. *Proc. Fifth Manitoba Conference on Numerical Math.* pp. 49–89.

26 Hellman, M. E. 1978. An overview of public key cryptography. *IEEE Communications Society Magazine* 16 (November): 24–32.

27 ———. 1980. A cryptanalytic time-memory tradeoff. *IEEE Trans. Info. Theory*. IT-26 (July).

28 Herlestam, T. 1978. Critical remarks on some public-key cryptosystems. *BIT* 18: 493–496.

29 Kahn, D. 1967. *The Codebreakers*. New York: Macmillan.

30 Kohnfelder, L. M. 1978. On the signature reblocking problem in public-key cryptosystems. *Comm. ACM* 21: 179.

31   Leung-Yan-Cheong, S. K. 1977. On a special class of wiretap channels. *IEEE Trans. Info. Theory*. IT-23: 625–627.

32   Leung-Yan-Cheong, S. K. and Hellman, M. E. 1978. The Gaussian wiretap channel. *IEEE Trans. Info. Theory* IT-24: 451–456.

33   Leung-Yan-Cheong, S. K. and Vacon, G. V. A method for private communication over a public channel, preprint.

34   MacWilliams, F. J. and Sloane, N. J. A. 1976. Pseudo-random sequences and arrays. *Proc. IEEE* 64: 1715–1729.

35   ———. 1977. *The Theory of Error-Correcting Codes*. New York: Elsevier.

36   McEliece, R. J. 1977. *The Theory of Information and Coding*. Reading, Mass.: Addison-Wesley.

37   ———. 1978. A public-key cryptosystem based on algebraic coding theory. *Deep Space Network Progress Report* 42–44. Pasadena: Jet Propulsion Labs (January) pp. 114–116.

38   Merkle, R. C. and Hellman, M. E. 1978. Hiding information and signatures in trapdoor knapsacks. *IEEE Trans Info. Theory* IT-24: 525–530.

39   Meyer, C. H. and Tuchman, W. L. 1972. Pseudorandom codes can be cracked. *Electronic Design* 20 (November 9): 74–76.

40   Morris, R. 1978. The Data Encryption Standard—retrospective and prospects. *IEEE Communications Society Magazine* 16 (November): 11–14.

41   Morris, R., Sloane, N. J. A. and Wyner, A. D. 1977. Assessment of the National Bureau of Standards Proposed Federal Data Encryption Standard. *Cryptologia* 1: 281–306.

42   Purdy, G. B. 1974. A high security log-in procedure. *Communications ACM* 17: 442–445.

43   Rivest, R. L. 1978. Remarks on a proposed cryptanalytic attack on the M.I.T. public-key cryptosystem. *Cryptologia* 2: 62–65.

44   Rivest, R. L., Shamir, A. and Adelman, L. M. 1978. A method for obtaining digital signatures and public-key cryptosystems. *Comm. ACM* 21: 120–126.

45   Shamir, A. 1978. A fast signature scheme. *Report TM-107*. Laboratory for Computer Science, M.I.T.

46   ———. 1979. Factoring numbers in $O(\log n)$ arithmetic steps. *Info. Processing Letters* 8: 28–31.

47   Shamir, A., Rivest, R. L. and Adleman, L. M. Mental Poker. *Intra.*, pp. 37–43.

48   Shamir, A. and Zippel, R. E. 1980. On the security of the Merkle-Hellman cryptographic scheme. *IEEE Trans. Info. Theory* IT-26 (May).

49   Shannon, C. E. 1949. Communication theory of secrecy systems. *Bell Syst. Tech. J*. 28: 656–715.

50 Simmons, G. J. and Norris, M. J. 1977. Preliminary comments on the M.I.T. public-key cryptosystem. *Cryptologia* 1: 406–414.

51 Smith, J. L. 1971. The design of Lucifer, a cryptographic device for data communications. *Report RC-3326*. Yorktown Heights, N.Y.: IBM Watson Research Center.

52 Solovay, R. and Strassen, V. 1977. A fast Monte-Carlo test for primality. *SIAM J. Computing* 6: 84–85 and 7 (1978): 18.

53 Sugarman, R. et al., 1979. On foiling computer crime. *IEEE Spectrum* 16 (July): 31–41.

54 Vernam, G. S. 1926. Cipher printing telegraph systems for secret wire and radio telegraphic communications. *J. AIEE* 45: 109–115.

55 Verriest, E. and Hellman, M. E. 1979. Convolutional encoding for Wyner's wiretrap channel. *IEEE Trans. Info. Theory*, IT-25: 234–237.

56 Wyner, A. D. 1975. The wire-tap channel. *Bell Syst. Tech. J.* 54: 1355–1387.

57 Yuval, G. 1979. How to swindle Rabin. *Cryptologia* 3: 187–189.

SOUTH ORANGE PUBLIC LIBRARY

3 9507 00063267 8

793.74

The Mathematical
Gardner

DATE DUE

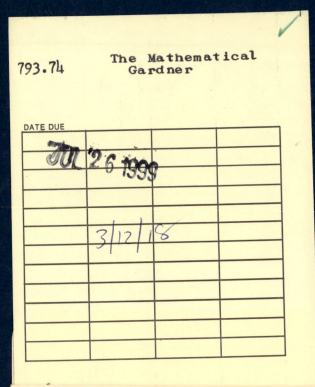